Scientific Management

FREDERICK WINSLOW TAYLOR

SCIENTIFIC MANAGEMENT

COMPRISING

Shop Management

The Principles of Scientific Management

Testimony Before the Special House Committee

BY

FREDERICK WINSLOW TAYLOR

With a Foreword by
HARLOW S. PERSON

CONSULTANT IN BUSINESS ECONOMICS AND MANAGEMENT;
FORMERLY PRESIDENT AND MANAGING DIRECTOR OF THE
TAYLOR SOCIETY, NEW YORK

GREENWOOD PRESS, PUBLISHERS
WESTPORT, CONNECTICUT

Library of Congress Cataloging in Publication Data

Taylor, Frederick Winslow, 1856-1915.
   Scientific management.

    1. Industrial management. 2. Factory management.
I. Title.
[T55.9.T38 1972]   658'.001   77-138133
ISBN 0-8371-5706-4

Copyright 1911 by Frederick W. Taylor, 1939 by Louise M. S. Taylor, 1947 by Harper & Brothers

All rights in this book are reserved. No part of the book may be reproduced in any manner whatsoever without written permission except in the case of brief quotations embodied in critical articles and reviews.

This edition originally published in 1947 by Harper & Brothers, Publishers, New York

Reprinted with the permission of Harper & Row, Publishers

Reprinted by Greenwood Press, Inc.

First Greenwood reprinting 1972
Second Greenwood reprinting 1974
Third Greenwood reprinting 1975
Fourth Greenwood reprinting 1976
Fifth Greenwood reprinting 1977

Library of Congress catalog card number 77-138133

ISBN 0-8371-5706-4

Printed in the United States of America

FOREWORD

By Harlow S. Person, Consultant in Business Economics and Management, New York; formerly President and Managing Director of the Taylor Society, New York.

IT IS a matter of significance that continuing demand for explanations of Scientific Management in Taylor's own words, earlier printings of which have for several years been unavailable, should induce the publishers to offer a new printing. It is of even greater significance that the publishers have decided to include under one cover "Shop Management," "The Principles of Scientific Management," and Taylor's testimony at "Hearings Before Social Committee of the House of Representatives to Investigate the Taylor and Other Systems of Shop Management." Prepared at different times for different audiences and under circumstances that inspired different emphases, a study of all three is essential to one who seeks understanding of the dominant force that has guided the development of twentieth century management.

Taylor's papers might well be classified as "occasional papers." He was in temperament, training and experience an engineer-executive, a doer. He was not interested in writing for its own sake, and, although he wrote painstakingly, he found the process laborious. Furthermore, he did not believe that management could be learned from reading or taught in the classroom; it had to be learned in the doing. The preparation of a formal, comprehensive treatise on Scientific Management would never have interested him. Each of his expositions was the result of a challenge of circumstances. That is why each represents a particular approach and emphasis.

"Shop Management" is a paper presented at the Saratoga, N. Y., meeting of the American Society of Mechanical Engineers in 1903. Because the audience was a group of engineer-executives, and because ASME was particular that all papers should be concise and free from what it then conceived to be extraneous matter, technique was emphasized and principles and social significance were touched lightly. The audience consisted chiefly of industrial executives in a position of authority to adopt and develop his technique, once they grasped the interrelation of details, and for that reason he emphasized the mechanist aspects.

"Principles of Scientific Management" was published during the early months of 1911. At that time circumstances were different. During the intervening years discussion of the technique had progressed and inevitably questions of principle were raised. The concept of Scientific Management had become controversial. Consequently in 1909 Taylor prepared a paper designed to emphasize principles and submitted it to the proper committee of ASME for consideration. This committee held it without action for nearly a year. During this period—the latter

FOREWORD

part of 1910—rate case hearings before the Interstate Commerce Commission in Washington had aroused an intense public interest in what during the hearings was for the first time labeled "Scientific Management." The press and monthly magazines found news value in the matter identified by this striking label. Special writers began to interview Taylor and his associates and to prepare special articles. Taylor felt that an authoritative statement, emphasizing aspects of public interest, was essential. Consequently he withdrew the paper from ASME. He published at his own expense, to meet professional requirements, an edition which he sent to all members of the society, and then authorized Harper and Brothers to print an edition for the public. As the title indicates, Taylor's emphasis was on principles, with enough of technique and of results for illustration.

Viewed in present-day perspective it was not an adequate presentation of principles. On the one hand, Taylor's mind was pretty much the opposite of the academic type of mind that thinks in terms of generalizations; he was interested in action and its immediate measurable results. On the other hand, at the time "Principles" was prepared there had been no external force to extract from him what capacity for generalization he possessed. That external force presented itself during the winter months of 1911-12 at hearings before a special committee of the House of Representatives, and the publisher has wisely included in the present volume Taylor's testimony at those hearings.

The appointment of this special committee was inspired by organized labor, which by this time was showing concern over the effect on its organization and procedures of measured individual productivity in even the best examples of Scientific Management, and especially over the use of its mechanisms (which Taylor said could be employed for good or bad) by unscrupulous employers and managers. Because Scientific Management had been developed in several arsenals of the army, labor asserted that the matter was one of Congressional concern, and the special committee was appointed.

It is the fact that generally committees appointed to "investigate" are not entirely objective, and sometimes not fair. Their appointment may be inspired by interests against the matter of investigation whose power of suggestion reaches through to influence the constitution of the committee. The committee with which we are here concerned is not free from criticism on that score. Yet its establishment was a public service insofar as its questioning inspired Taylor to utterances concerning philosophy, principles and technique that he would never have thought of writing into a professional paper. Some of them are eloquent as well as clarifying; for example, the famous passages concerning what Scientific Management is not, as well as what it is, beginning on page 26 of this edition.

Frederick Winslow Taylor was born in 1856 in a cultured and well-to-do but not wealthy family in Philadelphia. His parents desired that he enter the law and sent him to Phillips-Exeter Academy to prepare for Harvard entrance examinations. Although not a brilliant student, by seriousness of purpose and hard study he led his class at the Academy. But he paid the price of serious impairment of vision because of too much study by kerosene light. The doctors advised against Harvard and any career involving close study. So young Taylor returned to his parents' home uncertain as to his future activity.

Energetic, conscientious and restless, he looked for a career that would not call for too much reading. Accordingly in 1874 he began an apprenticeship as a pattern-maker and as a machinist in a small shop in Philadelphia. In 1878 he had become a journeyman machinist and journeyman pattern-maker. Attracted by the reputation of William Sellers, president and general manager of Midvale Steel Company, he applied for and secured a job at the works of that company. However, this first job was neither as machinist nor pattern-maker, but as an ordinary laborer. His energy and genius are manifest in the following promotions: within a period of eight years he progressed through the stages of ordinary laborer, time keeper, machinist, gang boss, foreman and assistant engineer to chief engineer of the works. By night study in absentia, his eyesight having improved, he earned the M.E. degree at Stevens Institute. In the course of his day-to-day work he developed and proved the value of that technique of management which he identified as the task system, which his associates termed the Taylor System, and everybody eventually designated as Scientific Management. The development of this technique came about in the following manner. When he was appointed gang boss he sought to increase the output by putting pressure on the men. A serious struggle between gang boss and workers ensued. Taylor finally won in the struggle, but the experience hurt him. He gave the matter thought and decided that the primary cause of such conflicts is that management, without knowing what is a proper day's work, tries to secure output by pressure. If management knew what is a proper day's work, it could then get output by demonstration. He decided by experiment to discover what was a proper day's work for every operation in the shop. His experiments along this line continued throughout his service with the Midvale Steel Company, then at the Bethlehem Steel Company, and later in various types of enterprises as consultant. Within a few years he had developed a technique of managing that in its factual basis and scope was more effective both in productivity and in good worker relations than any management elsewhere.

This new technique of managing involved two major elements. First, discovery by experiment of the best way of performing and the proper time for every operation and every component unit of an operation: in the light of the state of the art, the best material, tool, machine, manipulation of tool or machine, and the best flow of work and sequence of unit operations. These data were classified, indexed and lodged in the data files for use as new orders came along. Second, a new division of labor as between management and workers: the assignment to management of the responsibility for discovering these best ways of performing units of operations, and the further responsibility of planning operations and actually making available at the proper time and place, and in the proper quantity, the materials, tools, instructions and other facilities required by the workers. The great gains in productivity accruing from this technique of management come not from greater exertion on the part of workers (it is generally simplified and reduced) but from elimination of wastes—waste of workers' time and machine time through delays of misapplied effort, of failure in coordination of quantities, and so forth.

Taylor became a member of ASME in 1885, attended its meetings and listened with great interest to discussions of management, especially as stimulated by Henry R. Towne's paper "The Engineer as Economist," in 1886. But he became impatient of these discussions and of their controlling point of view. They were chiefly about premium and other differential wage systems, reflecting the point of view of what Taylor called the management of "initiative and incentive." In this type of management the manager tried through a premium or bonus to stimulate the workers' incentive to greater productivity by their own greater efforts. No thought here of what management itself could do to increase productivity and lighten labor's efforts.

Taylor decided, therefore, to present a paper describing his technique of management. Because wage systems were then the focal point of interest he tried to work a description of his technique of managing (which he considered extremely important) into a paper on a differential piece rate system with which he had experimented (and which he did not consider of great importance). In 1895 he presented his paper "A Piece Rate System." The piece rate feature of the paper was given attention and discussed; the technique of managing was ignored. This was a disappointment to the young man, but he took it philosophically. He decided that he had spoken before he was really prepared. He would wait patiently and at a later date present a paper on management after he had more experience and had assembled his material properly. Eight years later (1903) he presented "Shop Management." The members of ASME generally were disposed to brush this paper to one side; but a few of them—men of vision like Henry R. Towne—perceived its significance and before long it was the storm center of controversial discussion throughout the management world.

In the course of his testimony before the House committee Taylor was asked how many concerns used his system in its entirety. His reply was: "In its entirety—none; not one." Then in response to another question he went on to say that a great many used it substantially, to a greater or less degree. Were Mr. Taylor alive to respond to the same question in 1947—thirty-five years later—his reply would have to be essentially the same. Yet there is a continuing demand for his papers that calls for this new edition. What is the meaning of this paradox?

American industry, and industry in parts of western Europe, has been profoundly influenced by Scientific Management and is densely spotted with fragments of it. Practically every manufacturing establishment of stability has a planning room. Time study technicians are employed by the thousands. The sales programs, budgets and quotas of the best-managed marketing departments are in these devices utilizing the technique to a greater or less degree. General administrative schedules, budgets and standards have been inspired by Scientific Management. Modern cost accounting in terms of products, operations and processes would be impossible without it. Yet these are primarily mechanisms and they may carry with them in any particular organization little or none of the spirit of Scientific Management.

The most stirring part of Taylor's testimony before the House committee is that section in which he develops the thought that true Scientific Management requires a mental revolution on the parts both of management and of workers. They must accept the philosophy that, except for minor adjustments to keep different desirable products in balance, the interests of both and of society in the long run call for ever greater output of want-satisfying commodities. Output requires expenditure of human and material energies; therefore both workers and management should join in the search for discovery of the laws of least waste. They should join in these rearrangements which under division of labor are required to make these laws effective.

In the small plants with which Taylor was concerned in his active life these joint efforts came about informally; every worker was a participant observer in the development of standards. Taylor was never in a situation which called for consideration of formalized collective bargaining. Yet he did not disbelieve in collective bargaining as an institution, and since his day Scientific Management has in places been developed under collective bargaining auspices. While Taylor was not unsympathetic to bargaining whether the development of Scientific Management should be undertaken in an establishment, he was not tolerant of the concept that one might discover by bargaining a particular fact that lends itself rather to discovery by research and experiment. Recognition of the need of ever greater productivity, recognition of the necessity of discovering by scientific methods the laws governing the conservation of human and material energies in achieving the greater productivity, arrangements jointly by management and workers to give effect to these laws, and patience, and ever more patience—these were what Taylor considered the corner stones of true Scientific Management.

Therefore, true Scientific Management calls for a unifying point of view and a unity of interests and of efforts seldom present in a particular establishment. The directors must understand it in purpose and principle; that it is a matter of development, not installation; that it is in the nature of an investment the returns from which, though great, may be deferred; that the development takes time and patience. The active managers, all of them, must understand these things and have great skill in developing new standards to supersede obsolete standards, and in substituting the new for the old without interrupting orderly processing. And especially must management be skilled in aiding workers to understand the purpose and meaning of Scientific Management and in maintaining their confidence in the purpose and in the management. Taylor said in his testimony that it takes two to five years—more frequently five years—to develop Scientific Management in an enterprise. It must be planted, and cultivated and fertilized, and pruned and shaped, like a shrub or tree. It is not something to be bought and installed like a boiler or a machine.

It is because of a recognition or sensing of all these conditions to a development of Scientific Management "in its entirety" that there are so few examples of such developments. Directors and managers are inclined to be more opportunistic and make the most of mechanistic fragments of the technique.

The continuing and, of late, increasing demand for Taylor's papers seems to indicate that not only industrialists but students of social problems sense that they have failed to explore the values of Scientific Management as a great social force. It has been appraised generally in terms of its first emphasis; as a technique for conserving energy and increasing productivity by the use of scientific methods at the individual workplace. But since "Shop Management" was written nearly half a century ago, this technique of conservation has been applied to coordination of all the workplaces of great departments of huge enterprises, and in a few instances to coordination and conservation of the energies of entire enterprises. Some industrialists sense the fact that if we explore what is potential in Scientific Management with larger perspective, we may discover that the philosophy, principles and technique are applicable to conservation problems of entire nations, and perhaps of an entire world.

Never was the need greater for evaluation of every means of recovery from the vast wastes of war, of preservation of remaining human and physical energies, and of reorganization looking towards a new coordination of the surviving fragments of shattered economies. The very survival of democratic institutions may depend on a lifting of productivity to new degrees of adequacy which will rapidly eliminate starvation, establish a feeling of a greater economic security, and destroy impulses to follow false leaders along the paths of violence toward a totalitarian world.

Shop Management

Shop Management

BY

FREDERICK WINSLOW TAYLOR, M.E., Sc.D.

PAST PRESIDENT OF THE AMERICAN SOCIETY OF
MECHANICAL ENGINEERS

Author of "The Principles of Scientific Management"

WITH AN INTRODUCTION BY

HENRY R. TOWNE

LATE PRESIDENT OF THE YALE & TOWNE MFG. CO.

FOREWORD

BY HENRY R. TOWNE
PAST PRESIDENT, A.S.M.E.

Late President of the Yale and Towne Manufacturing Company

AS a fellow-worker with Dr. Taylor, in the field of industrial management, I have followed the development of his work, almost from its commencement, with constantly increasing admiration for the exceptional talent which he has brought to this new field of investigation, and with constantly increasing realization of the fundamental importance of the methods which he has initiated. The substitution of machinery for unaided human labor was the great industrial achievement of the nineteenth century. The new achievement to which Dr. Taylor points the way consists in elevating human labor itself to a higher plane of efficiency and of earning power.

In a paper entitled "The Engineer as an Economist," contributed to the *Proceedings* of The American Society of Mechanical Engineers in May, 1886, I made the following statements:

"The monogram of our national initials, which is the symbol for our monetary unit, the dollar, is almost as frequently conjoined to the figures of an engineer's calculations as are the symbols indicating feet, minutes, pounds, or gallons. The final issue of his work, in probably a majority of cases, resolves itself into a question of dollars and cents, of relative or absolute values. . . . To ensure the best results, the organization of productive labor must be directed and controlled by persons having not only good executive ability, and possessing the practical familiarity of a mechanic or engineer, with the goods produced and the processes employed, but having also, and equally, a practical knowledge of how to observe, record, analyze, and compare essential facts in relation to wages, supplies, expense accounts, and all else that enters into or affects the economy of production and the cost of the product."

As pertinent to the subject of industrial engineering, I will also quote the following from an address delivered by me, in February, 1905, to the graduating students of Purdue University:

"The *dollar* is the final term in almost every equation which arises in the practice of engineering in any or all of its branches, except qualifiedly as to military and naval engineering, where in some cases cost may be ignored. In other words, the true function of the engineer is, or should be, not only to determine how physical problems may be solved, but also how they may be solved most economically. For example, a railroad may have to be carried over a gorge or arroyo. Obviously it does not need an engineer to point out that this may be done by filling the chasm with earth, but only a bridge engineer is competent to determine whether it is cheaper to do this or to bridge it, and to design the bridge which will safely and most cheaply serve, the cost of which should be compared with that of an earth fill. Therefore the engineer is, by the nature of his vocation, an economist. His function is not only to design, but also so to design as to ensure the best economical result. He who designs an unsafe structure or an inoperative machine is a bad engineer; he who designs them so that they are safe and operative, but needlessly expensive, is a poor engineer, and, it may be remarked, usually earns poor pay; he who designs good work, which can be executed at a fair cost, is a sound and usually a successful engineer; he who does the best work at the lowest cost sooner or later stands at the top of his profession, and usually has the reward which this implies."

I avail of these quotations to emphasize the fact that industrial engineering, of which shop management is an integral and vital part, implies not merely the making of a given product, but the making of that product at the *lowest cost* consistent with the maintenance of the intended standard of quality. The attainment of this result is the object which Dr. Taylor has had in view during the many years through which he has pursued his studies and investigations. The methods explained and the rules laid down in the following monograph by him — probably the most valuable contribution yet made to the literature of industrial engineering — are intended to enable and to assist others engaged in this field of work to utilize and apply his methods to their several individual problems.

The monograph which is here republished was Dr. Taylor's first great contribution to industrial engineering, the second being the paper entitled "On the Art of Cutting Metals" (248 pages, with 24 insert folders covering illustrations and tables) which he presented as his Presidential Address to The American Society of Mechanical Engineers at its meeting in December, 1906, in the discussion of which at that meeting I made the following comments:

"Mr. Taylor's paper on 'The Art of Cutting Metals' is a masterpiece. Based on what is undoubtedly the longest, largest, and most exhaustive series of experiments ever conducted in this field, its summary of the conclusions deduced therefrom embodies the most important contribution to our knowledge of this subject which has ever been made. The subject itself relates to the foundation on which all of our metal-working industries are built.

"About sixty years ago American invention lifted one of the earliest and most universal of the manual arts from the plane on which it had stood from the dawn of civilization to the high level of modern mechanical industry. This was the achievement of the sewing-machine. About thirty years ago, American invention again took one of the oldest of the manual arts, that of writing, and brought it fairly within the scope of modern mechanical development. This was the achievement of the typewriting-machine. The art of forming and tempering metal tools undoubtedly is coeval with the passing of the stone age, and, therefore, in antiquity is at least as old, if indeed it does not outrank, the arts of sewing and writing. Like them it has remained almost unchanged from the beginning until nearly the present time. The work of Mr. Taylor and his associates has lifted it at once from the plane of empiricism and tradition to the high level of modern science, and apparently has gone far to reduce it almost to an exact science. In no other field of original research, that I can recall, has investigation, starting from so low a point, attained so high a level as the result of a single continued effort."

The investigations on which the report last referred to was based extended over a period of twenty-six years and involved the expenditure of some $200,000, the funds being contributed by ten industrial corporations. No other argument is needed to demonstrate Dr. Taylor's thoroughness and inexhaustible patience than the simple fact that he pursued these investigations continuously through that long period before deciding that he was ready and prepared to make known to the world his conclusions.

The conclusions embodied in Dr. Taylor's "Shop Management" constitute in effect the foundations for a new science — "The Science of Industrial Management." As in the case of constructive work the ideal engineer is he who does the best work at the lowest cost, so also, in the case of industrial operations, the best manager is he who so organizes the forces under his control that each individual shall work at his best efficiency and shall be compensated accordingly. Dr. Taylor has demonstrated conclusively that, to accomplish this, it is essential to segregate the *planning* of work from its *execution*; to employ for the former trained experts possessing the right mental equipment, and for the latter men having the right physical equipment for their respective tasks and being receptive of expert guidance in their performance. Under Dr. Taylor's leadership the combination of these elements has produced, in numberless cases, astonishing increments of output and of earnings per employé.

We are proud of the fact that the United States has led all other nations in the development of labor-saving machinery in almost every field of industry. Dr. Taylor has shown us methods whereby we can duplicate this achievement by vastly increasing the efficiency of human labor, and of accomplishing thereby a large increase in the wage-earning capacity of the workman, and a still larger decrease in the labor cost of his product.

The records of experience, and the principles deduced therefrom, set forth by Dr. Taylor in this book, should interest and appeal to all workers in the industrial field, employer and employé alike, for they point the way to increased efficiency and earning power for both. We are justly proud of the high wage rates which prevail throughout our country, and jealous of any interference with them by the products of the cheaper labor of other countries. To maintain this condition, to strengthen our control of home markets, and, above all, to broaden our opportunities in foreign markets where we must compete with the products of other industrial nations, we should welcome and encourage every influence tending to increase the efficiency of our productive processes. Dr. Taylor's contributions to this end are fundamental in character and immeasurable in ultimate effect. They concern organized industry in each and all of its infinite forms and manifestations. If intelligently and effectively utilized, they will greatly enhance the incomes of our wage-earners.

Believing profoundly in the truth of these statements, I express the hope that all who are concerned in our national industries, of every kind, will study and profit by the new science of Scientific Management, of which Dr. Taylor is concededly the leading investigator and exponent, and of which the basic principles are set forth in the following pages.

PREFACE

"SHOP MANAGEMENT" is a handbook for those interested in the management of industrial enterprises and in the production of goods. It was first published in 1903, under the auspices of The American Society of Mechanical Engineers, having been read at a meeting of that society held at Saratoga, N. Y., in June of that year.

The growing interest in scientific management on the part of the lay public has seemed to call for a new edition of this book. The demands upon the author's time have been such as to preclude his personally giving much attention to seeing the book through the press. No material changes in the text have been found necessary. At several points words have been added to make the author's meaning clear to those with no technical knowledge of the subject. A number of inconsistencies as between the text and the tables and figures have been removed; some minor additions to the time-study data have been made; the illustrations have been redrawn or reset, and a comprehensive index appended. That part of the discussion of the monograph which took place at the meeting at which it was presented, and which seemed pertinent, has been worked in with the text.

"The Principles of Scientific Management," published uniform with this book, is simply an argument for Mr. Taylor's Philosophy of Human Labor, — an outline of the fundamental principles on which it rests. In "Shop Management," however, the effort is made to describe the organization and some of the mechanisms by means of which this philosophy and these principles can be made effective in the workshop, or on the market place.

Mr. Taylor has written "Shop Management" in such a way that everything in it should be intelligible to any one with a high school education. It is the general testimony, however, of those who have used the book in actual practice that, with each re-reading, a larger significance attaches to its industrial program.

We are indebted to Mr. Calvin W. Rice, the distinguished Secretary of The American Society of Mechanical Engineers, for his encouragement in bringing out this new edition of "Shop Management."

THE EDITOR.

MAY, 1911.

Shop Management

Shop Management

THROUGH his business in changing the methods of shop management, the writer has been brought into intimate contact over a period of years with the organization of manufacturing and industrial establishments, covering a large variety and range of product, and employing workmen in many of the leading trades.

In taking a broad view of the field of management, the two facts which appear most noteworthy are:

(*a*) What may be called the *great unevenness*, or lack of uniformity shown, even in our best run works, in the development of the several elements, which together constitute what is called *the management*.

(*b*) *The lack of apparent relation* between good shop management and the payment of dividends.

Although the day of trusts is here, still practically each of the component companies of the trusts was developed and built up largely through the energies and especial ability of some one or two men who were the master spirits in directing its growth. As a rule, this leader rose from a more or less humble position in one of the departments, say in the commercial or the manufacturing department, until he became the head of his particular section. Having shown especial ability in his line, he was for that reason made manager of the whole establishment.

In examining the organization of works of this class, it will frequently be found that the management of the particular department in which this master spirit has grown up towers to a high point of excellence, his success having been due to a thorough knowledge of all of the smallest requirements of his section, obtained through personal contact, and the gradual training of the men under him to their maximum efficiency.

The remaining departments, in which this man has had but little personal experience, will often present equally glaring examples of inefficiency. And this, mainly because management is not yet looked upon as an art, with laws as exact, and as clearly defined, for instance, as the fundamental principles of engineering, which demand long and careful thought and study. Management is still looked upon as a question of men, the old view being that if you have the right man the methods can be safely left to him.

The following, while rather an extreme case, may still be considered as a fairly typical illustration of the *unevenness of management*. It became desirable to combine two rival manufactories of chemicals. The great obstacle to this combination, however, and one which for several years had proved insurmountable, was that the two men, each of whom occupied the position of owner and manager of his company, thoroughly despised one another. One of these men had risen to the top of his works through the office at the commercial end, and the other had come up from a workman in the factory. Each one was sure that the other was a fool, if not worse. When they were finally combined it was found that each was right in his judgment of the other in a certain way. A comparison of their books showed that the manufacturer was producing his chemicals more than forty per cent. cheaper than his rival, while the business man made up the difference by insisting on maintaining the highest quality, and by his superiority in selling, buying, and the management of the commercial side of the business. A combination of the two, however, finally resulted in mutual respect, and saving the forty per cent. formerly lost by each man.

The second fact that has struck the writer as most noteworthy is that there is no apparent relation in many, if not most cases, between good shop management and the success or failure of the company, many unsuccessful companies having good shop management while the reverse is true of many which pay large dividends.

We, however, who are primarily interested in the shop, are apt to forget that success, instead of hinging upon shop management, depends in many cases mainly upon other elements, namely, — the location of the company, its financial strength and ability, the efficiency of its business and sales departments, its engineering ability, the superiority of its plant and equipment, or the protection afforded either by patents, combination, location or other partial monopoly.

And even in those cases in which the efficiency of shop management might play an important part it must be remembered that for success no company need be better organized than its competitors.

The most severe trial to which any system can be subjected is that of a business which is in keen competition over a large territory, and in which the labor cost of production forms a large element of the expense, and it is in such establishments that one would naturally expect to find the best type of management.

Yet it is an interesting fact that in several of the largest and most important classes of industries in this country shop practice is still twenty to thirty years behind what might be called modern management. Not only is no attempt made by them to do tonnage or piece work, but the oldest of old-fashioned day work is still in vogue under which one overworked foreman manages the men. The workmen in these shops are still herded in classes, all of those in a class being paid the same wages, regardless of their respective efficiency.

In these industries, however, although they are keenly competitive, the poor type of shop management does not interfere with dividends, since they are in this respect all equally bad.

It would appear, therefore, that as an index to the quality of shop management the earning of dividends is but a poor guide.

Any one who has the opportunity and takes the time to study the subject will see that neither good nor bad management is confined to any one system or type. He will find a few instances of good management containing all of the elements necessary for permanent prosperity for both employers and men under ordinary day work, the task system, piece work, contract work, the premium plan, the bonus system and the differential rate; and he will find a very much larger number of instances of bad management under these systems containing as they do the elements which lead to discord and ultimate loss and trouble for both sides.

If neither the prosperity of the company nor any particular type or system furnishes an index to proper management, what then is the touchstone which indicates good or bad management?

The art of management has been defined, "as knowing exactly what you want men to do, and then seeing that they do it in the best and cheapest way." No concise definition can fully describe an art, but the relations between employers and men form without question the most important part of this art. In considering the subject, therefore, until this part of the problem has been fully discussed, the other phases of the art may be left in the background.

The progress of many types of management is punctuated by a series of disputes, disagreements and compromises between employers and men, and each side spends more than a considerable portion of its time thinking and talking over the injustice which it receives at the hands of the other. All such types are out of the question, and need not be considered.

It is safe to say that no system or scheme of management should be considered which does not in the long run give satisfaction to both employer and employé, which does not make it apparent that their best interests are mutual, and which does not bring about such thorough and hearty coöperation that they can pull together instead of apart. It cannot be said that this condition has as yet been at all generally recognized as the necessary foundation for good management. On the contrary, it is still quite generally regarded as a fact by both sides that in many of the most vital matters the best interests of employers are necessarily opposed to those of the men. In fact, the two elements which we will all agree are most wanted on the one hand by the men and on the other hand by the employers are generally looked upon as antagonistic.

What the workmen want from their employers beyond anything else is high wages, and what employers want from their workmen most of all is a low labor cost of manufacture.

These two conditions are not diametrically opposed to one another as would appear at first glance. On the contrary, they can be made to go together in all classes of work, without exception, and in the writer's judgment the existence or absence of these two elements forms the best index to either good or bad management.

This book is written mainly with the object of advocating *high wages* and *low labor cost* as the foundation of the best management, of pointing out the general principles which render it possible to maintain these conditions even under the most trying circumstances, and of indicating the various steps which the writer thinks should be taken in changing from a poor system to a better type of management.

The condition of high wages and low labor cost is far from being accepted either by the average manager or the average workman as a practical working basis. It is safe to say that the majority of employers have a feeling of satisfaction when their workmen are receiving lower wages than those of their competitors. On the other hand very many workmen feel contented if they find themselves doing the same amount of work per day as other similar workmen do and yet are getting more pay for it. Employers and workmen alike should look upon both of these conditions with apprehension, as either of them are sure, in the long run, to lead to trouble and loss for both parties.

Through unusual personal influence and energy, or more frequently through especial conditions which are but temporary, such as dull times when there is a surplus of labor, a superintendent may succeed in getting men to work extra hard for ordinary wages. After the men, however, realize that this is the case and an opportunity comes for them to change these conditions, in their reaction against what they believe unjust treatment they are almost sure to lean so far in the other direction as to do an equally great injustice to their employer.

On the other hand, the men who use the opportunity offered by a scarcity of labor to exact wages higher than the average of their class, without doing more than the average work in return, are merely laying up trouble for themselves in the long run. They grow accustomed to a high rate of living and expenditure, and when the inevitable turn comes and they are either thrown out of employment or forced to accept low wages, they are the losers by the whole transaction.

The only condition which contains the elements of stability and permanent satisfaction is that in which both employer and employés are doing as well or better than their competitors are likely to do, and this in nine cases out of ten means high wages and low labor cost, and both parties should be equally anxious for these conditions to prevail. With them the employer can hold his own with his competitors at all times and secure sufficient work to keep his men busy even in dull times. Without them both parties may do well enough in busy times, but both parties are likely to suffer when work becomes scarce.

The possibility of coupling high wages with a low labor cost rests mainly upon the enormous difference between the amount of work which a first-class man can do under favorable circumstances and the work which is actually done by the average man.

That there is a difference between the average and the first-class man is known to all employers, but that the first-class man can do in most cases from two to four times as much as is done by an average man is known to but few, and is fully realized only by those who have made a thorough and scientific study of the possibilities of men.

The writer has found this enormous difference between the first-class and average man to exist in all of the trades and branches of labor which he has investigated, and these cover a large field, as he, together with several of his friends, has been engaged with more than usual opportunities for thirty years past in carefully and systematically studying this subject.

This difference in the output of first-class and average men is as little realized by the workmen as by their employers. The first-class men know that they can do more work than the average, but they have rarely made any careful study of the matter. And the writer has over and over again found them utterly incredulous when he informed them, after close observation and study, how much they were able to do. In fact, in most cases when first told that they are able to do two or three times as much as they have done they take it as a joke and will not believe that one is in earnest.

It must be distinctly understood that in referring to the possibilities of a first-class man the writer does not mean what he can do when on a spurt or when he is over-exerting himself, but what a good man can keep up for a long term of years without injury to his health. It is a pace under which men become happier and thrive.

The second and equally interesting fact upon which the possibility of coupling high wages with low labor cost rests, is that first-class men are not only willing but glad to work at their maximum speed, providing they are paid from 30 to 100 per cent. more than the average of their trade.

The exact percentage by which the wages must be increased in order to make them work to their maximum is not a subject to be theorized over, settled by boards of directors sitting in solemn conclave, nor voted upon by trades unions. It is a fact inherent in human nature and has only been determined through the slow and difficult process of trial and error.

The writer has found, for example, after making many mistakes above and below the proper mark, that to get the maximum output for ordinary shop work requiring neither especial brains, very close application, skill, nor extra hard work, such, for instance, as the more ordinary kinds of routine machine shop work, it is necessary to pay about 30 per cent. more than the average. For ordinary day labor requiring little brains or special skill, but calling for strength, severe bodily exertion, and fatigue, it is necessary to pay from 50 per cent. to 60 per cent. above the average. For work requiring especial skill or brains, coupled with close application, but without severe bodily exertion, such as the more difficult and delicate machinist's work, from 70 per cent. to 80 per cent. beyond the average. And for work requiring skill, brains, close application, strength, and severe bodily exertion, such, for instance, as that involved in operating a well run steam hammer doing miscellaneous work, from 80 per cent. to 100 per cent. beyond the average.

There are plenty of good men ready to do their best for the above percentages of increase, but if the endeavor is made to get the right men to work at this maximum for less than the above increase, it will be found that most of them will prefer their old rate of speed with the lower pay. After trying the high speed piece work for a while they will one after another throw up their jobs and return to the old day work conditions. Men will not work at their best unless assured a good liberal increase, which must be permanent.

It is the writer's judgment, on the other hand, that for their own good it is as important that workmen should not be very much over-paid, as it is that they should not be under-paid. If over-paid, many will work irregularly and tend to become more or less shiftless, extravagant, and dissipated. It does not do for most men to get rich too fast. The writer's observation, however, would lead him to the conclusion that most men tend to become more instead of less thrifty when they receive the proper increase for an extra hard day's work, as, for example, the percentages of increase referred to above. They live rather better, begin to save money, become more sober, and work more steadily. And this certainly forms one of the strongest reasons for advocating this type of management.

In referring to high wages and low labor cost as fundamental in good management, the writer is most desirous not to be misunderstood.

By high wages he means wages which are high only with relation to the average of the class to which the man belongs and which are paid only to those who do much more or better work than the average of their class. He would not for an instant advocate the use of a high-priced tradesman to do the work which could be done by a trained laborer or a lower-priced man. No one would think of using a fine trotter to draw a grocery wagon nor a Percheron to do the work of a little mule. No more should a mechanic be allowed to do work for which a trained laborer can be used, and the writer goes so far as to say that almost any job that is repeated over and over again, however great skill and dexterity it may require, providing there is enough of it to occupy a man throughout a considerable part of the year, should be done by a trained laborer and not by a mechanic. A man with only the intelligence of an average laborer can be taught to do the most difficult and delicate work if it is repeated enough times; and his lower mental caliber renders him more fit than the mechanic to stand the monotony of repetition. It would seem to be the duty of employers, therefore, both in their own interest and in that of their employés, to see that each workman is given as far as possible the highest class of work for which his brains and physique fit him. A man, however, whose mental caliber and education do not 'fit him to become a good mechanic (and that grade of man is the one referred to as belonging to the "laboring class"), when he is trained to do some few especial jobs, which were formerly done by mechanics, should not expect to be paid the wages of a mechanic. He should get more than the average laborer, but less than a mechanic; thus insuring high wages to the workman, and low labor cost to the employer, and in this way making it most apparent to both that their interests are mutual.

To summarize, then, what the aim in each establishment should be:

(a) That each workman should be given as far as possible the highest grade of work for which his ability and physique fit him.

(b) That each workman should be called upon to

turn out the maximum amount of work which a first-rate man of his class can do and thrive.

(c) That each workman, when he works at the best pace of a first-class man, should be paid from 30 per cent. to 100 per cent. according to the nature of the work which he does, beyond the average of his class.

And this means *high wages* and a *low labor cost*. These conditions not only serve the best interests of the employer, but they tend to raise each workman to the highest level which he is fitted to attain by making him use his best faculties, forcing him to become and remain ambitious and energetic, and giving him sufficient pay to live better than in the past.

Under these conditions the writer has seen many first-class men developed who otherwise would have remained second or third class all of their lives.

Is not the presence or absence of these conditions the best indication that any system of management is either well or badly applied? And in considering the relative merits of different types of management, is not that system the best which will establish these conditions with the greatest certainty, precision, and speed?

In comparing the management of manufacturing and engineering companies by this standard, it is surprising to see how far they fall short. Few of those which are best organized have attained even approximately the maximum output of first-class men.

Many of them are paying much higher prices per piece than are required to secure the maximum product; while owing to a bad system, lack of exact knowledge of the time required to do work, and mutual suspicion and misunderstanding between employers and men, the output per man is so small that the men receive little if any more than average wages, both sides being evidently the losers thereby.

The chief causes which produce this loss to both parties are: *First* (and by far the most important), the profound ignorance of employers and their foremen as to the time in which various kinds of work should be done, and this ignorance is shared largely by the workmen.

Second: The indifference of the employers and their ignorance as to the proper system of management to adopt and the method of applying it, and further their indifference as to the individual character, worth, and welfare of their men.

On the part of the men the greatest obstacle to the attainment of this standard is the slow pace which they adopt, or the loafing or "soldiering," marking time, as it is called.

This loafing or soldiering proceeds from two causes. First, from the natural instinct and tendency of men to take it easy, which may be called *natural soldiering*. Second, from more intricate second thought and reasoning caused by their relations with other men, which may be called *systematic soldiering*.

There is no question that the tendency of the average man (in all walks of life) is toward working at a slow, easy gait, and that it is only after a good deal of thought and observation on his part or as a result of example, conscience, or external pressure that he takes a more rapid pace.

There are, of course, men of unusual energy, vitality, and ambition who naturally choose the fastest gait, set up their own standards, and who will work hard, even though it may be against their best interests. But these few uncommon men only serve by affording a contrast to emphasize the tendency of the average.

This common tendency to "take it easy" is greatly increased by bringing a number of men together on similar work and at a uniform standard rate of pay by the day.

Under this plan the better men gradually but surely slow down their gait to that of the poorest and least efficient. When a naturally energetic man works for a few days beside a lazy one, the logic of the situation is unanswerable: "Why should I work hard when that lazy fellow gets the same pay that I do and does only half as much work?"

A careful time study of men working under these conditions will disclose facts which are ludicrous as well as pitiable.

To illustrate: The writer has timed a naturally energetic workman who, while going and coming from work, would walk at a speed of from three to four miles per hour, and not infrequently trot home after a day's work. On arriving at his work he would immediately slow down to a speed of about one mile an hour. When, for example, wheeling a loaded wheelbarrow he would go at a good fast pace even up hill in order to be as short a time as possible under load, and immediately on the return walk slow down to a mile an hour, improving every opportunity for delay short of actually sitting down. In order to be sure not to do more than his lazy neighbor he would actually tire himself in his effort to go slow.

These men were working under a foreman of good reputation and one highly thought of by his employer who, when his attention was called to this state of things, answered: "Well, I can keep them from sitting down, but the devil can't make them get a move on while they are at work."

The natural laziness of men is serious, but by far the greatest evil from which both workmen and employers are suffering is the *systematic soldiering* which is almost universal under all of the ordinary schemes of management and which results from a careful study on the part of the workmen of what they think will promote their best interests.

The writer was much interested recently to hear one small but experienced golf caddy boy of twelve explaining to a green caddy who had shown special energy and interest the necessity of going slow and lagging behind his man when he came up to the ball, showing him that since they were paid by the hour, the faster they went the less money they got, and finally telling him that if he went too fast the other boys would give him a licking.

This represents a type of systematic soldiering which is not, however, very serious, since it is done with the knowledge of the employer, who can quite easily break it up if he wishes.

The greater part of the *systematic soldiering*, however, is done by the men with the deliberate object of keeping their employers ignorant of how fast work can be done.

So universal is soldiering for this purpose, that hardly a competent workman can be found in a large establishment, whether he works by the day or on piece work, contract work or under any of the ordinary systems of compensating labor, who does not devote a considerable part of his time to studying just how slowly he can work and still convince his employer that he is going at a good pace.

The causes for this are, briefly, that practically all employers determine upon a maximum sum which they feel it is right for each of their classes of employés to earn per day, whether their men work by the day or piece.

Each workman soon finds out about what this figure is for his particular case, and he also realizes that when his employer is convinced that a man is capable of doing more work than he has done, he will find sooner or later some way of compelling him to do it with little or no increase of pay.

Employers derive their knowledge of how much of a given class of work can be done in a day from either their own experience, which has frequently grown hazy with age, from casual and unsystematic observation of their men, or at best from records which are kept, showing the quickest time in which each job has been done. In many cases the employer will feel almost certain that a given job can be done faster than it has been, but he rarely cares to take the drastic measures necessary to force men to do it in the quickest time, unless he has an actual record, proving conclusively how fast the work can be done.

It evidently becomes for each man's interest, then, to see that no job is done faster than it has been in the past. The younger and less experienced men are taught this by their elders, and all possible persuasion and social pressure is brought to bear upon the greedy and selfish men to keep them from making new records which result in temporarily increasing their wages, while all those who come after them are made to work harder for the same old pay.

Under the best day work of the ordinary type, when accurate records are kept of the amount of work done by each man and of his efficiency, and when each man's wages are raised as he improves, and those who fail to rise to a certain standard are discharged and a fresh supply of carefully selected men are given work in their places, both the natural loafing and systematic soldiering can be largely broken up. This can be done, however, only when the men are thoroughly convinced that there is no intention of establishing piece work even in the remote future, and it is next to impossible to make men believe this when the work is of such a nature that they believe piece work to be practicable. In most cases their fear of making a record which will be used as a basis for piece work will cause them to soldier as much as they dare.

It is, however, under piece work that the art of systematic soldiering is thoroughly developed. After a workman has had the price per piece of the work he is doing lowered two or three times as a result of his having worked harder and increased his output, he is likely to entirely lose sight of his employer's side of the case and to become imbued with a grim determination to have no more cuts if soldiering can prevent it. Unfortunately for the character of the workman, soldiering involves a deliberate attempt to mislead and deceive his employer, and thus upright and straight-forward workmen are compelled to become more or less hypocritical. The employer is soon looked upon as an antagonist, if not as an enemy, and the mutual confidence which should exist between a leader and his men, the enthusiasm, the feeling that they are all working for the same end and will share in the results, is entirely lacking.

The feeling of antagonism under the ordinary piecework system becomes in many cases so marked on the part of the men that any proposition made by their employers, however reasonable, is looked upon with suspicion. Soldiering becomes such a fixed habit that men will frequently take pains to restrict the product of machines which they are running when even a large increase in output would involve no more work on their part.

On work which is repeated over and over again and the volume of which is sufficient to permit it, the plan of making a contract with a competent workman to do a certain class of work and allowing him to employ his own men subject to strict limitations, is successful.

As a rule, the fewer the men employed by the contractor and the smaller the variety of the work, the greater will be the success under the contract system, the reason for this being that the contractor, under the spur of financial necessity, makes personally so close a study of the quickest time in which the work can be done that soldiering on the part of his men becomes difficult and the best of them teach laborers or lower-priced helpers to do the work formerly done by mechanics.

The objections to the contract system are that the machine tools used by the contractor are apt to deteriorate rapidly, his chief interest being to get a large output, whether the tools are properly cared for or not, and that through the ignorance and inexperience of the contractor in handling men, his employés are frequently unjustly treated.

These disadvantages are, however, more than counterbalanced by the comparative absence of soldiering on the part of the men.

The greatest objection to this system is the soldiering which the contractor himself does in many cases, so as to secure a good price for his next contract.

It is not at all unusual for a contractor to restrict the output of his own men and to refuse to adopt improvements in machines, appliances, or methods while in the midst of a contract, knowing that his next contract price will be lowered in direct proportion to the profits which he has made and the improvements introduced.

Under the contract system, however, the relations between employers and men are much more agreeable and normal than under piece work, and it is to be regretted that owing to the nature of the work done in most shops this system is not more generally applicable.

The writer quotes as follows from his paper on "A Piece Rate System," read in 1895, before The American Society of Mechanical Engineers:

"Coöperation, or profit sharing, has entered the mind of every student of the subject as one of the possible and most attractive solutions of the problem; and there have been certain instances, both in England and France, of at least a partial success of coöperative experiments.

"So far as I know, however, these trials have been made either in small towns, remote from the manufacturing centers, or in industries which in many respects are not subject to ordinary manufacturing conditions.

"Coöperative experiments have failed, and, I think, are generally destined to fail, for several reasons, the first and most important of which is, that no form of coöperation has yet been devised in which each individual is allowed free scope for his personal ambition. Personal ambition always has been and will remain a more powerful incentive to exertion than a desire for the general welfare. The few misplaced drones, who do the loafing and share equally in the profits with the rest, under coöperation are sure to drag the better men down toward their level.

"The second and almost equally strong reason for failure lies in the remoteness of the reward. The average workman (I don't say all men) cannot look forward to a profit which is six months or a year away. The nice time which they are sure to have to-day, if they take things easily, proves more attractive than hard work, with a possible reward to be shared with others six months later.

"Other and formidable difficulties in the path of coöperation are, the equitable division of the profits, and the fact that, while workmen are always ready to share the profits, they are neither able nor willing to share the losses. Further than this, in many cases, it is neither right nor just that they should share either in the profits or the losses, since these may be due in great part to causes entirely beyond their influence or control, and to which they do not contribute."

Of all the ordinary systems of management in use (in which no accurate scientific study of the time problem is undertaken, and no carefully measured tasks are assigned to the men which must be accomplished in a given time) the best is the plan fundamentally originated by Mr. Henry R. Towne, and improved and made practical by Mr. F. A. Halsey. This plan is described in papers read by Mr. Towne before The American Society of Mechanical Engineers in 1886, and by Mr. Halsey in 1891, and has since been criticised and ably defended in a series of articles appearing in the "American Machinist."

The Towne-Halsey plan consists in recording the quickest time in which a job has been done, and fixing this as a standard. If the workman succeeds in doing the job in a shorter time, he is still paid his same wages per hour for the time he works on the job, and in addition is given a premium for having worked faster, consisting of from one-quarter to one-half the difference between the wages earned and the wages originally paid when the job was done in standard time. Mr. Halsey recommends the payment of one-third of the difference as the best premium for most cases. The difference between this system and ordinary piece work is that the workman on piece work gets the whole of the difference between the actual time of a job and the standard time, while under the Towne-Halsey plan he gets only a fraction of this difference.

It is not unusual to hear the Towne-Halsey plan referred to as practically the same as piece work. This is far from the truth, for while the difference between the two does not appear to a casual observer to be great, and the general principles of the two seem to be the same, still we all know that success or failure in many cases hinges upon small differences.

In the writer's judgment, the Towne-Halsey plan is a great invention, and, like many other great inventions, its value lies in its simplicity.

This plan has already been successfully adopted by a large number of establishments, and has resulted in giving higher wages to many workmen, accompanied by a lower labor cost to the employer, and at the same time materially improving their relations by lessening the feeling of antagonism between the two.

This system is successful because it diminishes soldiering, and this rests entirely upon the fact that since the workman only receives say one-third of the increase in pay that he would get under corresponding conditions on piece work, there is not the same temptation for the employer to cut prices.

After this system has been in operation for a year or two, if no cuts in prices have been made, the tendency of the men to soldier on that portion of the work which is being done under the system is diminished, although it does not entirely cease. On the other hand, the tendency of the men to soldier on new work which is started, and on such portions as are still done on day work, is even greater under the Towne-Halsey plan than under piece work.

To illustrate: Workmen, like the rest of mankind, are more strongly influenced by object lessons than by theories. The effect on men of such an object lesson as the following will be apparent. Suppose that two men, named respectively Smart and Honest, are at work by the day and receive the same pay, say 20 cents per hour. Each of these men is given a new piece of work which could be done in one hour. Smart does his job in four hours (and it is by no means unusual for men to soldier to this extent). Honest does his in one and one-half hours.

Now, when these two jobs start on this basis under the Towne-Halsey plan and are ultimately done in one hour each, Smart receives for his job 20 cents per hour + a premium of $\frac{60}{3}$ = 20 cents = *a total of 40 cents*. Honest receives for his job 20 cents per hour + a premium of $\frac{10}{3}$ = $3\frac{1}{3}$ cents = *a total of $23\frac{1}{3}$ cents*.

Most of the men in the shop will follow the example of Smart rather than that of Honest and will "soldier" to the extent of three or four hundred per cent. if allowed to do so.

The Towne-Halsey system shares with ordinary piece work then, the greatest evil of the latter, namely that its very foundation rests upon deceit, and under both of these systems there is necessarily, as we have seen, a great lack of justice and equality in the starting-point of different jobs.

Some of the rates will have resulted from records obtained when a first-class man was working close to his maximum speed, while others will be based on the performance of a poor man at one-third or one-quarter speed.

The injustice of the very foundation of the system is thus forced upon the workman every day of his life, and no man, however kindly disposed he may be toward his employer, can fail to resent this and be seriously influenced by it in his work. These systems are, therefore, of necessity slow and irregular in their operation in reducing costs. They "drift" gradually toward an increased output, but under them the attainment of the maximum output of a first-class man is almost impossible.

Objection has been made to the use of the word "drifting" in this connection. It is used absolutely without any intention of slurring the Towne-Halsey system or in the least detracting from its true merit.

It appears to me, however, that "drifting" very accurately describes it, for the reason that the management, having turned over the entire control of the speed problem to the men, the latter being influenced by their prejudices and whims, drift sometimes in one direction and sometimes in another; but on the whole, sooner or later, under the stimulus of the premium, move toward a higher rate of speed. This drifting, accompanied as it is by the irregularity and uncertainty both as to the final result which will be attained and as to how long it will take to reach this end, is in marked contrast to the distinct goal which is always kept in plain sight of both parties under task management, and the clear-cut directions which leave no doubt as to the means which are to be employed nor the time in which the work must be done; and these elements constitute the fundamental difference between the two systems. Mr. Halsey, in objecting to the use of the word "drifting" as describing his system, has referred to the use of his system in England in connection with a "rate-fixing" or planning department, and quotes as follows from his paper to show that he contemplated control of the speed of the work by the management:

"On contract work undertaken for the first time the method is the same except that the premium is based on the estimated time for the execution of the work."

In making this claim Mr. Halsey appears to have entirely lost sight of the real essence of the two plans. It is task management which is in use in England, not the Towne-Halsey system; and in the above quotation Mr. Halsey describes not his system but a type of task management, in which the men are paid a premium for carrying out the directions given them by the management.

There is no doubt that there is more or less confusion in the minds of many of those who have read about the task management and the Towne-Halsey system. This extends also to those who are actually using and working under these systems. This is practically true in England, where in some cases task management is actually being used under the name of the "Premium Plan." It would therefore seem desirable to indicate once again and in a little different way the essential difference between the two.

The one element which the Towne-Halsey system and task management have in common is that both recognize the all-important fact that workmen cannot be induced to work extra hard without receiving extra pay. Under both systems the men who succeed are daily and automatically, as it were, paid an extra premium. The payment of this daily premium forms such a characteristic feature in both systems, and so radically differentiates these systems from those which were in use before, that people are apt to look upon this one element as the essence of both systems and so fail to recognize the more important, underlying principles upon which the success of each of them is based.

In their essence, with the one exception of the payment of a daily premium, the systems stand at the two opposite extremes in the field of management; and it is owing to the distinctly radical, though opposite, positions taken by them that each one owes its success; and it seems to me a matter of importance that this should be understood. In any executive work which involves the coöperation of two different men or parties, where both parties have anything like equal power or voice in its direction, there is almost sure to be a certain amount of bickering, quarreling, and vacillation, and the success of the enterprise suffers accordingly. If, however, either one of the parties has the entire direction, the enterprise will progress consistently and probably harmoniously, even although the wrong one of the two parties may be in control.

Broadly speaking, in the field of management there are two parties — the superintendents, etc., on one side and the men on the other, and the main questions at issue are the speed and accuracy with which the work shall be done. Up to the time that task management was introduced in the Midvale Steel Works, it can be fairly said that under the old systems of management the men and the management had about equal weight in deciding how fast the work should be done. Shop records showing the quickest time in which each job had been done and more or less shrewd guessing being the means on which the management depended for bargaining with and coercing the men; and deliberate soldiering for the purpose of misinforming the management being the weapon used by the men in self-defense. Under the old system the incentive was entirely lacking which is needed to induce men to coöperate heartily with the management in increasing the speed with which work is turned out. It is chiefly due, under the old systems, to this divided control of the speed with which the work shall be done that such an amount of bickering, quarreling, and often hard feeling exists between the two sides.

The essence of task management lies in the fact that the control of the speed problem rests entirely with the management; and, on the other hand, the true strength of the Towne-Halsey system rests upon the fact that under it the question of speed is settled entirely by the men without interference on the part of the management. Thus in both cases, though from diametrically opposite causes, there is undivided control, and this is the chief element needed for harmony.

The writer has seen many jobs successfully nursed in several of our large and well managed establishments under these drifting systems, for a term of ten to fifteen years, at from one-third to one-quarter speed. The workmen, in the meanwhile, apparently enjoyed the confidence of their employers, and in many cases the employers not only suspected the deceit, but felt quite sure of it.

The great defect, then, common to all the ordinary systems of management (including the Towne-Halsey system, the best of this class) is that their starting-point, their very foundation, rests upon ignorance and deceit, and that throughout their whole course in the one element which is most vital both to employer and workmen, namely, the speed at which work is done, they are allowed to drift instead of being intelligently directed and controlled.

The writer has found, through an experience of thirty years, covering a large variety in manufactures, as well as in the building trades, structural and engineering work, that it is not only practicable but comparatively easy to obtain, through a systematic and scientific time study, exact information as to how much of any given kind of work either a first-class or an average man can do in a day, and with this information as a foundation, he has over and over again seen the fact demonstrated that workmen of all classes are not only willing, but glad to give up all idea of soldiering, and devote all of their energies to turning out the maximum work possible, providing they are sure of a suitable permanent reward.

With accurate time knowledge as a basis, surprisingly large results can be obtained under any scheme of management from day work up; there is no question that even ordinary day work resting upon this foundation will give greater satisfaction than any of the systems in common use, standing as they do upon soldiering as a basis.

To many of the readers of this book both the fundamental objects to be aimed at, namely, *high wages with low labor cost*, and the means advocated by the writer for attaining this end; namely, *accurate time study*, will appear so theoretical and so far outside of the range of their personal observation and experience that it would seem desirable, before proceeding farther, to give a brief illustration of what has been accomplished in this line.

The writer chooses from among a large variety of trades to which these principles have been applied, the yard labor handling raw materials in the works of the Bethlehem Steel Company at South Bethlehem, Pa., not because the results attained there have been greater than in many other instances, but because the case is so elementary that the results are evidently due to no other cause than thorough time study as a basis, followed by the application of a few simple principles with which all of us are familiar.

In almost all of the other more complicated cases

the large increase in output is due partly to the actual physical changes, either in the machines or small tools and appliances, which a preliminary time study almost always shows to be necessary, so that for purposes of illustration the simple case chosen is the better, although the gain made in the more complicated cases is none the less legitimately due to the system.

Up to the spring of the year 1899, all of the materials in the yard of the Bethlehem Steel Company had been handled by gangs of men working by the day, and under the foremanship of men who had themselves formerly worked at similar work as laborers. Their management was about as good as the average of similar work, although it was bad; all of the men being paid the ruling wages of laborers in this section of the country, namely, $1.15 per day, the only means of encouraging or disciplining them being either talking to them or discharging them; occasionally, however, a man was selected from among these men and given a better class of work with slightly higher wages in some of the companies' shops, and this had the effect of slightly stimulating them. From four to six hundred men were employed on this class of work throughout the year.

The work of these men consisted mainly of unloading from railway cars and shoveling on to piles, and from these piles again loading as required, the raw materials used in running three blast furnaces and seven large open-hearth furnaces, such as ore of various kinds, varying from fine, gravelly ore to that which comes in large lumps, coke, limestone, special pig, sand, etc., unloading hard and soft coal for boilers gas-producers, etc., and also for storage and again loading the stored coal as required for use, loading the pig-iron produced at the furnaces for shipment, for storage, and for local use, and handling billets, etc., produced by the rolling mills. The work covered a large variety as laboring work goes, and it was not usual to keep a man continuously at the same class of work.

Before undertaking the management of these men, the writer was informed that they were steady workers, but slow and phlegmatic, and that nothing would induce them to work fast.

The first step was to place an intelligent, college-educated man in charge of progress in this line. This man had not before handled this class of labor, although he understood managing workmen. He was not familiar with the methods pursued by the writer, but was soon taught the art of determining how much work a first-class man can do in a day. This was done by timing with a stop watch a first-class man while he was working fast. The best way to do this, in fact almost the only way in which the timing can be done with certainty, is to divide the man's work into its elements and time each element separately. For example, in the case of a man loading pig-iron on to a car, the elements should be: (*a*) picking up the pig from the ground or pile (time in hundredths of a minute); (*b*) walking with it on a level (time per foot walked); (*c*) walking with it up an incline to car (time per foot walked); (*d*) throwing the pig down (time in hundredths of a minute), or laying it on a pile (time in hundredths of a minute); (e) walking back empty to get a load (time per foot walked).

In case of important elements which were to enter into a number of rates, a large number of observations were taken when practicable on different first-class men, and at different times, and they were averaged.

The most difficult elements to time and decide upon in this, as in most cases, are the percentage of the day required for rest, and the time to allow for accidental or unavoidable delays.

In the case of the yard labor at Bethlehem, each class of work was studied as above, each element being timed separately, and, in addition, a record was kept in many cases of the total amount of work done by the man in a day. The record of the gross work of the man (who is being timed) is, in most cases, not necessary after the observer is skilled in his work. As the Bethlehem time observer was new to this work, the gross time was useful in checking his detailed observations and so gradually educating him and giving him confidence in the new methods.

The writer had so many other duties that his personal help was confined to teaching the proper methods and approving the details of the various changes which were in all cases outlined in written reports before being carried out.

As soon as a careful study had been made of the time elements entering into one class of work, a single first-class workman was picked out and started on ordinary piece work on this job. His task required him to do between *three and one-half* and *four times* as much work in a day as had been done in the past on an average.

Between twelve and thirteen tons of pig-iron per man had been carried from a pile on the ground, up an inclined plank, and loaded on to a gondola car by the average pig-iron handler while working by the day. The men in doing this work had worked in gangs of from five to twenty men.

The man selected from one of these gangs to make the first start under the writer's system was called upon to load on piece work from forty-five to forty-eight tons (2,240 lbs. each) per day.

He regarded this task as an entirely fair one, and earned on an average, from the start, $1.85 per day, which was 60 per cent. more than he had been paid by the day. This man happened to be considerably lighter than the average good workman at this class of work. He weighed about 130 pounds. He proved, however, to be especially well suited to this job, and was kept at it steadily throughout the time that the writer was in Bethlehem, and some years later was still at the same work.

Being the first piece work started in the works, it excited considerable opposition, both on the part of the workmen and of several of the leading men in the town, their opposition being based mainly on the old fallacy that if piece work proved successful a great many men would be thrown out of work, and that thereby not only the workmen but the whole town would suffer.

One after another of the new men who were started singly on this job were either persuaded or intimidated into giving it up. In many cases they were given other work by those interested in preventing piece work, at wages higher than the ruling wages. In the meantime, however, the first man who started on the work earned steadily $1.85 per day, and this object lesson gradually wore out the concerted opposition, which ceased rather suddenly after about two months. From this time on there was no difficulty in getting plenty of good men who were anxious to start on piece work, and the difficulty lay in making with sufficient rapidity the accurate time study of the elementary operations or "unit times" which forms the foundation of this kind of piece work.

Throughout the introduction of piece work, when after a thorough time study a new section of the work was started, one man only was put on each new job, and not more than one man was allowed to work at it until he had demonstrated that the task set was a fair one by earning an average of $1.85 per day. After a few sections of the work had been started in this way, the complaint on the part of the better workmen was that they were not allowed to go on to piece work fast enough.

It required about two years to transfer practically all of the yard labor from day to piece work. And the larger part of the transfer was made during the last six months of this time.

As stated above, the greater part of the time was taken up in studying "unit times," and this time study was greatly delayed by having successively the two leading men who had been trained to the work leave because they were offered much larger salaries elsewhere. The study of "unit times" for the yard labor took practically the time of two trained men for two years. Throughout this time the day and piece workers were under entirely separate and distinct management. The original foremen continued to manage the day work, and day and piece workers were never allowed to work together. Gradually the day work gang was diminished and the piece workers were increased as one section of work after another was transformed from the former to the latter.

Two elements which were important to the success of this work should be noted:

First, on the morning following each day's work, each workman was given a slip of paper informing him in detail just how much work he had done the day before, and the amount he had earned. This enabled him to measure his performance against his earnings while the details were fresh in his mind. Without this there would have been great dissatisfaction among those who failed to climb up to the task asked of them, and many would have gradually fallen off in their performance.

Second, whenever it was practicable, each man's work was measured by itself. Only when absolutely necessary was the work of two men measured up together and the price divided between them, and then care was taken to select two men of as nearly as possible the same capacity. Only on few occasions, and then upon special permission, signed by the writer, were more than two men allowed to work on gang work, dividing their earnings between them. Gang work almost invariably results in a falling off in earnings and consequent dissatisfaction.

An interesting illustration of the desirability of individual piece work instead of gang work came to our attention at Bethlehem. Several of the best piece workers among the Bethlehem yard laborers were informed by their friends that a much higher price per ton was paid for shoveling ore in another works than the rate given at Bethlehem. After talking the matter over with the writer he advised them to go to the other works, which they accordingly did. In about a month they were all back at work in Bethlehem again, having found that at the other works they were obliged to work with a gang of men instead of on individual piece work, and that the rest of the gang worked so slowly that in spite of the high price paid per ton they earned much less than at Bethlehem.

Table 1, on page 54, gives a summary of the work done by the piece-work laborers in handling raw materials, such as ores, anthracite and bituminous coal, coke, pig-iron, sand, limestone, cinder, scale, ashes, etc., in the works of the Bethlehem Steel Company, during the year ending April 30, 1900. This work consisted mainly in loading and unloading cars on arrival or departure from the works, and for local transportation, and was done entirely by hand, *i.e.*, without the use of cranes or other machinery.

The greater part of the credit for making the accurate time study and actually managing the men on this work should be given to Mr. A. B. Wadleigh, the writer's assistant in this section at that time.

	Piece Work	Day Work
Number of tons (2,240 lbs. per ton) handled on piece work during the year ending April 30, 1901	924,040$\frac{13}{100}$	
Total cost of handling 924,040$\frac{13}{100}$ tons including the piece work wages paid the men, and in addition all incidental day labor used.........................	$30,797.78	
Former cost of handling the same number of tons of similar materials on day work....	....	$67,215.47
Net saving in handling 924,040$\frac{13}{100}$ tons of materials, effected in one year through substituting piece work for day work	$36,417.69	
Average cost for handling a ton (2,240 lbs.) on piece and day work.................	$0.033	$0.072
Average earnings per day, per man	[1] $1.88	$1.15
Average number of tons handled per day per man.............................	[2] 57	16

TABLE 1. — SHOWING RELATIVE COST OF YARD LABOR UNDER TASK PIECE WORK AND OLD STYLE DAY WORK.

When the writer left the steel works, the Bethlehem piece workers were the finest body of picked laborers that he has ever seen together. They were practically all first-class men, because in each case the task which they were called upon to perform was such that only a first-class man could do it. The tasks were all purposely made so severe that not more

[1] It was our intention to fix piece work rates which should enable first-class workmen to average about 60 per cent. more than they had been earning on day work, namely $1.85 per day. A year's average shows them to have earned $1.88 per day, or three cents per man per day more than we expected — an error of 1$\frac{6}{10}$ per cent.

[2] The piece workers handled on an average 3$\frac{56}{100}$ times as many tons per day as the day workers.

than one out of five laborers (perhaps even a smaller percentage than this) could keep up.

It was clearly understood by each newcomer as he went to work that unless he was able to average at least $1.85 per day he would have to make way for another man who could do so. As a result, first-class men from all over that part of the country, who were in most cases earning from $1.05 to $1.15 per day, were anxious to try their hands at earning $1.85 per day. If they succeeded they were naturally contented, and if they failed they left, sorry that they were unable to maintain the proper pace, but with no hard feelings either toward the system or the management. Throughout the time that the writer was there, labor was as scarce and as difficult to get as it ever has been in the history of this country, and yet there was always a surplus of first-class men ready to leave other jobs and try their hand at Bethlehem piece work.

Perhaps the most notable difference between these men and ordinary piece workers lay in their changed mental attitude toward their employers and their work, and in the total absence of soldiering on their part. The ordinary piece worker would have spent a considerable part of his time in deciding just how much his employer would allow him to earn without cutting prices and in then trying to come as close as possible to this figure, while carefully guarding each job so as to keep the management from finding out how fast it really could be done. These men, however, were faced with a new but very simple and straightforward proposition, namely, am I a first-class laborer or not? Each man felt that if he belonged in the first class all he had to do was to work at his best and he would be paid sixty per cent. more than he had been paid in the past. Each piece work price was accepted by the men without question. They never bargained over nor complained about rates, and there was no occasion to do so, since they were all equally fair, and called for almost exactly the same amount of work and fatigue per dollar of wages.

A careful inquiry into the condition of these men when away from work developed the fact that out of the whole gang only two were said to be drinking men. This does not, of course, imply that many of them did not take an occasional drink. The fact is that a steady drinker would find it almost impossible to keep up with the pace which was set, so that they were practically all sober. Many if not most of them were saving money, and they all lived better than they had before. The results attained under this system were most satisfactory both to employer and workmen, and show in a convincing way the possibility of uniting high wages with a low labor cost.

This is virtually a labor union of first-class men, who are united together to secure the extra high wages, which belong to them by right and which in this case are begrudged them by none, and which will be theirs through dull times as well as periods of activity. Such a union commands the unqualified admiration and respect of all classes of the community; the respect equally of workmen, employers, political economists, and philanthropists. There are no dues for membership, since all of the expenses are paid by the company. The employers act as officers of the Union, to enforce its rules and keep its records, since the interests of the company are identical and bound up with those of the men. It is never necessary to plead with, or persuade men to join this Union, since the employers themselves organize it free of cost; the best workmen in the community are always anxious to belong to it. The feature most to be regretted about it is that the membership is limited.

The words "labor union" are, however, unfortunately so closely associated in the minds of most people with the idea of disagreement and strife between employers and men that it seems almost incongruous to apply them to this case. Is not this, however, the ideal "labor union," with character and special ability of a high order as the only qualifications for membership.

It is a curious fact that with the people to whom the writer has described this system, the first feeling, particularly among those more philanthropically inclined, is one of pity for the inferior workmen who lost their jobs in order to make way for the first-class men. This sympathy is entirely misplaced. There was such a demand for labor at the time that no workman was obliged to be out of work for more than a day or two, and so the poor workmen were practically as well off as ever. The feeling, instead of being one of pity for the inferior workmen, should be one of congratulation and rejoicing that many first-class men — who through unfortunate circumstances had never had the opportunity of proving their worth — at last were given the chance to earn high wages and become prosperous.

What the writer wishes particularly to emphasize is that this whole system rests upon an accurate and scientific study of unit times, which is by far the most important element in scientific management. With it, greater and more permanent results can be attained even under ordinary day work or piece work than can be reached under any of the more elaborate systems without it.

In 1895 the writer read a paper before The American Society of Mechanical Engineers entitled "A Piece Rate System." His chief object in writing it was to advocate the study of unit times as the foundation of good management. Unfortunately, he at the same time described the "differential rate" system of piece work, which had been introduced by him in the Midvale Steel Works. Although he called attention to the fact that the latter was entirely of secondary importance, the differential rate was widely discussed in the journals of this country and abroad while practically nothing was said about the study of "unit times." Thirteen members of the Society discussed the piece rate system at length, and only two briefly referred to the study of the "unit times."

The writer most sincerely trusts that his leading object in writing this book will not be overlooked, and that *scientific time study* will receive the attention which it merits. Bearing in mind the Bethlehem yard labor as an illustration of the application of the study of unit times as the foundation of success in management, the following would seem to him a fair comparison of the older methods with the more modern plan.

For each job there is the quickest time in which it can be done by a first-class man. This time may be called the "quickest time," or the "standard time" for the job. Under all the ordinary systems, this "quickest time" is more or less completely shrouded in mist. In most cases, however, the workman is nearer to it and sees it more clearly than the employer.

Under ordinary piece work the management watch every indication given them by the workmen as to what the "quickest time" is for each job, and endeavor continually to force the men toward this "standard time," while the workmen constantly use every effort to prevent this from being done and to lead the management in the wrong direction. In spite of this conflict, however, the "standard time" is gradually approached.

Under the Towne-Halsey plan the management gives up all direct effort to reach this "quickest time," but offers mild inducements to the workmen to do so, and turns over the whole enterprise to them. The workmen, peacefully as far as the management is concerned, but with considerable pulling and hauling among themselves, and without the assistance of a trained guiding hand, drift gradually and slowly in the direction of the "standard time," but rarely approach it closely.

With accurate time study as a basis, the "quickest time" for each job is at all times in plain sight of both employers and workmen, and is reached with accuracy, precision, and speed, both sides pulling hard in the same direction under the uniform simple and just agreement that whenever a first-class man works his best he will receive from 30 to 100 per cent. more than the average of his trade.

Probably a majority of the attempts that are made to radically change the organization of manufacturing companies result in a loss of money to the company, failure to bring about the change sought for, and a return to practically the original organization. The reason for this being that there are but few employers who look upon management as an art, and that they go at a difficult task without either having understood or appreciated the time required for organization or its cost, the troubles to be met with, or the obstacles to be overcome, and without having studied the means to be employed in doing so.

Before starting to make any changes in the organization of a company the following matters should be carefully considered: *First*, the importance of choosing the general type of management best suited to the particular case. *Second*, that in all cases money must be spent, and in many cases a great deal of money, before the changes are completed which result in lowering cost. *Third*, that it takes time to reach any result worth aiming at. *Fourth*, the importance of making changes in their proper order, and that unless the right steps are taken, and taken in their proper sequence, there is great danger from deterioration in the quality of the output and from serious troubles with the workmen, often resulting in strikes.

As to the type of management to be ultimately aimed at, before any changes whatever are made, it is necessary, or at least highly desirable, that the most careful consideration should be given to the type to be chosen; and once a scheme is decided upon it should be carried forward step by step without wavering or retrograding. Workmen will tolerate and even come to have great respect for one change after another made in logical sequence and according to a consistent plan. It is most demoralizing, however, to have to recall a step once taken, whatever may be the cause, and it makes any further changes doubly difficult.

The choice must be made between some of the types of management in common use, which the writer feels are properly designated by the word "drifting," and the more modern and scientific management based on an accurate knowledge of how long it should take to do the work. If, as is frequently the case, the managers of an enterprise find themselves so overwhelmed with other departments of the business that they can give but little thought to the management of the shop, then some one of the various "drifting" schemes should be adopted; and of these the writer believes the Towne-Halsey plan to be the best, since it drifts safely and peacefully though slowly in the right direction; yet under it the best results can never be reached. The fact, however, that managers are in this way overwhelmed by their work is the best proof that there is something radically wrong with the plan of their organization and in self defense they should take immediate steps toward a more thorough study of the art.

It is not at all generally realized that whatever system may be used, — providing a business is complex in its nature — the building up of an efficient organization is necessarily slow and sometimes very expensive. Almost all of the directors of manufacturing companies appreciate the economy of a thoroughly modern, up-to-date, and efficient plant, and are willing to pay for it. Very few of them, however, realize that the best organization, whatever its cost may be, is in many cases even more important than the plant; nor do they clearly realize that no kind of an efficient organization can be built up without spending money. The spending of money for good machinery appeals to them because they can see machines after they are bought; but putting money into anything so invisible, intangible, and to the average man so indefinite, as an organization seems almost like throwing it away.

There is no question that when the work to be done is at all complicated, a good organization with a poor plant will give better results than the best plant with a poor organization. One of the most successful manufacturers in this country was asked recently by a number of financiers whether he thought that the difference between one style of organization and another amounted to much providing the company had an up-to-date plant properly located. His answer was, "If I had to choose now between abandoning my present organization and burning down all of my plants which have cost me millions, I should choose the latter. My plants could be rebuilt in a short while with borrowed money, but I could hardly replace my organization in a generation."

Modern engineering can almost be called an exact science; each year removes it further from guess work and from rule-of-thumb methods and establishes it more firmly upon the foundation of fixed principles.

The writer feels that management is also destined to become more of an art, and that many of the elements which are now believed to be outside the field of exact knowledge will soon be standardized, tabulated, accepted, and used, as are now many of the elements of engineering. Management will be studied as an art and will rest upon well recognized, clearly defined, and fixed principles instead of depending upon more or less hazy ideas received from a limited observation of the few organizations with which the individual may have come in contact. There will, of course, be various successful types, and the application of the underlying principles must be modified to suit each particular case. The writer has already indicated that he thinks the first object in management is to unite high wages with a low labor cost. He believes that this object can be most easily attained by the application of the following principles:

(a) A LARGE DAILY TASK. — Each man in the establishment, high or low, should daily have a clearly defined task laid out before him. This task should not in the least degree be vague nor indefinite, but should be circumscribed carefully and completely, and should not be easy to accomplish.

(b) STANDARD CONDITIONS. — Each man's task should call for a full day's work, and at the same time the workman should be given such standardized conditions and appliances as will enable him to accomplish his task with certainty.

(c) HIGH PAY FOR SUCCESS. — He should be sure of large pay when he accomplishes his task.

(d) LOSS IN CASE OF FAILURE. — When he fails he should be sure that sooner or later he will be the loser by it.

When an establishment has reached an advanced state of organization, in many cases a fifth element should be added, namely: the task should be made so difficult that it can only be accomplished by a first-class man.

There is nothing new nor startling about any of these principles and yet it will be difficult to find a shop in which they are not daily violated over and over again. They call, however, for a greater departure from the ordinary types of organization than would at first appear. In the case, for instance, of a machine shop doing miscellaneous work, in order to assign daily to each man a carefully measured task, a special planning department is required to lay out all of the work at least one day ahead. All orders must be given to the men in detail in writing; and in order to lay out the next day's work and plan the entire progress of work through the shop, daily returns must be made by the men to the planning department in writing, showing just what has been done. Before each casting or forging arrives in the shop the exact route which it is to take from machine to machine should be laid out. An instruction card for each operation must be written out stating in detail just how each operation on every piece of work is to be done and the time required to do it, the drawing number, any special tools, jigs, or appliances required, etc. Before the four principles above referred to can be successfully applied it is also necessary in most shops to make important physical changes. All of the small details in the shop, which are usually regarded as of little importance and are left to be regulated according to the individual taste of the workman, or, at best, of the foreman, must be thoroughly and carefully standardized; such details, for instance, as the care and tightening of the belts; the exact shape and quality of each cutting tool; the establishment of a complete tool room from which properly ground tools, as well as jigs, templets, drawings, etc., are issued under a good check system, etc.; and as a matter of importance (in fact, as the foundation of scientific management) an accurate study of unit times must be made by one or more men connected with the planning department, and each machine tool must be standardized and a table or slide rule constructed for it showing how to run it to the best advantage.

At first view the running of a planning department, together with the other innovations, would appear to involve a large amount of additional work and expense, and the most natural question would be is whether the increased efficiency of the shop more than offsets this outlay? It must be borne in mind, however, that, with the exception of the study of unit times, there is hardly a single item of work done in the planning department which is not already being done in the shop. Establishing a planning department merely concentrates the planning and much other brainwork in a few men especially fitted for their task and trained in their especial lines, instead of having it done, as heretofore, in most cases by high priced mechanics, well fitted to work at their trades, but poorly trained for work more or less clerical in its nature.

There is a close analogy between the methods of modern engineering and this type of management. Engineering now centers in the drafting room as modern management does in the planning department. The new style engineering has all the appearance of complication and extravagance, with its multitude of drawings; the amount of study and work which is put into each detail; and its corps of draftsmen, all of whom would be sneered at by the old engineer as "non-producers." For the same reason, modern management, with its minute time study and a managing department in which each operation is carefully planned, with its many written orders and its apparent red tape, looks like a waste of money; while the ordinary management in which the planning is mainly done by the workmen themselves, with the help of one or two foremen, seems simple and economical in the extreme.

The writer, however, while still a young man, had all lingering doubt as to the value of a drafting room dispelled by seeing the chief engineer, the foreman of the machine shop, the foreman of the foundry, and one or two workmen, in one of our large and successful engineering establishments of the old school, stand over the cylinder of an engine which was being built, with chalk and dividers, and discuss for more than an hour the proper size and location of the studs for fastening on the cylinder head. This was simplicity, but not economy. About the same time he became thoroughly convinced of the necessity and economy of a planning department with time study, and with written instruction cards and returns. He saw over and over again a workman shut down his machine and hunt up the foreman to inquire, perhaps, what work to put into his machine next, and then chase around the shop to find it or to have a special tool or templet looked up or made. He saw workmen carefully nursing their jobs by the hour and doing next to nothing to avoid making a record, and he was even more forcibly convinced of the necessity for a change while he was still working as a machinist by being ordered by the other men to slow down to half speed under penalty of being thrown over the fence.

No one now doubts the economy of the drafting room, and the writer predicts that in a very few years from now no one will doubt the economy and necessity of the study of unit times and of the planning department.

Another point of analogy between modern engineering and modern management lies in the fact that modern engineering proceeds with comparative certainty to the design and construction of a machine or structure of the maximum efficiency with the minimum weight and cost of materials, while the old style engineering at best only approximated these results and then only after a series of breakdowns, involving the practical reconstruction of the machine and the lapse of a long period of time. The ordinary system of management, owing to the lack of exact information and precise methods, can only approximate to the desired standard of high wages accompanied by low labor cost and then only slowly, with marked irregularity in results, with continued opposition, and, in many cases, with danger from strikes. Modern management, on the other hand, proceeds slowly at first, but with directness and precision, step by step, and, after the first few object lessons, almost without opposition on the part of the men, to high wages and low labor cost; and as is of great importance, it assigns wages to the men which are uniformly fair. They are not demoralized, and their sense of justice offended by receiving wages which are sometimes too low and at other times entirely too high.

One of the marked advantages of scientific management lies in its freedom from strikes. The writer has never been opposed by a strike, although he has been engaged for a great part of his time since 1883 in introducing this type of management in different parts of the country and in a great variety of industries. The only case of which the writer can think in which a strike under this system might be unavoidable would be that in which most of the employés were members of a labor union, and of a union whose rules were so inflexible and whose members were so stubborn that they were unwilling to try any other system, even though it assured them larger wages than their own. The writer has seen, however, several times after the introduction of this system, the members of labor unions who were working under it leave the union in large numbers because they found that they could do better under the operation of the system than under the laws of the union.

There is no question that the average individual accomplishes the most when he either gives himself, or some one else assigns him, a definite task, namely, a given amount of work which he must do within a given time; and the more elementary the mind and character of the individual the more necessary does it become that each task shall extend over a short period of time only. No school teacher would think of telling children in a general way to study a certain book or subject. It is practically universal to assign each day a definite lesson beginning on one specified page and line and ending on another; and the best progress is made when the conditions are such that a definite study hour or period can be assigned in which the lesson must be learned. Most of us remain, through a great part of our lives, in this respect, grown-up children, and do our best only under pressure of a task of comparatively short duration.

Another and perhaps equally great advantage

of assigning a daily task as against ordinary piece work lies in the fact that the success of a good workman or the failure of a poor one is thereby daily and prominently called to the attention of the management. Many a poor workman might be willing to go along in a slipshod way under ordinary piece work, careless as to whether he fell off a little in his output or not. Very few of them, however, would be willing to record a daily failure to accomplish their task even if they were allowed to do so by their foreman; and also since on ordinary piece work the price alone is specified without limiting the time which the job is to take, a quite large falling off in output can in many cases occur without coming to the attention of the management at all. It is for these reasons that the writer has above indicated "a large daily task" for each man as the first of four principles which should be included in the best type of management.

It is evident, however, that it is useless to assign a task unless at the same time adequate measures are taken to enforce its accomplishment. As Artemus Ward says, "I can call the spirits from the windy deep, but damn 'em they won't come!" It is to compel the completion of the daily task then that two of the other principles are required, namely, "high pay for success" and "loss in case of failure." The advantage of Mr. H. L. Gantt's system of "task work with a bonus," and the writer's "differential rate piece work" over the other systems lies in the fact that with each of these the men automatically and daily receive either an extra reward in case of complete success, or a distinct loss in case they fall off even a little.

The four principles above referred to can be successfully applied either under day work, piece work, task work with a bonus, or differential rate piece work, and each of these systems has its own especial conditions under which it is to be preferred to either of the other three. In no case, however, should an attempt be made to apply these principles unless accurate and thorough time study has previously been made of every item entering into the day's task.

They should be applied under day work only when a number of miscellaneous jobs have to be done day after day, none of which can occupy the entire time of a man throughout the whole of a day and when the time required to do each of these small jobs is likely to vary somewhat each day. In this case a number of these jobs can be grouped into a daily task which should be assigned, if practicable, to one man, possibly even to two or three, but rarely to a gang of men of any size. To illustrate: In a small boiler house in which there is no storage room for coal, the work of wheeling the coal to the fireman, wheeling out the ashes, helping clean fires and keeping the boiler room and the outside of the boilers clean can be made into the daily task for a man, and if these items do not sum up into a full day's work, on the average, other duties can be added until a proper task is assured. Or, the various details of sweeping, cleaning, and keeping a certain section of a shop floor windows, machines, etc., in order can be united to form a task. Or, in a small factory which turns out a uniform product and in uniform quantities day after day, supplying raw materials to certain parts of the factory and removing finished product from others may be coupled with other definite duties to form a task. The task should call for a large day's work, and the man should be paid more than the usual day's pay so that the position will be sought for by first-class, ambitious men. Clerical work can very properly be done by the task in this way, although when there is enough of it, piece work at so much per entry is to be preferred.

In all cases a clear cut, definite inspection of the task is desirable at least once a day and sometimes twice. When a shop is not running at night, a good time for this inspection is at seven o'clock in the morning, for instance. The inspector should daily sign a printed card, stating that he has inspected the work done by ——, and enumerating the various items of the task. The card should state that the workman has satisfactorily performed his task, "except the following items," which should be enumerated in detail.

When men are working on task work by the day they should be made to start to work at the regular starting hour. They should, however, have no regular time for leaving. As soon as the task is finished they should be allowed to go home; and, on the other hand, they should be made to stay at work until their task is done, even if it lasts into the night, no deduction being made for shorter hours nor extra pay allowed for overtime. It is both inhuman and unwise to ask a man, working on task work, to stay in the shop after his task is finished "to maintain the discipline of the shop," as is frequently done. It only tends to make men eye servants.

An amusing instance of the value of task work with freedom to leave when the task is done was given the writer by his friend, Mr. Chas. D. Rogers, for many years superintendent of the American Screw Works, of Providence, R. I., one of the greatest mechanical geniuses and most resourceful managers that this country has produced, but a man who, owing to his great modesty, has never been fully appreciated outside of those who know him well. Mr. Rogers tried several modifications of day and piece work in an unsuccessful endeavor to get the children who were engaged in sorting over the very small screws to do a fair day's work. He finally met with great success by assigning to each child a fair day's task and allowing him to go home and play as soon as his task was done. Each child's playtime was his own and highly prized while the greater part of his wages went to his parents.

Piece work embodying the task idea can be used to advantage when there is enough work of the same general character to keep a number of men busy regularly; such work, for instance, as the Bethlehem yard labor previously described, or the work of bicycle ball inspection referred to later on. In piece work of this class the task idea should always be maintained by keeping it clearly before each man that his average daily earnings must amount to a given high sum (as in the case of the Bethlehem laborers, $1.85 per day), and that failure to average this amount will surely result in his being laid off. It must be remembered that on plain piece work the less competent workmen will always bring what influence and pressure they can to cause the best men to slow down towards their level and that the task idea is needed to counteract this influence. Where the labor market is large enough to secure in a reasonable time enough strictly first-class men, the piece work rates should be fixed on such a basis that only a first-class man working at his best can earn the average amount called for. This figure should be, in the case of first-class men as stated above, from 30 per cent. to 100 per cent. beyond the wages usually paid. The task idea is emphasized with this style of piece work by two things — the high wages and the laying off, after a reasonable trial, of incompetent men; and for the success of the system, the number of men employed on practically the same class of work should be large enough for the workmen quite often to have the object lesson of seeing men laid off for failing to earn high wages and others substituted in their places.

There are comparatively few machine shops, or even manufacturing establishments, in which the work is so uniform in its nature as to employ enough men on the same grade of work and in sufficiently close contact to one another to render piece work preferable to the other systems. In the great majority of cases the work is so miscellaneous in its nature as to call for the employment of workmen varying greatly in their natural ability and attainments, all the way, for instance, from the ordinary laborer, through the trained laborer, helper, rough machinist, fitter, machine hand, to the highly skilled special or all-round mechanic. And while in a large establishment there may be often enough men of the same grade to warrant the adoption of piece work with the task idea, yet, even in this case, they are generally so scattered in different parts of the shop that laying off one of their number for incompetence does not reach the others with sufficient force to impress them with the necessity of keeping up with their task.

It is evident then that, in the great majority of cases, the four leading principles in management can be best applied through either task work with a bonus or the differential piece rate in spite of the slight additional clerical work and the increased difficulty in planning ahead incident to these systems of paying wages. Three of the principles of management given above, namely, (a) a large daily task, (b) high pay for success, and (c) loss in case of failure form the very essence of both of these systems and act as a daily stimulant for the men. The fourth principle of management is a necessary preliminary, since without having first thoroughly standardized all of the conditions surrounding work, neither of these two plans can be successfully applied.

In many cases the greatest good resulting from the application of these systems of paying wages is the indirect gain which comes from the enforced standardization of all details and conditions, large and small, surrounding the work. All of the ordinary systems can be and are almost always applied without adopting and maintaining thorough shop standards. But the task idea can not be carried out without them.

The differential rate piece work is rather simpler in its application than task work with bonus and is the more forceful of the two. It should be used wherever it is practicable, but in no case until after all the accompanying conditions have been perfected and completely standardized and a thorough time study has been made of all of the elements of the work. This system is particularly useful where the same kind of work is repeated day after day, and also whenever the maximum possible output is desired, which is almost always the case in the operation of expensive machinery or of a plant occupying valuable ground or a large building. It is more forceful than task work with a bonus because it not only pulls the man up from the top but pushes him equally hard from the bottom. Both of these systems give the workman a large extra reward when he accomplishes his full task within the given time. With the differential rate, if for any reason he fails to do his full task, he not only loses the large extra premium which is paid for complete success, but in addition he suffers the direct loss of the piece price for each piece by which he falls short. Failure under the task with a bonus system involves a corresponding loss of the extra premium or bonus, but the workman, since he is paid a given price per hour, receives his ordinary day's pay in case of failure and suffers no additional loss beyond that of the extra premium whether he may have fallen short of the task to the extent of one piece or a dozen.

In principle, these two systems appear to be almost identical, yet this small difference, the slightly milder nature of task work with a bonus, is sufficient to render it much more flexible and therefore applicable to a large number of cases in which the differential rate system cannot be used. Task work with a bonus was invented by Mr. H. L. Gantt, while he was assisting the writer in organizing the Bethlehem Steel Company. The possibilities of his system were immediately recognized by all of the leading men engaged on the work, and long before it would have been practicable to use the differential rate, work was started under this plan. It was successful from the start, and steadily grew in volume and in favor, and to-day is more extensively used than ever before.

Mr. Gantt's system is especially useful during the difficult and delicate period of transition from the slow pace of ordinary day work to the high speed which is the leading characteristic of good management. During this period of transition in the past, a time was always reached when a sudden long leap was taken from improved day work to some form of piece work; and in making this jump many good men inevitably fell and were lost from the procession. Mr. Gantt's system bridges over this difficult stretch and enables the workman to go smoothly and with gradually accelerated speed from the slower pace of improved day work to the high speed of the new system.

It does not appear that Mr. Gantt has recognized the full advantages to be derived through the proper application of his system during this period of transition, at any rate he has failed to point them out in his papers and to call the attention to the best method of applying his plan in such cases.

No workman can be expected to do a piece of work the first time as fast as he will later. It should also be recognized that it takes a certain time for men who have worked at the ordinary slow rate of speed to change to high speed. Mr. Gantt's plan can be adapted to meet both of these conditions by allowing the workman to take a longer time to do the job at first and yet earn his bonus; and later compelling him to finish the job in the quickest time in order to get the premium. In all cases it is of the utmost importance that each instruction card should state the *quickest time* in which the workman will ultimately be called upon to do the work. There will then be no temptation for the man to soldier since he will see that the management know accurately how fast the work can be done.

There is also a large class of work in addition to that of the period of transition to which task work with a bonus is especially adapted. The higher pressure of the differential rate is the stimulant required by the workman to maintain a high rate of speed and secure high wages while he has the steady swing that belongs to work which is repeated over and over again. When, however, the work is of such variety that each day presents an entirely new task, the pressure of the differential rate is sometimes too severe. The chances of failing to quite reach the task are greater in this class of work than in routine work; and in many such cases it is better, owing to the increased difficulties, that the workman should feel sure at least of his regular day's rate, which is secured him by Mr. Gantt's system in case he falls short of the full task. There is still another case of quite frequent occurrence in which the flexibility of Mr. Gantt's plan makes it the most desirable. In many establishments, particularly those doing an engineering business of considerable variety or engaged in constructing and erecting miscellaneous machinery, it is necessary to employ continuously a number of especially skilful and high-priced mechanics. The particular work for which these men are wanted comes, however, in many cases, at irregular intervals, and there are frequently quite long waits between their especial jobs. During such periods these men must be provided with work which is ordinarily done by less efficient, lower-priced men, and if a proper piece price has been fixed on this work it would naturally be a price suited to the less skilful men, and therefore too low for the men in question. The alternative is presented of trying to compel these especially skilled men to work for a lower price than they should receive, or of fixing a special higher piece price for the work. Fixing two prices for the same piece of work, one for the man who usually does it and a higher price for the higher grade man, always causes the greatest feeling of injustice and dissatisfaction in the man who is discriminated against. With Mr. Gantt's plan, the less skilled workman would recognize the justice of paying his more experienced companion regularly a higher rate of wages by the day, yet when they were both working on the same kind of work each man would receive the same extra bonus for doing the full day's task. Thus, with Mr. Gantt's system, the total day's pay of the higher classed man would be greater than that of the less skilled man, even when on the same work, and the latter would not begrudge it to him. We may say that the difference is one of sentiment, yet sentiment plays an important part in all of our lives; and sentiment is particularly strong in the workman when he believes a direct injustice is being done him.

Mr. James M. Dodge, the distinguished Past President of The American Society of Mechanical Engineers, has invented an ingenious system of piece work which is adapted to meet this very case, and which has especial advantages not possessed by any of the other plans.

It is clear, then, that in carrying out the task idea after the required knowledge has been obtained through a study of unit times, each of the four systems, (*a*) day work, (*b*) straight piece work, (*c*) task work with a bonus, and (*d*) differential piece work, has its especial field of usefulness, and that in every large establishment doing a variety of work all four of these plans can and should be used at the same time. Three of these systems were in use at the Bethlehem Steel Company when the writer left there, and the fourth would have soon been started if he had remained.

Before leaving this part of the book which has been devoted to pointing out the value of the daily task in management, it would seem desirable to give an illustration of the value of the differential rate piece work and also of the desirability of making each task as simple and short as practicable.

The writer quotes as follows from a paper entitled "A Piece Rate System," read by him before The American Society of Mechanical Engineers in 1895:

"The first case in which a differential rate was applied during the year 1884, furnishes a good illustration of what can be accomplished by it. A standard steel forging, many thousands of which are used each year, had for several years been turned at the rate of from four to five per day under the ordinary system of piece work, 50 cents per piece being the price paid for the work. After analyzing the job, and determining the shortest time required to do each of the elementary operations of which it was composed, and then summing up the total, the writer became convinced that it was possible to turn ten pieces a day. To finish the forgings at this rate, however, the machinists were obliged to work at their maximum pace from morning to night, and the lathes were run as fast as the tools would allow, and under a heavy feed. Ordinary tempered tools 1 inch by $1\frac{1}{2}$ inch, made of carbon tool steel, were used for this work.

"It will be appreciated that this was a big day's work, both for men and machines, when it is understood that it involved removing, with a single 16-inch lathe, having two saddles, an average of more than

800 lbs. of steel chips in ten hours. In place of the 50 cent rate, that they had been paid before, the men were given 35 cents per piece when they turned them at the speed of 10 per day; and when they produced less than ten they received only 25 cents per piece.

"It took considerable trouble to induce the men to turn at this high speed, since they did not at first fully appreciate that it was the intention of the firm to allow them to earn permanently at the rate of $3.50 per day. But from the day they first turned ten pieces to the present time, a period of more than ten years, the men who understood their work have scarcely failed a single day to turn at this rate. Throughout that time until the beginning of the recent fall in the scale of wages throughout the country, the rate was not cut.

"During this whole period, the competitors of the company never succeeded in averaging over half of this production per lathe, although they knew and even saw what was being done at Midvale. They, however, did not allow their men to earn from over $2.00 to $2.50 per day, and so never even approached the maximum output.

"The following table will show the economy of paying high wages under the differential rate in doing the above job:

"COST OF PRODUCTION PER LATHE PER DAY

Ordinary System of Piece Work	Differential Rate System
Man's wages $2.50	Man's wages $3.50
Machine cost 3.37	Machine cost 3.37
Total cost per day..... 5.87	Total cost per day..... 6.87
5 pieces produced;	10 pieces produced;
Cost per piece $1.17	Cost per piece $0.69

"The above result was mostly though not entirely due to the differential rate. The superior system of managing all of the small details of the shop counted for considerable."

The exceedingly dull times that began in July, 1893, and were accompanied by a great fall in prices, rendered it necessary to lower the wages of machinists throughout the country. The wages of the men in the Midvale Steel Works were reduced at this time, and the change was accepted by them as fair and just.

Throughout the works, however, the principle of the differential rate was maintained, and was, and is still, fully appreciated by both the management and men. Through some error at the time of the general reduction of wages in 1893, the differential rate on the particular job above referred to was removed, and a straight piece work rate of 25 cents per piece was substituted for it. The result of abandoning the differential proved to be the best possible demonstration of its value. Under straight piece work, the output immediately fell to between six and eight pieces per day, and remained at this figure for several years, although under the differential rate it had held throughout a long term of years steadily at ten per day.

When work is to be repeated many times, the time study should be minute and exact. Each job should be carefully subdivided into its elementary operations, and each of these unit times should receive the most thorough time study. In fixing the times for the tasks, and the piece work rates on jobs of this class, the job should be subdivided into a number of divisions, and a separate time and price assigned to each division rather than to assign a single time and price for the whole job. This should be done for several reasons, the most important of which is that the average workman, in order to maintain a rapid pace, should be given the opportunity of measuring his performance against the task set him at frequent intervals. Many men are incapable of looking very far ahead, but if they see a definite opportunity of earning so many cents by working hard for so many minutes, they will avail themselves of it.

As an illustration, the steel tires used on car wheels and locomotives were originally turned in the Midvale Steel Works on piece work, a single piece-work rate being paid for all of the work which could be done on a tire at a single setting. A fixed price was paid for this work, whether there was much or little metal to be removed, and on the average this price was fair to the men. The apparent advantage of fixing a fair average rate was, that it made rate-fixing exceedingly simple, and saved clerk work in the time, cost and record keeping.

A careful time study, however, convinced the writer that for the reasons given above most of the men failed to do their best. In place of the single rate and time for all of the work done at a setting, the writer subdivided tire-turning into a number of short operations, and fixed a proper time and price, varying for each small job, according to the amount of metal to be removed, and the hardness and diameter of the tire. The effect of this subdivision was to increase the output, with the same men, methods, and machines, at least thirty-three per cent.

As an illustration of the minuteness of this subdivision, an instruction card similar to the one used is reproduced in Figure 1 on the next page. (This card was about 7 inches long by 4 inches wide.)

The cost of the additional clerk work involved in this change was so insignificant that it practically did not affect the problem. This principle of short tasks in tire turning was introduced by the writer in the Midvale Steel Works in 1883 and is still in full use there, having survived the test of over twenty years' trial with a change of management.

In another establishment a differential rate was applied to tire turning, with operations subdivided in this way, by adding fifteen per cent. to the pay of each tire turner whenever his daily or weekly piece work earnings passed a given figure.

Another illustration of the application of this principle of measuring a man's performance against a given task at frequent intervals to an entirely different line of work may be of interest. For this purpose the writer chooses the manufacture of bicycle balls in the works of the Symonds Rolling Machine Company, in Fitchburg, Mass. All of the work done in this factory was subjected to an accurate time study, and then was changed from day to piece work, through the assistance of functional foremanship, etc. The particular operation to be described, however, is that of inspecting bicycle balls before

Machine shop ....
Order for ....Tires...........
Do work on Tire No ....
As follows and per blue print ....

	Templet	Size to be cut to	Depth of cut	Driving belt	Feed	Rate	Time this operation should take
Surface to be machined	.....	.....	.....	.....	.....	.....	....
Set tire on machine ready to turn....	.....	.....	.....	.....	.....	.....	....
Rough face front edge	.....	.....	.....	.....	.....	.....	....
Finish face front edge	.....	.....	.....	.....	.....	.....	....
Rough bore front...	.....	.....	.....	.....	.....	.....	....
Finish bore front...	.....	.....	.....	.....	.....	.....	....
Rough face front I. S.C..............	.....	.....	.....	.....	.....	.....	....
Cut out filled	.....	.....	.....	.....	.....	.....	....
Rough bore front I. S.F..............	.....	.....	.....	.....	.....	.....	....
Rough face back edge	.....	.....	.....	.....	.....	.....	....
Finish face back edge	.....	.....	.....	.....	.....	.....	....
Finish bore back ...	.....	.....	.....	.....	.....	.....	....
Rough bore back ...	.....	.....	.....	.....	.....	.....	....
Rough face back I. S.F..............	.....	.....	.....	.....	.....	.....	....
Cut out filled	.....	.....	.....	.....	.....	.....	....
Cut recess	.....	.....	.....	.....	.....	.....	....
Rough turn thread..	.....	.....	.....	.....	.....	.....	....
Finish turn thread..	.....	.....	.....	.....	.....	.....	....
Rough turn flange ..	.....	.....	.....	.....	.....	.....	....
Finish turn edge....	.....	.....	.....	.....	.....	.....	....
Clean fillet of flange.	.....	.....	.....	.....	.....	.....	....
Remove tire from machine and clean face plate	.....	.....	.....	.....	.....	.....	....

FIGURE 1. — TIRE-TURNING INSTRUCTION CARD

they were finally boxed for shipment. Many millions of these balls were inspected annually. When the writer undertook to systematize this work, the factory had been running for eight or ten years on ordinary day work, so that the various employés were "old hands," and skilled at their jobs. The work of inspection was done entirely by girls — about one hundred and twenty being employed at it — all on day work.

This work consisted briefly in placing a row of small polished steel balls on the back of the left hand, in the crease between two of the fingers pressed together, and while they were rolled over and over, with the aid of a magnet held in the right hand, they were minutely examined in a strong light, and the defective balls picked out and thrown into especial boxes. Four kinds of defects were looked for — dented, soft, scratched, and fire cracked — and they were mostly so minute as to be invisible to an eye not especially trained to this work. It required the closest attention and concentration. The girls had worked on day work for years, ten and one-half hours per day, with a Saturday half-holiday.

The first move before in any way stimulating them toward a larger output was to insure against a falling off in quality. This was accomplished through over-inspection. Four of the most trustworthy girls were given each a lot of balls which had been examined the day before by one of the regular inspectors. The number identifying the lot having been changed by the foreman so that none of the over-inspectors knew whose work they were examining. In addition, one of the lots inspected by the four over-inspectors was examined on the following day by the chief inspector, selected on account of her accuracy and integrity.

An effective expedient was adopted for checking the honesty and accuracy of the over-inspection. Every two or three days a lot of balls was especially prepared by the foreman, who counted out a definite number of perfect balls, and added a recorded number of defective balls of each kind. The inspectors had no means of distinguishing this lot from the regular commercial lots. And in this way all temptation to slight their work or make false returns was removed.

After insuring in this way against deterioration in quality, effective means were at once adopted to increase the output. Improved day work was substituted for the old slipshod method. An accurate daily record, both as to quantity and quality, was kept for each inspector. In a comparatively short time this enabled the foreman to stir the ambition of all the inspectors by increasing the wages of those who turned out a large quantity and good quality, at the same time lowering the pay of those who fell short, and discharging others who proved to be incorrigibly slow or careless. An accurate time study was made through the use of a stop watch and record blanks, to determine how fast each kind of inspection should be done. This showed that the girls spent a considerable part of their time in partial idleness, talking and half working, or in actually doing nothing.

Talking while at work was stopped by seating them far apart. The hours of work were shortened from $10\frac{1}{2}$ per day, first to $9\frac{1}{2}$, and latter to $8\frac{1}{2}$; a Saturday half holiday being given them even with the shorter hours. Two recesses of ten minutes each were given them, in the middle of the morning and afternoon, during which they were expected to leave their seats, and were allowed to talk.

The shorter hours and improved conditions made it possible for the girls to really work steadily, instead of pretending to do so. Piece work was then introduced, a differential rate being paid, not for an increase in output, but for greater accuracy in the inspection; the lots inspected by the over-inspectors forming the basis for the payment of the differential. The work of each girl was measured every hour, and they were all informed whether they were keeping up with their tasks, or how far they had fallen short; and an assistant was sent by the foreman to encourage those who were falling behind, and help them to catch up.

The principle of measuring the performance of each workman against a standard at frequent intervals, of keeping them informed as to their progress, and of sending an assistant to help those who were falling down, was carried out throughout the works, and proved to be most useful.

The final results of the improved system in the inspecting department were as follows:

(*a*) Thirty-five girls did the work formerly done by one hundred and twenty.

(*b*) The girls averaged from $6.50 to $9.00 per week instead of $3.50 to $4.50, as formerly.

(*c*) They worked only $8\frac{1}{2}$ hours per day, with

Saturday a half-holiday, while they had formerly worked 10½ hours per day.

(d) An accurate comparison of the balls which were inspected under the old system of day work with those done under piece work, with over-inspection, showed that, in spite of the large increase in output per girl, there were 58 per cent. more defective balls left in the product as sold under day work than under piece work. In other words, the accuracy of inspection under piece work was one-third greater than that under day work.

That thirty-five girls were able to do the work which formerly required about one hundred and twenty is due, not only to the improvement in the work of each girl, owing to better methods, but to the weeding out of the lazy and unpromising candidates, and the substitution of more ambitious individuals.

A more interesting illustration of the effect of the improved conditions and treatment is shown in the following comparison. Records were kept of the work of ten girls, all "old hands," and good inspectors, and the improvement made by these skilled hands is undoubtedly entirely due to better management. All of these girls throughout the period of comparison were engaged on the same kind of work, viz.: inspecting bicycle balls, three-sixteenths of an inch in diameter.

The work of organization began in March, and although the records for the first three months were not entirely clear, the increased output due to better day work amounted undoubtedly to about 33 per cent. The increase per day from June on day work, to July on piece work, the hours each month being $10\tfrac{1}{2}$ per day, was 37 per cent. This increase was due to the introduction of piece work. The increase per day from July to August (the length of working days in July being $10\tfrac{1}{2}$ hours, and in August $9\tfrac{1}{2}$ hours, both months piece work) was 33 per cent.

The increase from August to September (the length of working day in August being $9\tfrac{1}{2}$ hours, and in September $8\tfrac{1}{2}$ hours) was 0.08 per cent. This means that the girls did practically the same amount of work per day in September, in $8\tfrac{1}{2}$ hours, that they did in August in $9\tfrac{1}{2}$ hours.

To summarize: the same ten girls did on an average each day in September, on piece work, when only working $8\tfrac{1}{2}$ hours per day, 2.42 times as much, or nearly two and one-half times as much, in a day (not per hour, the increase per hour was of course much greater) as they had done when working on day work in March with a working day of $10\tfrac{1}{2}$ hours. They earned $6.50 to $9.00 per week on piece work, while they had only earned $3.50 to $4.50 on day work. The accuracy of inspection under piece work was one-third greater than under day work.

The time study for this work was done by my friend, Sanford E. Thompson, C. E., who also had the actual management of the girls throughout the period of transition. At this time Mr. H. L. Gantt was general superintendent of the company, and the work of systematizing was under the general direction of the writer.

It is, of course, evident that the nature of the

organizations required to manage different types of business must vary to an enormous extent, from the simple tonnage works (with its uniform product, which is best managed by a single strong man who carries all of the details in his head and who, with a few comparatively cheap assistants, pushes the enterprise through to success) to the large machine works, doing a miscellaneous business, with its intricate organization, in which the work of any one man necessarily counts for but little.

It is this great difference in the type of the organization required that so frequently renders managers who have been eminently successful in one line utter failures when they undertake the direction of works of a different kind. This is particularly true of men successful in tonnage work who are placed in charge of shops involving much greater detail.

In selecting an organization for illustration, it would seem best to choose one of the most elaborate. The manner in which this can be simplified to suit a less intricate case will readily suggest itself to any one interested in the subject. One of the most difficult works to organize is that of a large engineering establishment building miscellaneous machinery, and the writer has therefore chosen this for description.

Practically all of the shops of this class are organized upon what may be called the military plan. The orders from the general are transmitted through the colonels, majors, captains, lieutenants and non-commissioned officers to the men. In the same way the orders in industrial establishments go from the manager through superintendents, foremen of shops, assistant foremen and gang bosses to the men. In an establishment of this kind the duties of the foremen, gang bosses, etc., are so varied, and call for an amount of special information coupled with such a variety of natural ability, that only men of unusual qualities to start with, and who have had years of special training, can perform them in a satisfactory manner. It is because of the difficulty — almost the impossibility — of getting suitable foremen and gang bosses, more than for any other reason, that we so seldom hear of a miscellaneous machine works starting in on a large scale and meeting with much, if any, success for the first few years. This difficulty is not fully realized by the managers of the old well established companies, since their superintendents and assistants have grown up with the business, and have been gradually worked into and fitted for their especial duties through years of training and the process of natural selection. Even in these establishments, however, this difficulty has impressed itself upon the managers so forcibly that most of them have of late years spent thousands of dollars in re-grouping their machine tools for the purpose of making their foremanship more effective. The planers have been placed in one group, slotters in another, lathes in another, etc., so as to demand a smaller range of experience and less diversity of knowledge from their respective foremen.

For an establishment, then, of this kind, starting up on a large scale, it may be said to be an impossibility to get suitable superintendents and foremen.

The writer found this difficulty at first to be an almost insurmountable obstacle to his work in organizing manufacturing establishments; and after years of experience, overcoming the opposition of the heads of departments and the foremen and gang bosses, and training them to their new duties, still remains the greatest problem in organization. The writer has had comparatively little trouble in inducing workmen to change their ways and to increase their speed, providing the proper object lessons are presented to them, and time enough is allowed for these to produce their effect. It is rarely the case, however, that superintendents and foremen can find any reasons for changing their methods, which, as far as they can see, have been successful. And having, as a rule, obtained their positions owing to their unusual force of character, and being accustomed daily to rule other men, their opposition is generally effective.

In the writer's experience, almost all shops are under-officered. Invariably the number of leading men employed is not sufficient to do the work economically. Under the military type of organization, the foreman is held responsible for the successful running of the entire shop, and when we measure his duties by the standard of the four leading principles of management above referred to, it becomes apparent that in his case these conditions are as far as possible from being fulfilled. His duties may be briefly enumerated in the following way. He must lay out the work for the whole shop, see that each piece of work goes in the proper order to the right machine, and that the man at the machine knows just what is to be done and how he is to do it. He must see that the work is not slighted, and that it is done fast, and all the while he must look ahead a month or so, either to provide more men to do the work or more work for the men to do. He must constantly discipline the men and readjust their wages, and in addition to this must fix piece work prices and supervise the timekeeping.

The first of the four leading principles in management calls for a clearly defined and circumscribed task. Evidently the foreman's duties are in no way clearly circumscribed. It is left each day entirely to his judgment what small part of the mass of duties before him it is most important for him to attend to, and he staggers along under this fraction of the work for which he is responsible, leaving the balance to be done in many cases as the gang bosses and workmen see fit. The second principle calls for such conditions that the daily task can always be accomplished. The conditions in his case are always such that it is impossible for him to do it all, and he never even makes a pretence of fulfilling his entire task. The third and fourth principles call for high pay in case the task is successfully done, and low pay in case of failure. The failure to realize the first two conditions, however, renders the application of the last two out of the question.

The foreman usually endeavors to lighten his burdens by delegating his duties to the various assistant foremen or gang bosses in charge of lathes, planers, milling machines, vise work, etc. Each of these men is then called upon to perform duties of almost as great variety as those of the foreman himself. The difficulty in obtaining in one man the variety of special information and the different mental and moral qualities necessary to perform all of the duties demanded of those men has been clearly summarized in the following list of the nine qualities which go to make up a well rounded man:

Brains.
Education.
Special or technical knowledge; manual dexterity or strength.
Tact.
Energy.
Grit.
Honesty.
Judgment or common sense and
Good health.

Plenty of men who possess only three of the above qualities can be hired at any time for laborers' wages. Add four of these qualities together and you get a higher priced man. The man combining five of these qualities begins to be hard to find, and those with six, seven, and eight are almost impossible to get. Having this fact in mind, let us go over the duties which a gang boss in charge, say, of lathes or planers, is called upon to perform, and note the knowledge and qualities which they call for.

First. He must be a good machinist — and this alone calls for years of special training, and limits the choice to a comparatively small class of men.

Second. He must be able to read drawings readily,

and have sufficient imagination to see the work in its finished state clearly before him. This calls for at least a certain amount of brains and education.

Third. He must plan ahead and see that the right jigs, clamps, and appliances, as well as proper cutting tools, are on hand, and are used to set the work correctly in the machine and cut the metal at the right speed and feed. This calls for the ability to concentrate the mind upon a multitude of small details, and take pains with little, uninteresting things.

Fourth. He must see that each man keeps his machine clean and in good order. This calls for the example of a man who is naturally neat and orderly himself.

Fifth. He must see that each man turns out work of the proper quality. This calls for the conservative judgment and the honesty which are the qualities of a good inspector.

Sixth. He must see that the men under him work steadily and fast. To accomplish this he should himself be a hustler, a man of energy, ready to pitch in and infuse life into his men by working faster than they do, and this quality is rarely combined with the painstaking care, the neatness and the conservative judgment demanded as the third, fourth, and fifth requirements of a gang boss.

Seventh. He must constantly look ahead over the whole field of work and see that the parts go to the machines in their proper sequence, and that the right job gets to each machine.

Eighth. He must, at least in a general way, supervise the timekeeping and fix piece work rates. Both the seventh and eighth duties call for a certain amount of clerical work and ability, and this class of work is almost always repugnant to the man suited to active executive work, and difficult for him to do; and the rate-fixing alone requires the whole time and careful study of a man especially suited to its minute detail.

Ninth. He must discipline the men under him, and readjust their wages; and these duties call for judgment, tact, and judicial fairness.

It is evident, then, that the duties which the ordinary gang boss is called upon to perform would demand of him a large proportion of the nine attributes mentioned above; and if such a man could be found he should be made manager or superintendent of a works instead of gang boss. However, bearing in mind the fact that plenty of men can be had who combine four or five of these attributes, it becomes evident that the work of management should be so subdivided that the various positions can be filled by men of this caliber, and a great part of the art of management undoubtedly lies in planning the work in this way. This can, in the judgment of the writer, be best accomplished by *abandoning the military type of organization* and introducing two broad and sweeping changes in the art of management:

(*a*) As far as possible the workmen, as well as the gang bosses and foremen, should be entirely relieved of the work of planning, and of all work which is more or less clerical in its nature. All possible brain work should be removed from the shop and centered in the planning or laying-out department, leaving for the foremen and gang bosses work strictly executive in its nature. Their duties should be to see that the operations planned and directed from the planning room are promptly carried out in the shop. Their time should be spent with the men, teaching them to think ahead, and leading and instructing them in their work.

(b) Throughout the whole field of management the military type of organization should be abandoned, and what may be called the "functional type" substituted in its place. "Functional management" consists in so dividing the work of management that each man from the assistant superintendent down shall have as few functions as possible to perform. If practicable the work of each man in the management should be confined to the performance of a single leading function.

Under the ordinary or military type the workmen are divided into groups. The men in each group receive their orders from one man only, the foreman or gang boss of that group. This man is the single agent through which the various functions of the management are brought into contact with the men. Certainly the most marked outward characteristic of functional management lies in the fact that each workman, instead of coming in direct contact with the management at one point only, namely, through his gang boss, receives his daily orders and help directly from eight different bosses, each of whom performs his own particular function. Four of these bosses are in the planning room and of these three send their orders to and receive their returns from the men, usually in writing. Four others are in the shop and personally help the men in their work, each boss helping in his own particular line or function only. Some of these bosses come in contact with each man only once or twice a day and then for a few minutes perhaps, while others are with the men all the time, and help each man frequently. The functions of one or two of these bosses require them to come in contact with each workman for so short a time each day that they can perform their particular duties perhaps for all of the men in the shop, and in their line they manage the entire shop. Other bosses are called upon to help their men so much and so often that each boss can perform his function for but a few men, and in this particular line a number of bosses are required, all performing the same function but each having his particular group of men to help. Thus the grouping of the men in the shop is entirely changed, each workman belonging to eight different groups according to the particular functional boss whom he happens to be working under at the moment.

The following is a brief description of the duties of the four types of executive functional bosses which the writer has found it profitable to use in the active work of the shop: (1) gang bosses, (2) speed bosses, (3) inspectors, and (4) repair bosses.

The gang boss has charge of the preparation of all work up to the time that the piece is set in the machine. It is his duty to see that every man under him has at all times at least one piece of work ahead at his machine, with all the jigs, templets, drawings, driving mechanism, sling chains, etc., ready to go into his machine as soon as the piece he is actually working on is done. The gang boss must show his men how to set their work in their machines in the quickest time, and see that they do it. He is responsible for the work being accurately and quickly set, and should be not only able but willing to pitch in himself and show the men how to set the work in record time.

The speed boss must see that the proper cutting tools are used for each piece of work, that the work is properly driven, that the cuts are started in the right part of the piece, and that the best speeds and feeds and depth of cut are used. His work begins only after the piece is in the lathe or planer, and ends when the actual machining ends. The speed boss must not only advise his men how best to do this work, but he must see that they do it in the quickest time, and that they use the speeds and feeds and depth of cut as directed on the instruction card. In many cases he is called upon to demonstrate that the work can be done in the specified time by doing it himself in the presence of his men.

The inspector is responsible for the quality of the work, and both the workmen and speed bosses must see that the work is all finished to suit him. This man can, of course, do his work best if he is a master of the art of finishing work both well and quickly.

The repair boss sees that each workman keeps his machine clean, free from rust and scratches, and that he oils and treats it properly, and that all of the standards established for the care and maintenance of the machines and their accessories are rigidly maintained, such as care of belts and shifters, cleanliness of floor around machines, and orderly piling and disposition of work.

The following is an outline of the duties of the four functional bosses who are located in the planning room, and who in their various functions represent the department in its connection with the men. The first three of these send their directions to and receive their returns from the men, mainly in writing. These four representatives of the planning department are, the (1) order of work and route clerk, (2) instruction card clerk, (3) time and cost clerk, and (4) shop disciplinarian.

Order of Work and Route Clerk. After the route clerk in the planning department has laid out the exact route which each piece of work is to travel through the shop from machine to machine in order that it may be finished at the time it is needed for assembling, and the work done in the most economical way, the order of work clerk daily writes lists instructing the workmen and also all of the executive shop bosses as to the exact order in which the work is to be done by each class of machines or men, and these lists constitute the chief means for directing the workmen in this particular function.

Instruction Card Clerks. The "instruction card," as its name indicates, is the chief means employed by the planning department for instructing both the executive bosses and the men in all of the details of their work. It tells them briefly the general and detail drawing to refer to, the piece number and the cost order number to charge the work to, the special jigs, fixtures, or tools to use, where to start each cut, the exact depth of each cut, and how many cuts to take, the speed and feed to be used for each cut, and the time within which each operation must be finished. It also informs them as to the piece rate, the differential rate, or the premium to be paid for completing the task within the specified time (according to the system employed); and further, when necessary, refers them by name to the man who will give them especial directions. This instruction card is filled in by one or more members of the planning department, according to the nature and complication of the instructions, and bears the same relation to the planning room that the drawing does to the drafting room. The man who sends it into the shop and who, in case difficulties are met with in carrying out the instructions, sees that the proper man sweeps these difficulties away, is called the instruction card foreman.

Time and Cost Clerk. This man sends to the men through the "time ticket" all the information they need for recording their time and the cost of the work, and secures proper returns from them. He refers these for entry to the cost and time record clerks in the planning room.

Shop Disciplinarian. In case of insubordination or impudence, repeated failure to do their duty, lateness or unexcused absence, the shop disciplinarian takes the workman or bosses in hand and applies the proper remedy. He sees that a complete record of each man's virtues and defects is kept. This man should also have much to do with readjusting the wages of the workmen. At the very least, he should invariably be consulted before any change is made. One of his important functions should be that of peace-maker.

Thus, under functional foremanship, we see that the work which, under the military type of organization, was done by the single gang boss, is subdivided among eight men: (1) route clerks, (2) instruction card clerks, (3) cost and time clerks, who plan and give directions from the planning room; (4) gang bosses, (5) speed bosses, (6) inspectors, (7) repair bosses, who show the men how to carry out their instructions, and see that the work is done at the proper speed; and (8) the shop disciplinarian, who performs this function for the entire establishment.

The greatest good resulting from this change is that it becomes possible in a comparatively short time to train bosses who can really and fully perform the functions demanded of them, while under the old system it took years to train men who were after all able to thoroughly perform only a portion of their duties. A glance at the nine qualities needed for a well rounded man and then at the duties of these functional foremen will show that each of these men requires but a limited number of the nine qualities in order to successfully fill his position; and that the special knowledge which he must acquire forms only a small part of that needed by the old style gang boss. The writer has seen men taken (some of them from the ranks of the workmen, others from the old style bosses and others from among the graduates of industrial schools, technical schools and colleges) and trained to become efficient functional foremen in from six to eighteen months. Thus it becomes possible with functional foremanship to thoroughly and completely equip even a new company starting on a large scale with competent officers in a reasonable time, which is entirely out of the question under the old system. Another great advantage resulting from functional or divided foremanship is that it becomes entirely practicable to apply the four leading principles of management to the bosses as well as to the workmen. Each foreman can have a task assigned him which is so accurately measured that he will be kept fully occupied and still will daily be able to perform his entire function. This renders it possible to pay him high wages when he is successful by giving him a premium similar to that offered the men and leave him with low pay when he fails.

The full possibilities of functional foremanship, however, will not have been realized until almost all of the machines in the shop are run by men who are of smaller calibre and attainments, and who are therefore cheaper than those required under the old system. The adoption of standard tools, appliances, and methods throughout the shop, the planning done in the planning room and the detailed instructions sent them from this department, added to the direct help received from the four executive bosses, permit the use of comparatively cheap men even on complicated work. Of the men in the machine shop of the Bethlehem Steel Company engaged in running the roughing machines, and who were working under the bonus system when the writer left them, about 95 per cent. were handy men trained up from laborers. And on the finishing machines, working on bonus, about 25 per cent. were handy men.

To fully understand the importance of the work which was being done by these former laborers, it must be borne in mind that a considerable part of their work was very large and expensive. The forgings which they were engaged in roughing and finishing weighed frequently many tons. Of course they were paid more than laborer's wages, though not as much as skilled machinists. The work in this shop was most miscellaneous in its nature.

Functional foremanship is already in limited use in many of the best managed shops. A number of managers have seen the practical good that arises from allowing two or three men especially trained in their particular lines to deal directly with the men instead of at second hand through the old style gang boss as a mouthpiece. So deep rooted, however, is the conviction that the very foundation of management rests in the military type as represented by the principle that no workman can work under two bosses at the same time, that all of the managers who are making limited use of the functional plan seem to feel it necessary to apologize for or explain away their use of it; as not really in this particular case being a violation of that principle. The writer has never yet found one, except among the works which he had assisted in organizing, who came out squarely and acknowledged that he was using functional foremanship because it was the right principle.

The writer introduced five of the elements of functional foremanship into the management of the small machine shop of the Midvale Steel Company of Philadelphia while he was foreman of that shop in 1882–1883: (1) the instruction card clerk, (2) the time clerk, (3) the inspector, (4) the gang boss, and (5) the shop disciplinarian. Each of these functional foremen dealt directly with the workmen instead of giving their orders through the gang boss. The dealings of the instruction card clerk and time clerk with the workmen were mostly in writing, and the writer himself performed the functions of shop disciplinarian, so that it was not until he introduced the inspector, with orders to go straight to the men instead of to the gang boss, that he appreciated the desirability of functional foremanship as a distinct principle in management. The prepossession in favor of the military type was so strong with the managers and owners of Midvale that it was not until years after functional foremanship was in continual use in this shop that he dared to advocate it to his superior officers as the correct principle.

Until very recently in his organization of works he has found it best to first introduce five or six of the elements of functional foremanship quietly, and get them running smoothly in a shop before calling attention to the principle involved. When the time for this announcement comes, it invariably acts as the proverbial red rag on the bull. It was some years later that the writer subdivided the duties of the "old gang boss" who spent his whole time with the men into the four functions of (1) speed boss, (2) repair boss, (3) inspector, and (4) gang boss, and it is the introduction of these four shop bosses directly helping the men (particularly that of the speed boss) in place of the single old boss, that has produced the greatest improvement in the shop.

When functional foremanship is introduced in a large shop, it is desirable that all of the bosses who are performing the same function should have their own foreman over them; for instance, the speed bosses should have a speed foreman over them, the gang bosses, a head gang boss; the inspectors, a chief inspector, etc., etc. The functions of these over-foremen are twofold. The first part of their work is to teach each of the bosses under them the exact nature of his duties, and at the start, also to nerve and brace them up to the point of insisting that the workmen shall carry out the orders exactly as specified on the instruction cards. This is a difficult task at first, as the workmen have been accustomed for years to do the details of the work to suit themselves, and many of them are intimate friends of the bosses and believe they know quite as much about their business as the latter. The second function of the over-foreman is to smooth out the difficulties which arise between the different types of bosses who in turn directly help the men. The speed boss, for instance, always follows after the gang boss on any particular job in taking charge of the workmen. In this way their respective duties come in contact edgeways, as it were, for a short time, and at the start there is sure to be more or less friction between the two. If two of these bosses meet with a difficulty which they cannot settle, they send for their respective overforemen, who are usually able to straighten it out. In case the latter are unable to agree on the remedy, the case is referred by them to the assistant superintendent, whose duties, for a certain time at least, may consist largely in arbitrating such difficulties and thus establishing the unwritten code of laws by which the shop is governed. This serves as one example of what is called the "exception principle" in management, which is referred to later.

Before leaving this portion of the subject the writer wishes to call attention to the analogy which functional foremanship bears to the management of a large, up-to-date school. In such a school the children are each day successively taken in hand by one teacher after another who is trained in his particular specialty, and they are in many cases disciplined by a man particularly trained in this function. The old style, one teacher to a class plan is entirely out of date.

The writer has found that better results are attained by placing the planning department in one office, situated, of course, as close to the center of the shop or shops as practicable, rather than by locating its members in different places according to their duties. This department performs more or less the functions of a clearing house. In doing their various duties, its members must exchange information frequently, and since they send their orders to and receive their returns from the men in the shop, principally in writing, simplicity calls for the use, when possible, of a single piece of paper for each job for conveying the instructions of the different members of the planning room to the men and another similar paper for receiving the returns from the men to the department. Writing out these orders and acting promptly on receipt of the returns and recording same requires the members of the department to be close together. The large machine shop of the Bethlehem Steel Company was more than a quarter of a mile long, and this was successfully run from a single planning room situated close to it. The manager, superintendent, and their assistants should, of course, have their offices adjacent to the planning room and, if practicable, the drafting room should be near at hand, thus bringing all of the planning and purely brain work of the establishment close together. The advantages of this concentration were found to be so great at Bethlehem that the general offices of the company, which were formerly located in the business part of the town, about a mile and a half away, were moved into the middle of the works adjacent to the planning room.

The shop, and indeed the whole works, should be managed, not by the manager, superintendent, or foreman, but by the planning department. The daily routine of running the entire works should be carried on by the various functional elements of this department, so that, in theory at least, the works could run smoothly even if the manager, superintendent and their assistants outside the planning room were all to be away for a month at a time.

The following are the leading functions of the planning department:

(a) The complete analysis of all orders for machines or work taken by the company.

(b) Time study for all work done by hand throughout the works, including that done in setting the work in machines, and all bench, vise work and transportation, etc.

(c) Time study for all operations done by the various machines.

(d) The balance of all materials, raw materials, stores and finished parts, and the balance of the work ahead for each class of machines and workmen.

(e) The analysis of all inquiries for new work received in the sales department and promises for time of delivery.

(f) The cost of all items manufactured with complete expense analysis and complete monthly comparative cost and expense exhibits.

(g) The pay department.

(h) The mnemonic symbol system for identification of parts and for charges.

(i) Information bureau.

(j) Standards.

(k) Maintenance of system and plant, and use of the tickler.

(l) Messenger system and post office delivery.

(m) Employment bureau.

(n) Shop disciplinarian.

(o) A mutual accident insurance association.

(p) Rush order department.

(q) Improvement of system or plant. These

several functions may be discribed more in detail as follows:

(*a*) THE COMPLETE ANALYSIS OF ALL ORDERS FOR MACHINES OR WORK TAKEN BY THE COMPANY.

This analysis should indicate the designing and drafting required, the machines or parts to be purchased and all data needed by the purchasing agent, and as soon as the necessary drawings and information come from the drafting room the lists of patterns, castings and forgings to be made, together with all instructions for making them, including general and detail drawing, piece number, the mnemonic symbol belonging to each piece (as referred to under (*h*) below) a complete analysis of the successive operations to be done on each piece, and the exact route which each piece is to travel from place to place in the works.

(*b*) TIME STUDY FOR ALL WORK DONE BY HAND THROUGHOUT THE WORKS, INCLUDING THAT DONE IN SETTING THE WORK IN MACHINES, AND ALL BENCH AND VISE WORK, AND TRANSPORTATION, ETC.

This information for each particular operation should be obtained by summing up the various unit times of which it consists. To do this, of course, requires the men performing this function to keep continually posted as to the best methods and appliances to use, and also to frequently consult with and receive advice from the executive gang bosses who carry out this work in the shop, and from the man in the department of standards and maintenance of plant (*j*) beneath. The actual study of unit times, of course, forms the greater part of the work of this section of the planning room.

(*c*) TIME STUDY FOR ALL OPERATIONS DONE BY THE VARIOUS MACHINES.

This information is best obtained from slide rules, one of which is made for each machine tool or class of machine tools throughout the works; one, for instance, for small lathes of the same type, one for planers of same type, etc. These slide rules show the best way to machine each piece and enable detailed directions to be given the workman as to how many cuts to take, where to start each cut, both for roughing out work and finishing it, the depth of the cut, the best feed and speed, and the exact time required to do each operation.

The information obtained through function (*b*), together with that obtained through (*c*) afford the basis for fixing the proper piece rate, differential rate or the bonus to be paid, according to the system employed.

(*d*) THE BALANCE OF ALL MATERIALS, RAW MATERIALS, STORES AND FINISHED PARTS, AND THE NUMBER OF DAYS' WORK AHEAD FOR EACH CLASS OF MACHINES AND WORKMEN.

Returns showing all receipts, as well as the issue of all raw materials, stores, partly finished work, and completed parts and machines, repair parts, etc., daily pass through the balance clerk, and each item of which there have been issues or receipts, or which has been appropriated to the use of a machine about to be manufactured, is daily balanced. Thus the balance clerk can see that the required stocks of materials are kept on hand by notifying at once the purchasing agent or other proper party when the amount on hand falls below the prescribed figure. The balance clerk should also keep a complete running balance of the hours of work ahead for each class of machines and workmen, receiving for this purpose daily from (a), (b), and (c) above statements of the hours of new work entered, and from the inspectors and daily time cards a statement of the work as it is finished. He should keep the manager and sales department posted through daily or weekly condensed reports as to the number of days of work ahead for each department, and thus enable them to obviate either a congestion or scarcity of work.

(e) THE ANALYSIS OF ALL INQUIRIES FOR NEW WORK RECEIVED IN THE SALES DEPARTMENT AND PROMISES AS TO TIME OF DELIVERY.

The man or men in the planning room who perform the duties indicated at (a) above should consult with (b) and (c) and obtain from them approximately the time required to do the work inquired for, and from (d) the days of work ahead for the various machines and departments, and inform the sales department as to the probable time required to do the work and the earliest date of delivery.

(f) THE COST OF ALL ITEMS MANUFACTURED, WITH COMPLETE EXPENSE ANALYSIS AND COMPLETE MONTHLY COMPARATIVE COST AND EXPENSE EXHIBITS.

The books of the company should be closed once a month and balanced as completely as they usually are at the end of the year, and the exact cost of each article of merchandise finished during the previous month should be entered on a comparative cost sheet. The expense exhibit should also be a comparative sheet. The cost account should be a completely balanced account, and not a memorandum account as it generally is. All the expenses of the establishment, direct and indirect, including the administration and sales expense, should be charged to the cost of the product which is to be sold.

(g) THE PAY DEPARTMENT.

The pay department should include not only a record of the time and wages and piece work earnings of each man, and his weekly or monthly payment, but the entire supervision of the arrival and departure of the men from the works and the various checks needed to insure against error or cheating. It is desirable that some one of the "exception systems" of time keeping should be used.

(h) THE MNEMONIC SYMBOL SYSTEM FOR IDENTIFICATION OF PARTS AND FOR CHARGES.

Some one of the mnemonic symbol systems should be used instead of numbering the parts or orders for identifying the various articles of manufacture, as well as the operations to be performed on each piece and the various expense charges of the establishment. This becomes a matter of great importance when written directions are sent from the planning room to the men, and the men make their returns in writing. The clerical work and chances for error are thereby greatly diminished.

(*i*) INFORMATION BUREAU.

The information bureau should include catalogues of drawings (providing the drafting room is close enough to the planning room) as well as all records and reports for the whole establishment. The art of properly indexing information is by no means a simple one, and as far as possible it should be centred in one man.

(*j*) STANDARDS.

The adoption and maintenance of standard tools, fixtures, and appliances down to the smallest item throughout the works and office, as well as the adoption of standard methods of doing all operations which are repeated, is a matter of importance, so that under similar conditions the same appliances and methods shall be used throughout the plant. This is an absolutely necessary preliminary to success in assigning daily tasks which are fair and which can be carried out with certainty.

(*k*) MAINTENANCE OF SYSTEM AND PLANT, AND USE OF THE TICKLER.

One of the most important functions of the planning room is that of the maintenance of the entire system, and of standard methods and appliances throughout the establishment, including the planning room itself. An elaborate time table should be made out showing daily the time when and place where each report is due, which is necessary to carry on the work and to maintain the system. It should be the duty of the member of the planning room in charge of this function to find out at each time through the day when reports are due, whether they have been received, and if not, to keep bothering the man who is behind hand until he has done his duty. Almost all of the reports, etc., going in and out of the planning room can be made to pass through this man. As a mechanical aid to him in performing his function the tickler is invaluable. The best type of tickler is one which has a portfolio for each day in the year, large enough to insert all reminders and even quite large instruction cards and reports without folding. In maintaining methods and appliances, notices should be placed in the tickler in advance, to come out at proper intervals throughout the year for the inspection of each element of the system and the inspection and overhauling of all standards as well as the examination and repairs at stated intervals of parts of machines, boilers, engines, belts, etc., likely to wear out or give trouble, thus preventing breakdowns and delays. One tickler can be used for the entire works and is preferable to a number of individual ticklers. Each man can remind himself of his various small routine duties to be performed either daily or weekly, etc., and which might be otherwise overlooked, by sending small reminders, written on slips of paper, to be placed in the tickler and returned to him at the proper time. Both the tickler and a thoroughly systematized messenger service should be immediately adjacent to this man in the planning room, if not directly under his management.

The proper execution of this function of the planning room will relieve the superintendent of some of the most vexatious and time-consuming of his duties, and at the same time the work will be done more thoroughly and cheaper than if he does it himself. By the adoption of standards and the use of instruction cards for overhauling machinery, etc., and the use of a tickler as above described, the writer reduced the repair force of the Midvale Steel Works to one-third its size while he was in the position of master mechanic. There was no planning department, however, in the works at that time.

(*l*) MESSENGER SYSTEM AND POST OFFICE DELIVERY.

The messenger system should be thoroughly organized and records kept showing which of the boys are the most efficient. This should afford one of the best opportunities for selecting boys fit to be taught trades, as apprentices or otherwise.

There should be a regular half hourly post office delivery system for collecting and distributing routine reports and records and messages in no especial hurry throughout the works.

(*m*) EMPLOYMENT BUREAU.

The selection of the men who are employed to fill vacancies or new positions should receive the most careful thought and attention and should be under the supervision of a competent man who will inquire into the experience and especial fitness and character of applicants and keep constantly revised lists of men suitable for the various positions in the shop. In this section of the planning room an individual record of each of the men in the works can well be kept showing his punctuality, absence without excuse, violation of shop rules, spoiled work or damage to machines or tools, as well as his skill at various kinds of work; average earnings, and other good qualities for the use of this department as well as the shop disciplinarian.

(*n*) THE SHOP DISCIPLINARIAN.

This man may well be closely associated with the employment bureau and, if the works is not too large, the two functions can be performed by the same man. The knowledge of character and of the qualities needed for various positions acquired in disciplining the men should be useful in selecting them for employment. This man should, of course, consult constantly with the various foremen and bosses, both in his function as disciplinarian and in the employment of men.

(*o*) A MUTUAL ACCIDENT INSURANCE ASSOCIATION.

A mutual accident insurance association should be established, to which the company contributes as well as the men. The object of this association is twofold: first, the relief of men who are injured, and second, an opportunity of returning to the workmen all fines which are imposed upon them in disciplining them, and for damage to company's property or work spoiled.

(*p*) RUSH ORDER DEPARTMENT.

Hurrying through parts which have been spoiled or have developed defects, and also special repair orders for customers, should receive the attention of one man.

(*q*) IMPROVEMENT OF SYSTEM OR PLANT.

One man should be especially charged with the work of improvement in the system and in the running of the plant.

The type of organization described in the foregoing paragraphs has such an appearance of complication and there are so many new positions outlined in the planning room which do not exist even in a well managed establishment of the old school, that it seems desirable to again call attention to the fact that, with the exception of the study of unit times and one or two minor functions, each item of work which is performed in the planning room with the superficial appearance of great complication must also be performed by the workmen in the shop under the old type of management, with its single cheap foreman and the appearance of great simplicity. In the first case, however, the work is done by an especially trained body of men who work together like a smoothly running machine, and in the second by a much larger number of men very poorly trained and ill-fitted for this work, and each of whom while doing it is taken away from some other job for which he is well trained. The work which is now done by one sewing machine, intricate in its appearance, was formerly done by a number of women with no apparatus beyond a simple needle and thread.

There is no question that the cost of production is lowered by separating the work of planning and the brain work as much as possible from the manual labor. When this is done, however, it is evident that the brain workers must be given sufficient work to keep them fully busy all the time. They must not be allowed to stand around for a considerable part of their time waiting for their particular kind of work to come along, as is so frequently the case.

The belief is almost universal among manufacturers that for economy the number of brain workers, or non-producers, as they are called, should be as small as possible in proportion to the number of producers, *i.e.*, those who actually work with their hands. An examination of the most successful establishments will, however, show that the reverse is true. A number of years ago the writer made a careful study of the proportion of producers to non-producers in three of the largest and most successful companies in the world, who were engaged in doing the same work in a general way. One of these companies was in France, one in Germany, and one in the United States. Being to a certain extent rivals in business and situated in different countries, naturally neither one had anything to do with the management of the other. In the course of his investigation, the writer found that the managers had never even taken the trouble to ascertain the exact proportion of non-producers to producers in their respective works; so that the organization of each company was an entirely independent evolution.

By "non-producers" the writer means such employés as all of the general officers, the clerks, foremen, gang bosses, watchmen, messenger boys, draftsmen, salesmen, etc.; and by "producers," only those who actually work with their hands.

In the French and German works there was found to be in each case one non-producer to between six and seven producers, and in the American works one non-producer to about seven producers. The writer found that in the case of another works, doing the same kind of business and whose management was notoriously bad, the proportion of non-producers to producers was one non-producer to about eleven producers. These companies all had large forges, foundries, rolling mills and machine shops turning out a miscellaneous product, much of which was machined. They turned out a highly wrought, elaborate and exact finished product, and did an extensive engineering and miscellaneous machine construction business.

In the case of a company doing a manufacturing business with a uniform and simple product for the maximum economy, the number of producers to each non-producer would of course be larger. No manager need feel alarmed then when he sees the number of non-producers increasing in proportion to producers, providing the non-producers are busy all of their time, and providing, of course, that in each case they are doing efficient work.

It would seem almost unnecessary to dwell upon the desirability of standardizing, not only all of the tools, appliances and implements throughout the works and office, but also the methods to be used in the multitude of small operations which are repeated day after day. There are many good managers of the old school, however, who feel that this standardization is not only unnecessary but that it is undesirable, their principal reason being that it is better to allow each workman to develop his individuality by choosing the particular implements and methods which suit him best. And there is considerable weight in this contention when the scheme of management is to allow each workman to do the work as he pleases and hold him responsible for results. Unfortunately, in ninety-nine out of a hundred such cases only the first part of this plan is carried out. The workman chooses his own methods and implements, but is not held in any strict sense accountable unless the quality of the work is so poor or the quantity turned out is so small as to almost amount to a scandal. In the type of management advocated by the writer, this complete standardization of all details and methods is not only desirable but absolutely indispensable as a preliminary to specifying the time in which each operation shall be done, and then insisting that it shall be done within the time allowed.

Neglecting to take the time and trouble to thoroughly standardize all of such methods and details is one of the chief causes for setbacks and failure in introducing this system. Much better results can be attained, even if poor standards be adopted, than can be reached if some of a given class of implements are the best of their kind while others are poor. It is uniformity that is required. Better have them uniformly second class than mainly first with some second and some third class thrown in at random. In the latter case the workmen will almost always adopt the pace which conforms to the third class instead of the first or second. In fact, however, it is not a matter involving any great expense or time to select in each case standard implements which shall be nearly the best or the best of their kinds. The writer has never failed to make enormous gains in the economy of running by the adoption of standards.

It was in the course of making a series of experiments with various air hardening tool steels with a view to adopting a standard for the Bethlehem works that Mr. J. Maunsel White, together with the writer, discovered the Taylor-White process of treating tool steel, which marks a distinct improvement in the art. The fact that this improvement was made not by manufacturers of tool steel, but in the course of the adoption of standards, shows both the necessity and fruitfulness of methodical and careful investigation in the choice of much neglected details. The economy to be gained through the adoption of uniform standards is hardly realized at all by the managers of this country. No better illustration of this fact is needed than that of the present condition of the cutting tools used throughout the machine shops of the United States. Hardly a shop can be found in which tools made from a dozen different qualities of steel are not used side by side, in many cases with little or no means of telling one make from another; and in addition, the shape of the cutting edge of the tool is in most cases left to the fancy of each individual workman. When one realizes that the cutting speed of the best treated air hardening steel is for a given depth of cut, feed and quality of metal being cut, say sixty feet per minute, while with the same shaped tool made from the best carbon tool steel and with the same conditions, the cutting speed will be only twelve feet per minute, it becomes apparent how little the necessity for rigid standards is appreciated.

Let us take another illustration. The machines of the country are still driven by belting. The motor drive, while it is coming, is still in the future. There is not one establishment in one hundred that does not leave the care and tightening of the belts to the judgment of the individual who runs the machine, although it is well known to all who have given any study to the subject that the most skilled machinist cannot properly tighten a belt without the use of belt clamps fitted with spring balances to properly register the tension. And the writer showed in a paper entitled "Notes on Belting" presented to The American Society of Mechanical Engineers in 1893, giving the results of an experiment tried on all of the belts in a machine shop and extending through nine years, in which every detail of the care and tightening and tension of each belt was recorded, that belts properly cared for according to a standard method by a trained laborer would average twice the pulling power and only a fraction of the interruptions to manufacture of those tightened according to the usual methods. The loss now going on throughout the country from failure to adopt and maintain standards for all small details is simply enormous. It is, however, a good sign for the future that a firm such as Messrs. Dodge & Day of Philadelphia, who are making a specialty of standardizing machine shop details, find their time fully occupied.

What may be called the "exception principle" in management is coming more and more into use, although, like many of the other elements of this art, it is used in isolated cases, and in most instances without recognizing it as a principle which should extend throughout the entire field. It is not an uncommon sight, though a sad one, to see the manager of a large business fairly swamped at his desk with an ocean of letters and reports, on each of which he thinks that he should put his initial or stamp. He feels that by having this mass of detail pass over his desk he is keeping in close touch with the entire business. The exception principle is directly the reverse of this. Under it the manager should receive only condensed, summarized, and *invariably* comparative reports, covering, however, all of the elements entering into the management, and even these summaries should all be carefully gone over by an assistant before they reach the manager, and have all of the exceptions to the past averages or to the standards pointed out, both the especially good and especially bad exceptions, thus giving him in a few minutes a full view of progress which is being made, or the reverse, and leaving him free to consider the broader lines of policy and to study the character and fitness of the important men under him. The exception principle can be applied in many ways, and the writer will endeavor to give some further illustrations of it later.

The writer has dwelt at length upon the desirability of concentrating as much as possible clerical and brain work in the planning department. There is, however, one such important exception to this rule that it would seem desirable to call attention to it. As already stated, the planning room gives its orders and instructions to the men mainly in writing and of necessity must also receive prompt and reliable written returns and reports which shall enable its members to issue orders for the next movement of each piece, lay out the work for each man for the following day, properly post the balance of work and materials accounts, enter the records on cost accounts and also enter the time and pay of each man on the pay sheet. There is no question that all of this information can be given both better and cheaper by the workman direct than through the intermediary of a walking time keeper, providing the proper instruction and report system has been introduced in the works with carefully ruled and printed instruction and return cards, and particularly providing a complete mnemonic system of symbols has been adopted so as to save the workmen the necessity of doing much writing. The principle to which the writer wishes to call particular attention is that the only way in which workmen can be induced to write out all of this information accurately and promptly is by having each man write his own time while on day work and pay when on piece work on the same card on which he is to enter the other desired information, and then refusing to enter his pay on the pay sheet until after all of the required information has been correctly given by him. Under this system as soon as a workman completes a job and at quitting time, whether the job is completed or not, he writes on a printed time card all of the information needed by the planning room in connection with that job, signs it and forwards it at once to the planning room. On arriving in the planning room each time card passes through the order of work or route clerk, the balance clerk, the cost clerk, etc., on its way to the pay sheet, and unless the workman has written the desired information the card is sent back to him, and he is apt to correct and return it promptly so as to have his pay entered up. The principle is clear that if one wishes to have routine clerical work done promptly and correctly it should somehow be attached to the pay card of the man who is to give it. This principle, of course, applies to the information desired from inspectors, gang bosses and others as well as workmen, and to reports required from various clerks. In the case of reports, a pay coupon can be attached to the report which will be detached and sent to the pay sheet as soon as the report has been found correct.

Before starting to make any radical changes leading toward an improvement in the system of management, it is desirable, and for ultimate success in most cases necessary, that the directors and the important owners of an enterprise shall be made to understand, at least in a general way, what is involved in the change. They should be informed of the leading objects which the new system aims at, such, for instance, as rendering mutual the interests of employer and employé through "high wages and low labor cost," the gradual selection and development of a body of first class picked workmen who will work extra hard and receive extra high wages and be dealt with individually instead of in masses. They should thoroughly understand that this can only be accomplished through the adoption of precise and exact methods, and having each smallest detail, both as to methods and appliances, carefully selected so as to be the best of its kind. They should understand the general philosophy of the system and should see that, as a whole, it must be in harmony with its few leading ideas, and that principles and details which are admirable in one type of management have no place whatever in another. They should be shown that it pays to employ an especial corps to introduce a new system just as it pays to employ especial designers and workmen to build a new plant; that, while a new system is being introduced, almost twice the number of foremen are required as are needed to run it after it is in; that all of this costs money, but that, unlike a new plant, returns begin to come in almost from the start from improved methods and appliances as they are introduced, and that in most cases the new system more than pays for itself as it goes along; that time, and a great deal of time, is involved in a radical change in management, and that in the case of a large works if they are incapable of looking ahead and patiently waiting for from two to four years, they had better leave things just as they are, since a change of system involves a change in the ideas, point of view and habits of many men with strong convictions and prejudices, and that this can only be brought about slowly and chiefly through a series of object lessons, each of which takes time, and through continued reasoning; and that for this reason, after deciding to adopt a given type, the necessary steps should be taken as fast as possible, one after another, for its introduction. The directors should be convinced that an increase in the proportion of non-producers to producers means increased economy and not red tape, providing the non-producers are kept busy at their respective functions. They should be prepared to lose some of their valuable men who cannot stand the change and also for the continued indignant protest of many of their old and trusted employés who can see nothing but extravagance in the new ways and ruin ahead. It is a matter of the first importance that, in addition to the directors of the company, all of those connected with the management should be given a broad and comprehensive view of the general objects to be attained and the means which will be employed. They should fully realize before starting on their work and should never lose sight of the fact that the great object of the new organization is to bring about two momentous changes in the men:

First. A complete revolution in their mental attitude toward their employers and their work.

Second. As a result of this change of feeling such an increase in their determination and physical activity, and such an improvement in the conditions under which the work is done as will result in many cases in their turning out from two to three times as much work as they have done in the past.

First, then, the men must be brought to see that the new system changes their employers from antagonists to friends who are working as hard as possible side by side with them, all pushing in the same direction and all helping to bring about such an increase in the output and to so cheapen the cost of production that the men will be paid permanently from thirty to one hundred per cent. more than they have earned in the past, and that there will still be a good profit left over for the company. At first workmen cannot see why, if they do twice as much work as they have done, they should not receive twice the wages. When the matter is properly explained to them and they have time to think it over, they will see that in most cases the increase in output is quite as much due to the improved appliances and methods, to the maintenance of standards and to the great help which they receive from the men over them as to their own harder work. They will realize that the company must pay for the introduction of the improved system, which costs thousands of dollars, and also the salaries of the additional foremen and of the clerks, etc., in the planning room as well as tool room and other expenses and that, in addition, the company is entitled to an increased profit quite as much as the men are. All but a few of them will come to understand in a general way that under the new order of things they are coöperating with their employers to make as great a saving as possible and that they will receive permanently their fair share of this gain. Then after the men acquiesce in the new order of things and are willing to do their part toward cheapening production, it will take time for them to change from their old easy-going ways to a higher rate of speed, and to learn to stay steadily at their work, think ahead and make every minute count. A certain percentage of them, with the best of intentions, will fail in this and find that they have no place in the new organization, while still others, and among them some of the best workers who are, however, either stupid or stubborn, can never be made to see that the new system is as good as the old; and these, too, must drop out. Let no one imagine, however, that this great change in the mental attitude of the men and the increase in their activity can be brought about by merely talking to them. Talking will be most useful — in fact indispensable — and no opportunity should be lost of explaining matters to them patiently, one man at a time, and giving them every chance to express their views.

Their real instruction, however, must come through a series of object lessons. They must be convinced that a great increase in speed is possible by seeing here and there a man among them increase his pace and double or treble his output. They must see this pace maintained until they are convinced that it is not a mere spurt; and, most important of all, they must see the men who "get there" in this way receive a proper increase in wages and become satisfied. It is only with these object lessons in plain sight that the new theories can be made to stick. It will be in presenting these object lessons and in smoothing away the difficulties so that the high speed can be maintained, and in assisting to form public opinion in the shop, that the great efficiency of functional foremanship under the direction of the planning room will first become apparent.

In reaching the final high rate of speed which shall be steadily maintained, the broad fact should be realized that the men must pass through several distinct phases, rising from one plane of efficiency to another until the final level is reached. First they must be taught to work under an improved system of day work. Each man must learn how to give up his own particular way of doing things, adapt his methods to the many new standards, and grow accustomed to receiving and obeying directions covering details, large and small, which in the past have been left to his individual judgment. At first the workmen can see nothing in all of this but red tape and impertinent interference, and time must be allowed them to recover from their irritation, not only at this, but at every stage in their upward march. If they have been classed together and paid uniform wages for each class, the better men should be singled out and given higher wages so that they shall distinctly recognize the fact that each man is to be paid according to his individual worth. After becoming accustomed to direction in minor matters, they must gradually learn to obey instructions as to the pace at which they are to work, and grasp the idea, first, that the planning department knows accurately how long each operation should take; and second, that sooner or later they will have to work at the required speed if they expect to prosper. After they are used to following the speed instructions given them, then one at a time they can be raised to the level of maintaining a rapid pace throughout the day. And it is not until this final step has been taken that the full measure of the value of the new system will be felt by the men through daily receiving larger wages, and by the company through a materially larger output and lower cost of production. It is evident, of course, that all of the workmen in the shop will not rise together from one level to another. Those engaged in certain lines of work will have reached their final high speed while others have barely taken the first step. The efforts of the new management should not be spread out thin over the whole shop. They should rather be focussed upon a few points, leaving the ninety and nine under the care of their former shepherds. After the efficiency of the men who are receiving special assistance and training has been raised to the desired level, the means for holding them there should be perfected, and they should never be allowed to lapse into their old ways. This will, of course, be accomplished in the most permanent way and rendered almost automatic, either through introducing task work with a bonus or the differential rate.

Before taking any steps toward changing methods the manager should realize that at no time during the introduction of the system should any broad, sweeping changes be made which seriously affect a large number of the workmen. It would be preposterous, for instance, in going from day to piece work to start a large number of men on piece work at the same time. Throughout the early stages of organization each change made should affect one workman only, and after the single man affected has become used to the new order of things, then change one man after another from the old system to the new, slowly at first, and rapidly as public opinion in the shop swings around under the influence of proper object lessons. Throughout a considerable part of the time, then, there will be two distinct systems of management in operation in the same shop; and in many cases it is desirable to have the men working under the new system managed by an entirely different set of foremen, etc., from those under the old.

The first step, after deciding upon the type of organization, should be the selection of a competent man to take charge of the introduction of the new system. The manager should think himself fortunate if he can get such a man at almost any price, since the task is a difficult and thankless one and but few men can be found who possess the necessary information coupled with the knowledge of men, the nerve, and the tact required for success in this work. The manager should keep himself free as far as possible from all active part in the introduction of the new system. While changes are going on it will require his entire energies to see that there is no falling off in the efficiency of the old system and that the quality and quantity of the output is kept up. The mistake which is usually made when a change in system is decided upon is that the manager and his principal assistants undertake to make all of the improvements themselves during their spare time, with the common result that weeks, months, and years go by without anything great being accomplished. The respective duties of the manager and the man in charge of improvement, and the limits of the authority of the latter should be clearly defined and agreed upon, always bearing in mind that responsibility should invariably be accompanied by its corresponding measure of authority.

The worst mistake that can be made is to refer to any part of the system as being "on trial." Once a given step is decided upon, all parties must be made to understand that it will go whether any one around the place likes it or not. In making changes in system the things that are given a "fair trial" fail, while the things that "must go," go all right.

To decide where to begin is a perplexing and bewildering problem which faces the reorganizer in management when he arrives in a large establishment. In making this decision, as in taking each subsequent step, the most important consideration, which should always be first in the mind of the reformer, is "what effect will this step have upon the workmen?" Through some means (it would almost appear some especial sense) the workman seems to scent the approach of a reformer even before his arrival in town. Their suspicions are thoroughly aroused, and they are on the alert for sweeping changes which are to be against their interests and which they are prepared to oppose from the start. Through generations of bitter experiences working men as a class have learned to look upon all change as antagonistic to their best interests. They do not ask the object of the change, but oppose it simply as *change*. The first changes, therefore, should be such as to allay the suspicions of the men and convince them by actual contact that the reforms are after all rather harmless and are only such as will ultimately be of benefit to all concerned. Such improvements then as directly affect the workmen least should be started first. At the same time it must be remembered that the whole operation is of necessity so slow that the new system should be started at as many points as possible, and constantly pushed as hard as possible. In the metal working plant which we are using for purposes of illustration a start can be made at once along all of the following lines:

First. The introduction of standards throughout the works and office.

Second. The scientific study of unit times on several different kinds of work.

Third. A complete analysis of the pulling, feeding power and the proper speeding of the various machine tools throughout the place with a view of making a slide rule for properly running each machine.

Fourth. The work of establishing the system of time cards by means of which ultimately all of the desired information will be conveyed from the men to the planning room.

Fifth. Overhauling the stores issuing and receiving system so as to establish a complete running balance of materials.

Sixth. Ruling and printing the various blanks that will be required for shop returns and reports, time cards, instruction cards, expense sheets, cost sheets, pay sheet, and balance records; storeroom; tickler; and maintenance of standards, system, and plant, etc.; and starting such functions of the planning room as do not directly affect the men.

If the works is a large one, the man in charge of introducing the system should appoint a special assistant in charge of each of the above functions just as an engineer designing a new plant would start a number of draftsmen to work upon the various elements of construction. Several of these assistants will be brought into close contact with the men, who will in this way gradually get used to seeing changes going on and their suspicion, both of the new men and the methods, will have been allayed to such an extent before any changes which seriously affect them are made, that little or no determined opposition on their part need be anticipated. The most important and difficult task of the organizer will be that of selecting and training the various functional foremen who are to lead and instruct the workmen, and his success will be measured principally by his ability to mold and reach these men. They cannot be found, they must be made. They must be instructed in their new functions largely, in the beginning at least, by the organizer himself; and this instruction, to be effective, should be mainly in actually doing the work. Explanation and theory will go a little way, but actual doing is needed to carry conviction. To illustrate: For nearly two and one-half years in the large shop of the Bethlehem Steel Company, one speed boss after another was instructed in the art of cutting metals fast on a large motor-driven lathe which was especially fitted to run at any desired speed within a very wide range. The work done in this machine was entirely connected, either with the study of cutting tools or the instruction of speed bosses. It was most interesting to see these men, principally either former gang bosses or the best workmen, gradually change from their attitude of determined and positive opposition to that in most cases of enthusiasm for, and earnest support of, the new methods. It was actually running the lathe themselves according to the new method and under the most positive and definite orders that produced the effect. The writer himself ran the lathe and instructed the first few bosses. It required from three weeks to two months for each man. Perhaps the most important part of the gang boss's and foreman's education lies in teaching them to promptly obey orders and instructions received not only from the superintendent or some official high in the company, but from any member of the planning room whose especial function it is to direct the rest of the works in his particular line; and it may be accepted as an unquestioned fact that no gang boss is fit to direct his men until after he has learned to promptly obey instructions received from any proper source, whether he likes his instructions and the instructor or not, and even although he may be convinced that he knows a much better way of doing the work. The first step is for each man to learn to obey the laws as they exist, and next, if the laws are wrong, to have them reformed in the proper way.

In starting to organize even a comparatively small shop, containing say from 75 to 100 men, it is best to begin by training in the full number of functional foremen, one for each function, since it must be remembered that about two out of three of those who are taught this work either leave of their own accord or prove unsatisfactory; and in addition, while both the workmen and bosses are adjusting themselves to their new duties, there are needed fully twice the number of bosses as are required to carry on the work after it is fully systematized.

Unfortunately, there is no means of selecting in advance those out of a number of candidates for a given work who are likely to prove successful. Many of those who appear to have all of the desired qualities, and who talk and appear the best, will turn out utter failures, while on the other hand, some of the most unlikely men rise to the top. The fact is, that the more attractive qualities of good manners, education, and even special training and skill, which are more apparent on the surface, count for less in an executive position than the grit, determination and bulldog endurance and tenacity that knows no defeat and comes up smiling to be knocked down over and over again.

The two qualities which count most for success in this kind of executive work are grit and what may be called "constructive imagination" — the faculty which enables a man to use the few facts that are stored in his mind in getting around the obstacles that oppose him, and in building up something useful in spite of them; and unfortunately, the presence of these qualities, together with honesty and common sense, can only be proved through an actual trial at executive work. As we all know, success at college or in the technical school does not indicate the presence of these qualities, even though the man may have worked hard. Mainly, it would seem, because the work of obtaining an education is principally that of absorption and assimilation; while that of active practical life is principally the direct reverse, namely, that of giving out.

In selecting men to be tried as foremen, or in fact for any position throughout the place, from the day laborer up, one of two different types of men should be chosen, according to the nature of the work to be done. For one class of work, men should be selected who are too good for the job; and for the other class of work, men who are barely good enough.

If the work is of a routine nature, in which the same operations are likely to be done over and over again, with no great variety, and in which there is no apparent prospect of a radical change being made, perhaps through a term of years, even though the work itself may be complicated in its nature, a man should be selected whose abilities are barely equal to the task. Time and training will fit him for his work, and since he will be better paid than in the past, and will realize that he has been given the chance to make his abilities yield him the largest return — all of the elements for promoting contentment will be present; and those men who are blessed with cheerful dispositions will become satisfied and remain so. Of course, a considerable part of mankind is so born or educated that permanent contentment is out of the question. No one, however, should be influenced by the discontent of this class. On the other hand, if the work to be done is of great variety — particularly if improvements in methods are to be anticipated — throughout the period of active organization the men engaged in systematizing should be too good for their jobs. For such work, men should be selected whose mental caliber and attainments will fit them, ultimately at least, to command higher wages than can be afforded on the work which they are at. It will prove a wise policy to promote such men both to better positions and pay, when they have shown themselves capable of accomplishing results and the opportunity offers. The results which these high-class men will accomplish, and the comparatively short time which they will take in organizing, will much more than pay for the expense and trouble, later on, of training other men, cheaper and of less capacity, to take their places. In many cases, however, gang bosses and men will develop faster than new positions open for them. When this occurs, it will pay employers well to find them positions in other works, either with better pay, or larger opportunities; not only as a matter of kindly feeling and generosity toward their men, but even more with the object of promoting the best interests of their own establishments. For one man lost in this way, five will be stimulated to work to the very limit of their abilities, and will rise ultimately to take the place of the man who has gone, and the best class of men will apply for work where these methods prevail. But few employers, however, are sufficiently broad-minded to adopt this policy. They dread the trouble and temporary inconvenience incident to training in new men.

Mr. James M. Dodge, Chairman of the Board of the Link-Belt Company, is one of the few men with whom the writer is acquainted who has been led by his kindly instincts, as well as by a far-sighted policy, to treat his employés in this way; and this, together with the personal magnetism and influence which belong to men of his type, has done much to render his shop one of the model establishments of the country, certainly as far as the relations of employer and men are concerned. On the other hand, this policy of promoting men and finding them new positions has its limits. No worse mistake can be made than that of allowing an establishment to be looked upon as a training school, to be used mainly for the education of many of its employés. All employés should bear in mind that each shop exists, first, last, and all the time, for the purpose of paying dividends to its owners. They should have patience, and never lose sight of this fact. And no man should expect promotion until after he has trained his successor to take his place. The writer is quite sure that in his own case, as a young man, no one element was of such assistance to him in obtaining new opportunities as the practice of invariably training another man to fill his position before asking for advancement.

The first of the functional foremen to be brought into actual contact with the men should be the inspector; and the whole system of inspection, with its proper safeguards, should be in smooth and successful operation before any steps are taken toward stimulating the men to a larger output; otherwise an increase in quantity will probably be accompanied by a falling off in quality.

Next choose for the application of the two principal functional foremen, viz., the speed boss and the gang boss, that portion of the work in which there is the largest need of, and opportunity for, making a gain. It is of the utmost importance that the first combined application of time study, slide rules, instruction cards, functional foremanship, and a premium for a large daily task should prove a success both for the workmen and for the company, and for this reason a simple class of work should be chosen for a start. The entire efforts of the new management should be centered on one point, and continue there until unqualified success has been attained.

When once this gain has been made, a peg should be put in which shall keep it from sliding back in the least; and it is here that the task idea with a time limit for each job will be found most useful. Under ordinary piece work, or the Towne-Halsey plan, the men are likely at any time to slide back a considerable distance without having it particularly noticed either by them or the management. With the task idea, the first falling off is instantly felt by the workman through the loss of his day's bonus, or his differential rate, and is thereby also forcibly brought to the attention of the management.

There is one rather natural difficulty which arises when the functional foremanship is first introduced. Men who were formerly either gang bosses, or foremen, are usually chosen as functional foremen, and these men, when they find their duties restricted to their particular functions, while they formerly were called upon to do everything, at first feel dissatisfied. They think that their field of usefulness is being greatly contracted. This is, however, a theoretical difficulty, which disappears when they really get into the full swing of their new positions. In fact the new position demands an amount of special information, forethought, and a clear-cut, definite responsibility that they have never even approximated in the past, and which is amply sufficient to keep all of their best faculties and energies alive and fully occupied. It is the experience of the writer that there is a great commercial demand for men with this sort of definite knowledge, who are used to accepting real responsibility and getting results; so that the training in their new duties renders them more instead of less valuable.

As a rule, the writer has found that those who were growling the most, and were loudest in asserting that they ought to be doing the whole thing, were only one-half or one-quarter performing their own particular functions. This desire to do every one's else work in addition to their own generally disappears when they are held to strict account in their particular line, and are given enough work to keep them hustling.

There are many people who will disapprove of the whole scheme of a planning department to do the thinking for the men, as well as a number of foremen to assist and lead each man in his work, on the ground that this does not tend to promote independence, self-reliance, and originality in the individual. Those holding this view, however, must take exception to the whole trend of modern industrial development; and it appears to the writer that they overlook the real facts in the case.

It is true, for instance, that the planning room, and functional foremanship, render it possible for an intelligent laborer or helper in time to do much of the work now done by a machinist. Is not this a good thing for the laborer and helper? He is given a higher class of work, which tends to develop him and gives him better wages. In the sympathy for the machinist the case of the laborer is overlooked. This sympathy for the machinist is, however, wasted, since the machinist, with the aid of the new system, will rise to a higher class of work which he was unable to do in the past, and in addition, divided or functional foremanship will call for a larger number of men in this class, so that men, who must otherwise have remained machinists all their lives, will have the opportunity of rising to a foremanship.

The demand for men of originality and brains was never so great as it is now, and the modern subdivision of labor, instead of dwarfing men, enables them all along the line to rise to a higher plane of efficiency, involving at the same time more brain work and less monotony. The type of man who was formerly a day laborer and digging dirt is now for instance making shoes in a shoe factory. The dirt handling is done by Italians or Hungarians.

After the planning room with functional foremanship has accomplished its most difficult task, of teaching the men how to do a full day's work themselves, and also how to get it out of their machines steadily, then, if desired, the number of non-producers can be diminished, preferably, by giving each type of functional foreman more to do in his specialty; or in the case of a very small shop, by combining two different functions in the same man. The former expedient is, however, much to be preferred to the latter. There need never be any worry about what is to become of those engaged in systematizing after the period of active organization is over. The difficulty will still remain even with functional foremanship, that of getting enough good men to fill the positions, and the demand for competent gang bosses will always be so great that no good boss need look for a job.

Of all the farces in management the greatest is that of an establishment organized along well planned lines, with all of the elements needed for success, and yet which fails to get either output or economy. There must be some man or men present in the organization who will not mistake the form for the essence, and who will have brains enough to find out those of their employés who "get there," and nerve enough to make it unpleasant for those who fail, as well as to reward those who succeed. No system can do away with the need of real men. Both system and good men are needed, and after introducing the best system, success will be in proportion to the ability, consistency, and respected authority of the management.

In a book of this sort, it would be manifestly impossible to discuss at any length all of the details which go toward making the system a success. Some of them are of such importance as to render at least a brief reference to them necessary. And first among these comes the study of unit times.

This, as already explained, is the most important element of the system advocated by the writer. Without it, the definite, clear-cut directions given to the workman, and the assigning of a full, yet just, daily task, with its premium for success, would be impossible; and the arch without the keystone would fall to the ground.

In 1883, while foreman of the machine shop of the Midvale Steel Company of Philadelphia, it occurred to the writer that it was simpler to time with a stop watch each of the elements of the various kinds of work done in the place, and then find the quickest time in which each job could be done by summing up the total times of its component parts, than it was to search through the time records of former jobs and guess at the proper time and price. After practising this method of time study himself for about a year, as well as circumstances would permit, it became evident that the system was a success.

The writer then established the time-study and rate-fixing department, which has given out piece work prices in the place ever since.

This department far more than paid for itself from the very start; but it was several years before the full benefits of the system were felt, owing to the fact that the best methods of making and recording time observations, as well as of determining the maximum capacity of each of the machines in the place, and of making working tables and time tables, were not at first adopted.

It has been the writer's experience that the difficulties of scientific time study are underestimated at first, and greatly overestimated after actually trying the work for two or three months. The average manager who decides to undertake the study of unit times in his works fails at first to realize that he is starting a new art or trade. He understands, for instance, the difficulties which he would meet with in establishing a drafting room, and would look for but small results at first, if he were to give a bright man the task of making drawings, who had never worked in a drafting room, and who was not even familiar with drafting implements and methods, but he entirely underestimates the difficulties of this new trade.

The art of studying unit times is quite as important and as difficult as that of the draftsman. It should be undertaken seriously, and looked upon as a profession. It has its own peculiar implements and methods, without the use and understanding of which progress will necessarily be slow, and in the absence of which there will be more failures than successes scored at first.

When, on the other hand, an energetic, determined man goes at time study as if it were his life's work, with the determination to succeed, the results which he can secure are little short of astounding. The difficulties of the task will be felt at once and so strongly by any one who undertakes it, that it seems important to encourage the beginner by giving at least one illustration of what has been accomplished.

Mr. Sanford E. Thompson, C. E., started in 1896 with but small help from the writer, except as far as the implements and methods are concerned, to study the time required to do all kinds of work in the building trades. In six years he has made a complete study of eight of the most important trades — excavation, masonry (including sewer-work and paving), carpentry, concrete and cement work, lathing and plastering, slating and roofing and rock quarrying. He took every stop watch observation himself and then, with the aid of two comparatively cheap assistants, worked up and tabulated all of his data ready for the printer. The magnitude of this undertaking will be appreciated when it is understood that the tables and descriptive matter for one of these trades alone take up about 250 pages. Mr. Thompson and the writer are both engineers, but neither of us was especially familiar with the above trades, and this work could not have been accomplished in a lifetime without the study of elementary units with a stop watch.

In the course of this work, Mr. Thompson has de-

OPERATION *Wheel-barrow Excavation* DATE *March 10* 189

DEPARTMENT *Construction*

MEN *Mike Flaherty*

MATERIALS *Sand requiring no pick. Hard clay in bank.*

IMPLEMENTS *No 3 shovel. Contractors wooden barrow*

CONDITIONS *Day work for contractor.*

By previous observation an average barrow load of sand is 2.32 cu ft meas in cut.

" " " " clay " 2.15 " " " " "

OP	TIME	AV	NO SHVLS	OP	TIME	AV	NO SHVLS	OP	TIME	AV	NO SHVLS	OP	TIME	AV	NO SHVLS
a	1.37	1.37	15	a	1.12	1.12	12	d	1.86	—	11				
b	1.56	0.19	—	b	1.39	0.27	—	d	1.81	—	13				
c	1.82	0.26	—	c	1.58	0.19	—	d	2.14	—	16				
d	1.97	0.15	—	d	1.70	0.12	—	d	1.98	—	14				
e	2.27	0.30		e	1.92	0.22									
f	2.36	0.09		f	2.05	0.13									
a	1.24	1.24	13	a	1.23	1.23	13								
b	1.36	0.12		b	1.38	0.15									
c	1.59	0.23		c	1.60	0.22									
d	1.83	0.24		d	1.78	0.18									
e	2.08	0.25		e	2.05	0.27									
f	2.33	0.25		f	2.23	0.18									

COMPLETE-OPERATIONS

TIME		TOTAL TIME IN MIN	TOTAL TIME PICKING IN MIN	TOTAL SHOVLG + WHLNG IN MIN	TIME PER BARROW IN MIN
7 AM	*Commenced loading sand.*				
9.02	*43 loads wheeled 50 ft.*	122		122	2.84
9.50	*Picking hard clay*	48			
11.39	*29 loads clay wheeled 50 ft.*	109			
11.46	*Picking clay again*	7	55		1.67
12.01	*4 loads clay wheeled 50 ft.*	15		124	3.76
		301			

NOTE *Comparison of "Detail" with "Complete" operations shows that about 27% of the total time was taken in rest and other necessary delays.*

About the same quantity loose as at the start.

DETAIL-OPERATIONS	NO. OF OBS	TIME PER BARROW IN MIN	TIME PER SHOVEL IN MIN	NO SHOVEL PER BARROW	TIME WHELNG 100 FT IN MIN
a Filling barrow with sand	4	1.240	0.094	13.2	
b Starting	4	0.182			
c Wheeling full 50 ft.	4	0.225			0.450
d Dumping and turning	4	0.172			
e Returning empty 50 ft.	4	0.260			0.520
f Dropping barrow & starting to shovel	4	0.162			
g		2.241			
a Filling barrow with clay	4	1.908	0.104	13.5	

OBSERVER *James Monroe.*

FIGURE 2. — TIME STUDY NOTE SHEET

veloped what are in many respects the best implements[1] in use, and with his permission some of them will be described. The blank form or note sheet used by Mr. Thompson, shown in Fig. 2 (see page 151), contains essentially:

(1) Space for the description of the work and notes in regard to it.

(2) A place for recording the total time of complete operations — that is, the gross time including all necessary delays, for doing a whole job or large portions of it.

(3) Lines for setting down the "detail operations," or "units" into which any piece of work may be divided, followed by columns for entering the averages obtained from the observations.

(4) Squares for recording the readings of the stop watch when observing the times of these elements. If these squares are filled, additional records can be entered on the back. The size of the sheets, which should be of best quality ledger paper, is $8\frac{3}{4}$ inches wide by 7 inches long, and by folding in the center they can be conveniently carried in the pocket, or placed in a case (see Fig. 3, page 153) containing one or more stop watches.

This case, or "watch book," is another device of Mr. Thompson's. It consists of a frame work, containing concealed in it one, two, or three watches, whose stop and start movements can be operated by pressing with the fingers of the left hand upon the proper portion of the cover of the note-book without the knowledge of the workman who is being

[1] Information about time study apparatus may be obtained from Sanford E. Thompson, Newton Highlands, Mass.

observed. The frame is bound in a leather case resembling a pocket note-book, and has a place for the note sheets described.

The writer does not believe at all in the policy of spying upon the workman when taking time observations for the purpose of time study. If the men observed are to be *ultimately affected by the re-*

FIGURE 3. — WATCH BOOK FOR TIME STUDY

sults of these observations, it is generally best to come out openly, and let them know that they are being timed, and what the object of the timing is. There are many cases, however, in which telling the workman that he was being timed in a minute way would only result in a row, and in defeating the whole object of the timing; particularly when only a few time units are to be studied on one man's work, and when this man will not be personally affected by the results of the observations. In these cases, the watch book of Mr. Thompson, holding the watches in the cover, is especially useful. A good deal of judgment is required to know when to time openly, or the reverse.

The operation selected for illustration on the note sheet shown in Fig. 2, page 151, is the excavation of earth with wheelbarrows, and the values given are fair averages of actual contract work where the wheelbarrow man fills his own barrow. It is obvious that similar methods of analyzing and recording may be applied to work ranging from unloading coal to skilled labor on fine machine tools.

The method of using the note sheets for timing a workman is as follows:

After entering the necessary descriptive matter at the top of the sheet, divide the operation to be timed into its elementary units, and write these units one after another under the heading "Detail Operations." If the job is long and complicated, it may be analyzed while the timing is going on, and the elementary units entered then instead of beforehand. In wheelbarrow work as illustrated in the example shown on the note sheet, the elementary units consist of "filling barrow," "starting" (which includes throwing down shovel and lifting handles of barrow), "wheeling full," etc. These units might have been further subdivided — the first one into time for loading one shovelful, or still further into the time for filling and the time for emptying each shovelful. The letters a, b, c, etc., which are printed, are simply for convenience in designating the elements.

We are now ready for the stop watch, which, to save clerical work, should be provided with a decimal dial similar to that shown in Fig. 4. The

FIGURE 4. — STOP WATCH WITH DECIMAL FACE

method of using this and recording the times depends upon the character of the time observations. In all cases, however, the stop watch times are recorded in the columns headed "Time" at the top of the right-hand half of the note sheet. These columns are the only place on the face of the sheet where stop watch readings are to be entered. If more space is required for these times, they should be entered on the back of the sheet. The rest of the figures (except those on the left-hand side of the note sheet, which may be taken from an ordinary timepiece) are the results of calculation, and may be made in the office by any clerk.

As has been stated, the method of recording the stop watch observations depends upon the work which is being observed. If the operation consists of the same element repeated over and over, the time of each may be set down separately; or, if the element is very small, the total time of, say, ten may be entered as a fraction, with the time for all ten observations as the numerator, and the number of observations for the denominator.

In the illustration given on the note sheet, Fig. 2, the operation consists of a series of elements. In such a case, the letters designating each elementary unit are entered under the columns "Op.," the stop watch is thrown to zero, and started as the man commences to work. As each new division of the operation (that is, as each elementary unit or unit time) is begun, the time is recorded. During any special delay the watch may be stopped, and started again from the same point, although, as a rule, Mr. Thompson advocates allowing the watch to run continuously, and enters the time of such a stop, designating it for convenience by the letter "Y."

In the case we are considering, two kinds of materials were handled — sand and clay. The time of each of the unit times, except the "filling," is the same for both sand and clay; hence, if we have sufficient observations on either one of the materials, the only element of the other which requires to be timed is the loading. This illustrates one of the merits of the elementary system.

The column "Av." is filled from the preceding column. The figures thus found are the actual net times of the different unit times. These unit times are averaged and entered in the "Time" column, on the lower half of the right-hand page, preceded, in the "No." column, by the number of observations which have been taken of each unit. These times, combined and compared with the gross times on the left-hand page, will determine the percentage lost in resting and other necessary delays. A convenient method for obtaining the time of an operation, like picking, in which the quantity is difficult to measure, is suggested by the records on the left-hand page.

The percentage of the time taken in rest and other necessary delays, which is noted on the sheet as, in this case, about 27 per cent., is obtained by a comparison of the average net "time per barrow" on the right with the "time per barrow" on the left. The latter is the quotient of the total time shoveling and wheeling divided by the number of loads wheeled.

It must be remembered that the example given is simply for illustration. To obtain accurate average times, for any item of work under specified conditions, it is necessary to take observations upon a number of men, each of whom is at work under conditions which are comparable. The total number of observations which should be taken of any one elementary unit depends upon its variableness, and also upon its frequency of occurrence in a day's work.

An expert observer can, on many kinds of work, time two or three men at the same time with the same watch, or he can operate two or three watches — one for each man. A note sheet can contain only a comparatively few observations. It is not convenient to make it of larger size than the dimensions given, when a watch-book is to be used, although it is perfectly feasible to make the horizontal rulings 8 lines to the inch instead of 5 lines to the inch as on the sample sheet. There will have to be, in almost all cases, a large number of note sheets on the same subject. Some system must be arranged for collecting and tabulating these records. On Tables 2A and 2B (pages 160 and 161) is shown the form used for tabulating. The length should be either 17 or 22 inches. The height of the form is 11 inches. With these dimensions a form may be folded and filed with ordinary letter sheets ($8\frac{1}{2}$ inches by 11 inches). The ruling which has been found most convenient is for the vertical divisions 3 columns to $1\frac{1}{8}$ inches, while the horizontal lines are ruled 6 to the inch. The columns may, or may not, have printed headings.

The data from the note sheet in Fig. 2 (page 151) is copied on to the table for illustration. The first columns of the table are descriptive. The rest of them are arranged so as to include all of the unit times, with any other data which are to be averaged or used when studying the results. At the extreme right of the sheet the gross times, including rest and necessary delay, are recorded and the percentages of rest are calculated.

Formulæ are convenient for combining the elements. For simplicity, in the example of barrow excavation, each of the unit times may be designated by the same letters used on the note sheet (Fig. 2) although in practise each element can best be designated by the initial letters of the words describing it.

Let

a = time filling a barrow with any material.
b = time preparing to wheel.
c = time wheeling full barrow 100 feet.
d = time dumping and turning.
e = time returning 100 feet with empty barrow.
f = time dropping barrow and starting to shovel.
p = time loosening one cubic yard with the pick.
P = percentage of a day required to rest and necessary delays.
L = load of a barrow in cubic feet.
B = time per cubic yard picking, loading, and wheeling any given kind of earth to any given distance when the wheeler loads his own barrow.

Then

$$B = \left(p + \left[a + b + d + f + \frac{\text{distance hauled}}{100}(c+e) \right] \frac{27}{L} \right)(1+P) \quad . \quad . \quad (1)$$

This general formula for barrow work can be simplified by choosing average values for the constants, and substituting numerals for the letters

TABLE 2A

LOADING BARROW

NOTE SHEET	Department	Men	Implements	Description	Material	Capacity of a barrow, cu. ft.	No. shovels per barrow	Capacity of a shovel, cu. ft.	No. obs.	Time filling barrow, minutes	Time per shovel, minutes	Time per cu. ft., minutes	Remarks
3-10-03	Construction	Johnson	No. 3 shovel and contractor's wood barrow	Wheeler loading his own barrow	Clay		13.5		4	1.948	0.144		
		Flaherty	Same	Same	Sand		13.2		4	1.240	0.094		

STARTING

No. obs.	Time per barrow, minutes	Time per cu. ft., minutes
4	0.182	

WHEELING FULL

No. obs.	Distance wheeled, ft.	Total time wheeling barrow, minutes	Time per 100 ft. per barrow, minutes	Time per 100 ft. per cu. ft., minutes
4	50	0.225	0.450	

DUMPING

Remarks	No. obs.	Time per barrow, minutes	Time per cu. ft., minutes
Level	4	0.172	

TABLE 2B

	RETURNING EMPTY				GETTING READY TO SHOVEL			SUMMATION OF DETAIL OPERATIONS	
No. obs.	Distance wheeled	Total time wheeling barrow, minutes	Time per 100 ft. per barrow, minutes	Time per 100 ft. per cu. ft., minutes	Remarks	No. obs.	Time per barrow, minutes	Time per cu. ft., minutes	No. obs.	Distance (each way), ft.
4	50	0.260	0.520		Level	4	0.162		4	50

SUMMATION OF DETAIL OPERATIONS		COMPLETE OPERATIONS						
Total time per barrow, minutes	Total time per cu. ft., minutes	No. trips	Distance (each way), ft.	Total time of all trips, minutes	Time per barrow, minutes	Time per cu. ft., minutes	Rest and delay, per cent	Remarks
2.241		33	50	124	3.76			Time loosening clay not included
		43	50	122	2.84		27	Sand required no loosening

TABLES 2A AND 2B. — TIME STUDY ASSEMBLING SHEET
Showing method of collating results of studies on earth-work barrows

now representing them. Substituting the average values from the note sheet on Fig. 2 (page 151), our formula becomes:

$B = (p + [a + 0.18 + 0.17 + 0.16 +$

or $\dfrac{\text{distance hauled}}{100}(0.22 + 0.26)]\dfrac{27}{L})1.27,$

$B = \left(p + [a + 0.51 + (0.0048) \text{ distance hauled}]\dfrac{27}{L}\right)1.27 \quad . \quad (2)$

Formula 2 is applicable to any kind of earth hauled by men working at the speeds recorded on the note sheet to any distance.

For sand, still using the values given on the note sheet (Fig. 2):

$B = \left(0 + [1.24 + 0.51 + 0.0048 \text{ (distance hauled)}]\dfrac{27}{2.32}\right)1.27,$

or

$B = 25.86 + 0.071 \text{ (distance hauled)} \quad \ldots \ldots \quad (3)$

For a 50-foot haul:

$B = 25.86 + 0.071\ (50) = 29.4$ min. as the time for one man to load and wheel one cubic yard of sand a distance of 50 feet.

In classes of work where the percentage of rest varies with the different elements of an operation it is most convenient to correct all of the elementary times by the proper percentages before combining them. Sometimes after having constructed a general formula, it may be solved by setting down the substitute numerical values in a vertical column for direct addition.

Table 3 (page 164) gives the times for throwing earth to different distances and different heights. It will be seen that for each special material the time for filling shovel remains the same regardless of the distance to which it is thrown. Each kind of material requires a different time for filling the shovel. The time throwing one shovelful, on the other hand, varies with the length of throw, but for any given distance it is the same for all of the earths. If the earth is of such a nature that it sticks to the shovel, this relation does not hold. For the elements of shoveling we have therefore:

s = time filling shovel and straightening up ready to throw.
t = time throwing one shovelful.
w = time walking one foot with loaded shovel.
w^1 = time returning one foot with empty shovel.
L = load of a shovel in cubic feet.
P = percentage of a day required for rest and necessary delays.
T = time for shoveling one cubic yard.

Our formula, then, for handling any earth after it is loosened, is:

$$T = \left([s + t + (w + w^1) \text{ distance carried}]\frac{27}{L}\right)(1 + P).$$

Where the material is simply thrown without walking, the formula becomes:

$$T = \left((s + t)\frac{27}{L}\right)(1 + P).$$

If weights are used instead of volumes:

Material	Throw: Vertical	Throw: Horizontal	Length of walk	Actual Continuous Work with no Allowance for Rest or other Stops: Time to fill shovel	Time to throw shovelful	Time walking with load	Time of back walk	Total time of complete operation	Volume of shovelful	Weight of shovelful	No. shovelfuls per minute	No. cubic yards per hour	No. pounds per hour	Per cent. of rest	Allowing for Rests and other Necessary Stops: No. shovelfuls per minute	No. cubic yards per hour	No. pounds per hour
	Feet	*Feet*	*Feet*	*Min.*	*Min.*	*Min.*	*Min.*	*Min.*	*Cu. ft.*	*Lbs.*	*Shovels*	*Cu. yds.*	*Lbs.*	*P.c.*	*Shovels*	*Yards*	*Lbs.*
Sand, or Sandy loam	4	5	...	0.073	0.031	......	.....	0.104	0.16	16	9.6	3.4	9,230	30	7.4	2.6	7,100
	6	5	...	0.073	0.043	......	.....	0.110	0.14	14	8.6	2.7	7,250	30	6.6	2.0	5,580
	8	5	...	0.073	0.056	......	.....	0.129	0.11	11	7.8	1.9	5,120	30	6.0	1.4	3,940
	4	7½	...	0.073	0.043	......	.....	0.116	0.14	14	8.6	2.7	7,250	30	6.6	2.1	5,580
	6	7½	...	0.073	0.056	......	.....	0.129	0.12	12	7.8	2.1	5,590	30	6.0	1.6	4,300
	4	10	...	0.073	0.058	......	.....	0.131	0.13	13	7.6	2.2	5,960	30	5.9	1.7	4,580
	6	10	...	0.073	0.076	......	.....	0.149	0.11	11	6.7	1.6	4,440	30	5.2	1.3	3,420
	...	...	20	0.073	0.020	0.080	0.080	0.253	0.20	20	4.0	1.8	4,750	5	3.8	1.7	4,520
	...	...	30	0.073	0.020	0.120	0.120	0.333	0.20	20	3.0	1.3	3,600	5	2.9	1.3	3,430
Loam, gravelly	4	5	...	0.092	0.031	......	.....	0.123	0.14	15.8	8.1	2.5	7,700	30	6.2	2.0	5,920
	6	5	...	0.092	0.043	......	.....	0.135	0.13	14.7	7.4	2.1	6,520	30	5.7	1.6	5,015
	8	5	...	0.092	0.056	......	.....	0.148	0.10	11.3	6.8	1.5	4,580	30	5.2	1.2	3,530
	4	7½	...	0.092	0.043	......	.....	0.135	0.13	14.7	7.4	2.1	6,510	30	5.7	1.6	5,010
	6	7½	...	0.092	0.056	......	.....	0.148	0.11	12.4	6.8	1.6	5,030	30	5.2	1.3	3,870
	4	10	...	0.092	0.058	......	.....	0.150	0.12	13.6	6.7	1.8	5,440	30	5.1	1.4	4,180
	6	10	...	0.092	0.076	......	.....	0.168	0.10	11.3	6.0	1.3	4,030	30	4.6	1.0	3,100
	...	...	20	0.092	0.020	0.080	0.080	0.272	0.19	21.5	3.7	1.6	4,750	5	3.5	1.5	4,520
	...	...	30	0.092	0.020	0.120	0.120	0.352	0.19	21.5	2.8	1.2	3,670	5	2.7	1.1	3,490
Gravel, medium	4	5	...	0.084	0.031	......	.....	0.115	0.12	17.0	8.7	2.3	8,870	30	6.7	1.8	6,820
	6	5	...	0.084	0.043	......	.....	0.127	0.10	14.2	7.9	1.8	6,720	30	6.0	1.3	5,170
	8	5	...	0.084	0.056	......	.....	0.140	0.08	11.4	7.1	1.3	4,880	30	5.5	1.0	3,750
	4	7½	...	0.084	0.043	......	.....	0.127	0.11	15.6	7.9	1.9	7,370	30	6.0	1.5	5,670
	6	7½	...	0.084	0.056	......	.....	0.140	0.09	12.8	7.1	1.4	5,480	30	5.5	1.1	4,220
	4	10	...	0.084	0.058	......	.....	0.142	0.10	14.2	7.0	1.6	6,000	30	5.4	1.2	4,620
	6	10	...	0.084	0.076	......	.....	0.160	0.08	11.4	6.2	1.1	4,270	30	4.8	0.8	3,280
	...	...	20	0.084	0.020	0.080	0.080	0.264	0.15	21.3	3.8	1.3	4,840	5	3.6	1.2	4,610
	...	...	30	0.084	0.020	0.120	0.120	0.344	0.15	21.3	2.9	1.0	3,720	5	2.8	0.9	3,540

TABLE 3.—SHOVELING EARTH IN AVERAGE CONTRACT WORK

EARTH PREVIOUSLY LOOSENED—VOLUMES ARE BASED ON MEASUREMENT IN CUT

Time shoveling one ton $= \left((s+t)\dfrac{\text{No. of lbs. in one ton}}{\text{weight of one shovelful}}\right)(1+P)$.

The writer has found the printed form shown on the insert, Fig. 5 (opposite page 166), useful in studying unit times in a certain class of the hand work done in a machine shop. This blank is fastened to a thin board held in the left hand and resting on the left arm of the observer. A stop watch is inserted in a small compartment attached to the back of the board at a point a little above its center, the face of the watch being seen from the front of the board through a small flap cut partly loose from the observation blank. While the watch is operated by the fingers of the left hand, the right hand of the operator is at all times free to enter the time observations on the blank. A pencil sketch of the work to be observed is made in the blank space on the upper left-hand portion of the sheet. In using this blank, of course, all attempt at secrecy is abandoned.

The mistake usually made by beginners is that of failing to note in sufficient detail the various conditions surrounding the job. It is not at first appreciated that the whole work of the time observer is useless if there is any doubt as to even one of these conditions. Such items, for instance, as the name of the man or men on the work, the number of helpers, and exact description of all of the implements used, even those which seem unimportant, such, for instance, as the diameter and length of bolts and the style of clamps used, the weight of the piece upon which work is being done, etc.

It is also desirable that, as soon as practicable

after taking a few complete sets of time observations, the operator should be given the opportunity of working up one or two sets at least by summing up the unit times and allowing the proper per cent. of rest, etc., and putting them into practical use, either by comparing his results with the actual time of a job which is known to be done in fast time, or by setting a time which a workman is to live up to.

The actual practical trial of the time student's work is most useful, both in teaching him the necessity of carefully noting the minutest details, and on the other hand convincing him of the practicability of the whole method, and in encouraging him in future work.

In making time observations, absolutely nothing should be left to the memory of the student. Every item, even those which appear self-evident, should be accurately recorded. The writer, and the assistant who immediately followed him, both made the mistake of not putting the results of much of their time study into use soon enough, so that many times observations which extended over a period of months were thrown away, in most instances because of failure to note some apparently unimportant detail.

It may be needless to state that when the results of time observations are first worked up, it will take far more time to pick out and add up the proper unit times, and allow the proper percentages of rest, etc., than it originally did for the workman to do the job. This fact need not disturb the operator, however. It will be evident that the slow time made at the start is due to his lack of experience,

FIGURE 5

delays. These elements can, however, be studied with about the same accuracy as the others.

Perhaps the greatest difficulty rests upon the fact that no two men work at exactly the same speed. The writer has found it best to take his time observations on first-class men only, when they can be found; and these men should be timed when working at their best. Having obtained the best time of a first-class man, it is a simple matter to determine the percentage which an average man will fall short of this maximum.

It is a good plan to pay a first-class man an extra price while his work is being timed. When workmen once understand that the time study is being made to enable them to earn higher wages, the writer has found them quite ready to help instead of hindering him in his work. The division of a given job into its proper elementary units, before beginning the time study, calls for considerable skill and good judgment. If the job to be observed is one which will be repeated over and over again, or if it is one of a series of similar jobs which form an important part of the standard work of an establishment, or of the trade which is being studied, then it is best to divide the job into elements which are rudimentary. In some cases this subdivision should be carried to a point which seems at first glance almost absurd.

For example, in the case of the study of the art of shoveling earths, referred to in Table 3, page 164, it will be seen that handling a shovelful of dirt is subdivided into,

$s =$ "Time filling shovel and straightening up ready to throw,"
and $t =$ "Time throwing one shovelful."

The first impression is that this minute subdivision of the work into elements, neither of which takes more than five or six seconds to perform, is little short of preposterous; yet if a rapid and thorough time study of the art of shoveling is to be made, this subdivision simplifies the work, and makes time study quicker and more thorough.

The reasons for this are twofold:

First. In the art of shoveling dirt, for instance, the study of fifty or sixty small elements, like those referred to above, will enable one to fix the exact time for many thousands of complete jobs of shoveling, constituting a very considerable proportion of the entire art.

Second. The study of single small elements is simpler, quicker, and more certain to be successful than that of a large number of elements combined. The greater the length of time involved in a single item of time study, the greater will be the likelihood of interruptions or accidents, which will render the results obtained by the observer questionable or even useless.

There is a considerable part of the work of most establishments that is not what may be called standard work, namely, that which is repeated many times. Such jobs as this can be divided for time study into groups, each of which contains several rudimentary elements. A division of this sort will be seen by referring to the data entered on face of note sheet, Fig. 2 (page 151).

In this case, instead of observing, first, the "time to fill a shovel," and then the time to "throw it into a wheelbarrow," etc., a number of these more rudimentary operations are grouped into the single operation of

a = "Time filling a wheelbarrow with any material."

This group of operations is thus studied as a whole.

Another illustration of the degree of subdivision which is desirable will be found by referring to the inserts, Fig. 5 (opposite page 166).

Where a general study is being made of the time required to do all kinds of hand work connected with and using machine tools, the items printed in detail should be timed singly.

When some special job, not to be repeated many times, is to be studied, then several elementary items can be grouped together and studied as a whole, in such groups for example as:

(a) Getting job ready to set.
(b) Setting work.
(c) Setting tool.
(d) Extra hand work.
(e) Removing work.

And in some cases even these groups can be further condensed.

An illustration of the time units which it is desirable to sum up and properly record and index for a certain kind of lathe work is given in Fig. 6.

THE MIDVALE STEEL CO.

Form D—124.   Machine Shop........................18..........

ESTIMATES FOR WORK ON LATHES

NAME..........................
Sketch.............Number...........
Order..............Weight...........
Metal..............Heat No. ....
Tensile Strength....Chem. Comp......
Per cent. of Stretch....................
HARDNESS, Class...................

OPERATIONS CONNECTED WITH PREPARING TO MACHINE WORK ON LATHES AND WITH REMOVING WORK TO FLOOR AFTER IT HAS BEEN MACHINED	
OPERATIONS	TIME IN MINUTES
Putting chain on, Work on Floor	
Putting chain on, Work on Centers	
Taking off chain, Work on Floor	
Taking off chain, Work on Centers	
Putting on Carrier	
Taking off "	
Lifting Work to Shears	
Getting Work on Centers	
Lifting Work from Centers to Floor	
Turning Work, end for end	
Adjusting Soda Water	
Stamping	
Center-punching	
Trying Trueness with Chalk	
" with Calipers	
" with Gauge	
Putting in Mandrel	
Taking out "	
Putting in Plug Centers	
Taking out " "	
Putting in False Centers	
Taking out " "	
Putting on Spiders	
Taking off "	
Putting on Follow Rest	
Taking off " "	
Putting on Face Plate	
Taking off " "	
Putting on Chuck	
Taking off "	
Laying out	
Changing Tools	
Putting in Packing	
Cut to Cut	
Learning what is to be done	
Considering how to Clamp	
Oiling up	
Cleaning Machine	
Changing Time Notes	
Changing Tools at Tool Room	
Shifting Work	
Putting on Former	
Taking off "	
Adjusting Feed	
" Speed	
" Poppet Head	
" Screw Cutting Gear	
SIGNED TOTAL	

OPERATIONS CONNECTED WITH MACHINING WORK ON LATHES

OPERATIONS	Speed	Feed	Cut	Tool	Inches	Minutes
Turning Feed In						
" " "						
" Hand Feed						
" " "						
Boring Feed In						
" " "						
" Hand Feed						
" " "						
Starting Cut						
" "						
Finishing Cut						
" "						
Fillet						
"						
Collar						
"						
Facing						
"						
Slicing						
"						
Nicking						
"						
Centering						
"						
Filling						
"						
Using Emery Cloth						
" " "						
TOTAL						

Machining—Two Heads Used
"   —One Head Used
Hand Work
Additional Allowance

TOTAL TIME
HIGH RATE
LOW RATE

Remarks

Time actually taken

FIGURE 6.—INSTRUCTION CARD FOR LATHE WORK

The writer has found that when some jobs are divided into their proper elements, certain of these elementary operations are so very small in time that it is difficult, if not impossible, to obtain accurate readings on the watch. In such cases, where the work consists of recurring cycles of elementary operations, that is, where a series of elementary operations is repeated over and over again, it is possible to take sets of observations on two or more of the successive elementary operations which occur in regular order, and from the times thus obtained to calculate the time of each element. An example of this is the work of loading pig iron on to bogies. The elementary operations or elements consist of:

(a) Picking up a pig.
(b) Walking with it to the bogie.
(c) Throwing or placing it on the bogie.
(d) Returning to the pile of pigs.

Here the length of time occupied in picking up the pig and throwing or placing it on the bogie is so small as to be difficult to time, but observations may be taken successively on the elements in sets of three. We may, in other words, take one set of observations upon the combined time of the three elements numbered 1, 2, 3; another set upon elements 2, 3, 4; another set upon elements, 3, 4, 1, and still another upon the set 4, 1, 2. By algebraic equations we may solve the values of each of the separate elements.

If we take a cycle consisting of five (5) elementary operations, a, b, c, d, e, and let observations be taken on three of them at a time, we have the equations:

$$a + b + c = A$$
$$b + c + d = B$$
$$c + d + e = C$$
$$d + e + a = D$$
$$e + a + b = E$$
$$A + B + C + D + E = S.$$

We may solve and obtain:

$$a = A + D - \tfrac{1}{3} S$$
$$b = B + E - \tfrac{1}{3} S$$
$$c = C + A - \tfrac{1}{3} S$$
$$d = D + B - \tfrac{1}{3} S$$
$$e = E + C - \tfrac{1}{3} S$$

The writer was surprised to find, however, that while in some cases these equations were readily solved, in others they were impossible of solution. My friend, Mr. Carl G. Barth, when the matter was referred to him, soon developed the fact that the number of elements of a cycle which may be observed together is subject to a mathematical law, which is expressed by him as follows:

The number of successive elements observed together must be prime to the total number of elements in the cycle.

Namely, the number of elements in any set must contain no factors; that is, must be divisible by no numbers which are contained in the total number of elements. The following table is, therefore, calculated by Mr. Barth showing how many operations may be observed together in various cases. The last column gives the number of observations in a set which will lead to the determination of the results with the minimum of labor.

174   SHOP MANAGEMENT

No. of Operations in the Cycle	No. of Operations that may be observed together	No. observed together that lead to a minimum of labor or is otherwise preferable
3	2	2
4	3	3
5	2, 3, or 4	3 or 4
6	5	5
7	2, 3, 4, 5, or 6	4 or 6
8	3, 5, or 7	5 or 7
9	2, 4, 5, 7, or 8	5 or 8
10	3, 7, or 9	7 or 9
11	2, 3, 4, 5, 6, 7, 8, 9, or 10	5 or 10
12	5, 7, or 11	7 or 11

When time study is undertaken in a systematic way, it becomes possible to do greater justice in many ways both to employers and workmen than has been done in the past. For example, we all know that the first time that even a skilled workman does a job it takes him a longer time than is required after he is familiar with his work, and used to a particular sequence of operations. The practised time student can not only figure out the time in which a piece of work should be done by a good man, after he has become familiar with this particular job through practice, but he should also be able to state how much more time would be required to do the same job when a good man goes at it for the first time; and this knowledge would make it possible to assign one time limit and price for new work, and a smaller time and price for the same job after being repeated, which is much more fair and just to both parties than the usual fixed price.

As the writer has said several times, the difference

between the best speed of a first-class man and the actual speed of the average man is very great. One of the most difficult pieces of work which must be faced by the man who is to set the daily tasks is to decide just how hard it is wise for him to make the task. Shall it be fixed for a first-class man, and if not, then at what point between the first-class and the average? One fact is clear, it should always be well above the performance of the average man, since men will invariably do better if a bonus is offered them than they have done without this incentive. The writer has, in almost all cases, solved this part of the problem by fixing a task which required a first-class man to do his best, and then offering a good round premium. When this high standard is set it takes longer to raise the men up to it. But it is surprising after all how rapidly they develop.

The precise point between the average and the first-class, which is selected for the task, should depend largely upon the labor market in which the works is situated. If the works were in a fine labor market, such, for instance, as that of Philadelphia, there is no question that the highest standard should be aimed at. If, on the other hand, the shop required a good deal of skilled labor, and was situated in a small country town, it might be wise to aim rather lower. There is a great difference in the labor markets of even some of the adjoining states in this country, and in one instance, in which the writer was aiming at a high standard in organizing a works, he found it necessary to import almost all of his men from a neighboring state before meeting with success.

Whether the bonus is given only when the work is done in the quickest time or at some point between this and the average time, *in all cases* the instruction card should state the best time in which the work can be done by a first-class man. There will then be no suspicion on the part of the men when a longer "bonus time" is allowed that the time student does not really know the possibilities of the case. For example, the instruction card might read:

Proper time . . . . 65 minutes
Bonus given first time job is done. 108 minutes

It is of the greatest importance that the man who has charge of assigning tasks should be perfectly straightforward in all of his dealings with the men. Neither in this nor in any other branch of the management should a man make any pretense of having more knowlege than he really possesses. He should impress the workmen with the fact that he is dead in earnest, and that he fully intends to know all about it some day; but he should make no claim to omniscience, and should always be ready to acknowledge and correct an error if he makes one. This combination of determination and fiankness establishes a sound and healthy relation between the management and men.

There is no class of work which cannot be profitably submitted to time study, by dividing it into its time elements, except such operations as take place in the head of the worker; and the writer has even seen a time study made of the speed of an average and first-class boy in solving problems in mathematics.

Clerk work can well be submitted to time study, and a daily task assigned in work of this class which at first appears to be very miscellaneous in its character.

One of the needs of modern management is that of literature on the subject of time study. The writer quotes as follows from his paper on "A Piece Rate System," written in 1895:

"Practically the greatest need felt in an establishment wishing to start a rate-fixing department is the lack of data as to the proper rate of speed at which work should be done. There are hundreds of operations which are common to most large establishments, yet each concern studies the speed problem for itself, and days of labor are wasted in what should be settled once for all, and recorded in a form which is available to all manufacturers.

"What is needed is a hand-book on the speed with which work can be done, similar to the elementary engineering handbooks. And the writer ventures to predict that such a book will before long be forthcoming. Such a book should describe the best method of making, recording, tabulating, and indexing time observations, since much time and effort are wasted by the adoption of inferior methods."

Unfortunately this prediction has not yet been realized. The writer's chief object in inducing Mr. Thompson to undertake a scientific time study of the various building trades and to join him in a publication of this work was to demonstrate on a large scale not only the desirability of accurate time study, but the efficiency and superiority of the method of studying elementary units as outlined above. He trusts that his object may be realized and that the publication of this book may be followed by similar works on other trades and more particularly on the details of machine shop practice, in which he is especially interested.

As a machine shop has been chosen to illustrate the application of such details of scientific management as time study, the planning department, functional foremanship, instruction cards, etc., the description would be far from complete without at least a brief reference to the methods employed in solving the time problem for machine tools.

The study of this subject involved the solution of four important problems:

First. The power required to cut different kinds of metals with tools of various shapes when using different depths of cut and coarseness of feed, and also the power required to feed the tool under varying conditions.

Second. An investigation of the laws governing the cutting of metals with tools, chiefly with the object of determining the effect upon the best cutting speed of each of the following variables:

(*a*) The quality of tool steel and treatment of tools (*i.e.*, in heating, forging, and tempering them).

(*b*) The shape of tool (*i.e.*, the curve or line of the cutting edge, the lip angle, and clearance angle).

(*c*) The duration of cut or the length of time the tool is required to last before being re-ground.

(*d*) The quality or hardness of the metal being cut (as to its effect on cutting speed).

(*e*) The depth of the cut.

(*f*) The thickness of the feed or shaving

(*g*) The effect on cutting speed of using water or other cooling medium on the tool.

Third. The best methods of analyzing the driving and feeding power of machine tools and, after considering their limitations as to speeds and feeds, of deciding upon the proper counter-shaft or other general driving speeds.

Fourth. After the study of the first, second, and third problems had resulted in the discovery of certain clearly defined laws, which were expressed by mathematical formulæ, the last and most difficult task of all lay in finding a means for solving the entire problem which should be so practical and simple as to enable an ordinary mechanic to answer quickly and accurately for each machine in the shop the question, "What driving speed, feed, and depth of cut will in each particular case do the work in the quickest time?"

In 1881, in the machine shop of the Midvale Steel Company, the writer began a systematic study of the laws involved in the first and second problems above referred to by devoting the entire time of a large vertical boring mill to this work, with special arrangements for varying the drive so as to obtain any desired speed. The needed uniformity of the metal was obtained by using large locomotive tires of known chemical composition, physical properties and hardness, weighing from 1,500 to 2,000 pounds.

For the greater part of the succeeding 22 years these experiments were carried on, first at Midvale and later in several other shops, under the general direction of the writer, by his friends and assistants, six machines having been at various times especially fitted up for this purpose.

The exact determination of these laws and their reduction to formulæ have proved a slow but most interesting problem; but by far the most difficult undertaking has been the development of the methods and finally the appliances (*i.e.*, slide rules) for making practical use of these laws after they were discovered.

In 1884 the writer succeeded in making a slow solution of this problem with the help of his friend, Mr. Geo. M. Sinclair, by indicating the values of these variables through curves and laying down one set of curves over another. Later my friend, Mr. H. L. Gantt, after devoting about $1\frac{1}{2}$ years exclusively to this work, obtained a much more rapid and simple solution. It was not, however, until 1900, in the works of the Bethlehem Steel Company, that Mr. Carl G. Barth, with the assistance of Mr. Gantt and a small amount of help from the writer, succeeded in developing a slide rule by means of which the entire problem can be accurately and quickly solved by any mechanic.

The difficulty from a mathematical standpoint of obtaining a rapid and accurate solution of this problem will be appreciated when it is remembered that twelve independent variables enter into each problem, and that a change in any of these will affect the answer.

The instruction card can be put to wide and varied use. It is to the art of management what the drawing is to engineering, and, like the latter, should vary in size and form according to the amount and variety of the information which it is to convey. In some cases it should consist of a pencil memorandum on a small piece of paper which will be sent directly to the man requiring the instructions, while in others it will be in the form of several pages of typewritten matter, properly varnished and mounted, and issued under the check or other record system, so that it can be used time after time. A description of an instruction card of this kind may be useful.

After the writer had become convinced of the economy of standard methods and appliances, and the desirability of relieving the men as far as possible from the necessity of doing the planning, while master mechanic at Midvale, he tried to get his assistant to write a complete instruction card for overhauling and cleaning the boilers at regular periods, to be sure that the inspection was complete, and that while the work was thoroughly done, the boilers should be out of use as short a time as possible, and also to have the various elements of this work done on piece work instead of by the day. His assistant, not having undertaken work of this kind before, failed at it, and the writer was forced to do it himself. He did all of the work of chipping, cleaning, and overhauling a set of boilers and at the same time made a careful time study of each of the elements of the work. This time study showed that a great part of the time was lost owing to the constrained position of the workman. Thick pads were made to fasten to the elbows, knees, and hips; special tools and appliances were made for the various details of the work; a complete list of the tools and implements was entered on the instruction card, each tool being stamped with its own number for identification, and all were issued from the tool room in a tool box so as to keep them together and save time. A separate piece work price was fixed for each of the elements of the job and a thorough inspection of each part of the work secured as it was completed.

The instruction card for this work filled several typewritten pages, and described in detail the order in which the operations should be done and the exact details of each man's work, with the number of each tool required, piece work prices, etc.

The whole scheme was much laughed at when it first went into use, but the trouble taken was fully justified, for the work was better done than ever before, and it cost only eleven dollars to completely overhaul a set of 300 H.P. boilers by this method, while the average cost of doing the same work on day work without an instruction card was sixty-two dollars.

Regarding the personal relations which should be maintained between employers and their men, the writer quotes the following paragraphs from a paper written in 1895. Additional experience has only served to confirm and strengthen these views; and although the greater part of this time, in his work of shop organization, has been devoted to the difficult and delicate task of inducing workmen to change their ways of doing things he has never been opposed by a strike.

"There has never been a strike by men working under this system, although it has been applied at the Midvale Steel Works for the past ten years; and the steel business has proved during this period the most fruitful field for labor organizations and strikes. And this notwithstanding the fact that the Midvale Company has never prevented its men from joining any labor organization. All of the best men in the company saw clearly that the success of a labor organization meant the lowering of their wages in order that the inferior men might earn more, and, of course, could not be persuaded to join.

"I attribute a great part of this success in avoiding strikes to the high wages which the best men were able to earn with the differential rates, and to the pleasant feeling fostered by this system; but this is by no means the whole cause. It has for years been the policy of that company to stimulate the personal ambition of every man in their employ by promoting them either in wages or position whenever they deserved it and the opportunity came.

"A careful record has been kept of each man's good points as well as his shortcomings, and one of the principal duties of each foreman was to make this careful study of his men so that substantial justice could be done to each. When men throughout an establishment are paid varying rates of daywork wages according to their individual worth, some being above and some below the average, it cannot be for the interest of those receiving high pay to join a union with the cheap men.

"No system of management, however good, should be applied in a wooden way. The proper personal relations should always be maintained between the employers and men; and even the prejudices of the workmen should be considered in dealing with them.

"The employer who goes through his works with kid gloves on, and is never known to dirty his hands or clothes, and who either talks to his men in a condescending or patronizing way, or else not at all, has no chance whatever of ascertaining their real thoughts or feelings.

Above all is it desirable that men should be talked to on their own level by those who are over them. Each man should be encouraged to discuss any trouble which he may have, either in the works or outside, with those over him. Men would far rather even be blamed by their bosses, especially if the 'tearing out' has a touch of human nature and feeling in it, than to be passed by day after day without a word, and with no more notice than if they were part of the machinery.

"The opportunity which each man should have of airing his mind freely, and having it out with his employers, is a safety-valve; and if the superintendents are reasonable men, and listen to and treat with respect what their men have to say, there is absolutely no reason for labor unions and strikes.

"It is not the large charities (however generous they may be) that are needed or appreciated by workmen so much as small acts of personal kindness and sympathy, which establish a bond of friendly feeling between them and their employers.

"The moral effect of this system on the men is marked. The feeling that substantial justice is being done them renders them on the whole much more manly, straightforward, and truthful. They work more cheerfully, and are more obliging to one another and their employers. They are not soured, as under the old system, by brooding over the injustice done them; and their spare minutes are not spent to the same extent in criticising their employers."

The writer has a profound respect for the working men of this country. He is proud to say that he has as many firm friends among them as among his other friends who were born in a different class, and he believes that quite as many men of fine character and ability are to be found among the former as in the latter. Being himself a college educated man, and having filled the various positions of foreman, master mechanic, chief draftsman, chief engineer, general superintendent, general manager, auditor, and head of the sales' department, on the one hand, and on the other hand having been for several years a workman, as apprentice, laborer, machinist, and gang boss, his sympathies are equally divided between the two classes.

He is firmly convinced that the best interests of workmen and their employers are the same; so that in his criticism of labor unions he feels that he is advocating the interests of both sides. The following paragraphs on this subject are quoted from the paper written in 1895 and above referred to:

"The author is far from taking the view held by

many manufacturers that labor unions are an almost unmitigated detriment to those who join them, as well as to employers and the general public.

"The labor unions — particularly the trades unions of England — have rendered a great service, not only to their members, but to the world, in shortening the hours of labor and in modifying the hardships and improving the conditions of wage workers.

"In the writer's judgment the system of treating with labor unions would seem to occupy a middle position among the various methods of adjusting the relations between employers and men.

"When employers herd their men together in classes, pay all of each class the same wages, and offer none of them any inducements to work harder or do better than the average, the only remedy for the men lies in combination; and frequently the only possible answer to encroachments on the part of their employers is a strike.

"This state of affairs is far from satisfactory to either employers or men, and the writer believes the system of regulating the wages and conditions of employment of whole classes of men by conference and agreement between the leaders of unions and manufacturers to be vastly inferior, both in its moral effect on the men and on the material interests of both parties, to the plan of stimulating each workman's ambition by paying him according to his individual worth, and without limiting him to the rate of work or pay of the average of his class."

The amount of work which a man should do in a day, what constitutes proper pay for this work, and the maximum number of hours per day which a man should work, together form the most important elements which are discussed between workmen and their employers. The writer has attempted to show that these matters can be much better determined by the expert time student than by either the union or a board of directors, and he firmly believes that in the future scientific time study will establish standards which will be accepted as fair by both sides.

There is no reason why labor unions should not be so constituted as to be a great help both to employers and men. Unfortunately, as they now exist they are in many, if not most, cases a hinderance to the prosperity of both.

The chief reasons for this would seem to be a failure on the part of the workmen to understand the broad principles which affect their best interests as well as those of their employers. It is undoubtedly true, however, that employers as a whole are not much better informed nor more interested in this matter than their workmen.

One of the unfortunate features of labor unions as they now exist is that the members look upon the dues which they pay to the union, and the time that they devote to it, as an investment which should bring them an annual return, and they feel that unless they succeed in getting either an increase in wages or shorter hours every year or so, the money which they pay into the union is wasted. The leaders of the unions realize this and, particularly if they are paid for their services, are apt to spend considerable of their time scaring up grievances whether they exist or not This naturally fosters antagonism instead of friendship between the two sides. There are, of course, marked exceptions to this rule; that of the Brotherhood of Locomotive Engineers being perhaps the most prominent.

The most serious of the delusions and fallacies under which workmen, and particularly those in many of the unions, are suffering is that it is for their interest to limit the amount of work which a man should do in a day.

There is no question that the greater the daily output of the average individual in a trade the greater will be the average wages earned in the trade, and that in the long run turning out a large amount of work each day will give them higher wages, steadier and more work, instead of throwing them out of work. The worst thing that a labor union can do for its members in the long run is to limit the amount of work which they allow each workman to do in a day. If their employers are in a competitive business, sooner or later those competitors whose workmen do not limit the output will take the trade away from them, and they will be thrown out of work. And in the meantime the small day's work which they have accustomed themselves to do demoralizes them, and instead of developing as men do when they use their strength and faculties to the utmost, and as men should do from year to year, they grow lazy, spend much of their time pitying themselves, and are less able to compete with other men. Forbidding their members to do more than a given amount of work in a day has been the greatest mistake made by the English trades unions. The whole of that country is suffering more or less from this error now. Their workmen are for this reason receiving lower wages than they might get, and in many cases the men, under the influence of this idea, have grown so slow that they would find it difficult to do a good day's work even if public opinion encouraged them in it.

In forcing their members to work slowly they use certain cant phrases which sound most plausible until their real meaning is analyzed. They continually use the expression, "Workmen should not be asked to do more than a fair day's work," which sounds right and just until we come to see how it is applied. The absurdity of its usual application would be apparent if we were to apply it to animals. Suppose a contractor had in his stable a miscellaneous collection of draft animals, including small donkeys, ponies, light horses, carriage horses and fine dray horses, and a law were to be made that no animal in the stable should be allowed to do more than "a fair day's work" for a donkey. The injustice of such a law would be apparent to every one. The trades unions, almost without an exception, admit all of those in the trade to membership — providing they pay their dues. And the difference between the first-class men and the poor ones is quite as great as that between fine dray horses and donkeys. In the case of horses this difference is well known to every one; with men, however, it is not at all generally recognized. When a labor union, under the cloak of the expression "a fair day's work," refuses to allow a first-class man to do any more work than a slow or inferior workman can do, its action is quite as absurd as limiting the work of a fine dray horse to that of a donkey would be.

Promotion, high wages, and, in some cases, shorter hours of work are the legitimate ambitions of a workman, but any scheme which curtails the output should be recognized as a device for lowering wages in the long run.

Any limit to the *maximum* wages which men are allowed to earn in a trade is equally injurious to their best interests. The "minimum wage" is the least harmful of the rules which are generally adopted by trades unions, though it frequently works an injustice to the better workmen. For example, the writer has been used to having his machinists earn all the way from $1.50 to seven and eight dollars per day, according to the individual worth of the men. Supposing a rule were made that no machinist should be paid less than $2.50 per day. It is evident that if an employer were forced to pay $2.50 per day to men who were only worth $1.50 or $1.75, in order to compete he would be obliged to lower the wages of those who in the past were getting more than $2.50, thus pulling down the better workers in order to raise up the poorer men. Men are not born equal, and any attempt to make them so is contrary to nature's laws and will fail.

Some of the labor unions have succeeded in persuading the people in parts of this country that there is something sacred in the cause of union labor and that, in the interest of this cause, the union should receive moral support whether it is right in any particular case or not.

Union labor is sacred just so long as its acts are fair and good, and it is damnable just as soon as its acts are bad. Its rights are precisely those of non-union labor, neither greater nor less. The boycott, the use of force or intimidation, and the oppression of non-union workmen by labor unions are damnable; these acts of tyranny are thoroughly un-American and will not be tolerated by the American people.

One of the most interesting and difficult problems connected with the art of management is how to persuade union men to do a full day's work if the union does not wish them to do it. I am glad of the opportunity of saying what I think on the matter, and of explaining somewhat in detail just how I should expect, in fact, how I have time after time induced union men to do a large day's work, quite as large as other men do.

In dealing with union men certain general principles should never be lost sight of. These principles are the proper ones to apply to all men, but in dealing with union men their application becomes all the more imperative.

First. One should be sure, beyond the smallest doubt, that what is demanded of the men is entirely just and can surely be accomplished. This certainty can only be reached by a minute and thorough time study.

Second. Exact and detailed directions should be given to the workman telling him, not in a general way but specifying in every small particular, just what he is to do and how he is to do it.

Third. It is of the utmost importance in starting to make a change that the energies of the management should be centered upon one single workman, and that no further attempt at improvement should be made until entire success has been secured in this case. Judgment should be used in selecting for a start work of such a character that the most clear cut and definite directions can be given regarding it, so that failure to carry out these directions will constitute direct disobedience of a single, straightforward order.

Fourth. In case the workman fails to carry out the order the management should be prepared to demonstrate that the work called for can be done by having some one connected with the management actually do it in the time called for.

The mistake which is usually made in dealing with union men, lies in giving an order which affects a number of workmen at the same time and in laying stress upon the increase in the output which is demanded instead of emphasizing one by one the details which the workman is to carry out in order to attain the desired result. In the first case a clear issue is raised: say that the man must turn out fifty per cent. more pieces than he has in the past, and therefore it will be assumed by most people that he must work fifty per cent. harder. In this issue the union is more than likely to have the sympathy of the general public, and they can logically take it up and fight upon it. If, however, the workman is given a series of plain, simple, and reasonable orders, and is offered a premium for carrying them out, the union will have a much more difficult task in defending the man who disobeys them. To illustrate: If we take the case of a complicated piece of machine work which is being done on a lathe or other machine tool, and the workman is called upon (under the old type of management) to increase his output by twenty-five or fifty per cent. there is opened a field of argument in which the assertion of the man, backed by the union, that the task is impossible or too hard, will have quite as much weight as that of the management. If, however, the management begins by analyzing in detail just how each section of the work should be done and then writes out complete instructions specifying the tools to be used in succession, the cone step on which the driving belt is to run, the depth of cut and the feed to be used, the exact manner in which the work is to be set in the machine, etc., and if before starting to make any change they have trained in as functional foremen several men who are particularly expert and well informed in their specialities, as, for instance, a speed boss, gang boss, and inspector; if you then place for example a speed boss alongside of that workman, with an instruction card clearly written out, stating what both the speed boss and the man whom he is instructing are to do, and that card says you are to use such and such a tool, put your driving belt on this cone, and use this feed on your machine, and if you do so you will get out the work in such and such a time, I can hardly conceive of a case in which a union could prevent the boss from ordering the man to put his driving belt just where he said and using just the feed that he said, and in doing that the workman can hardly fail to get the work out on time. No union would dare to say to the management of a works, you shall not run the machine with the belt on this or that cone step. They do not come down specifically in that way; they say, "You shall not work so fast," but they do not say, "You shall not use such and such a tool, or run with such a feed or at such a speed." However much they might like to do it, they do not dare to interfere specifically in this way. Now, when your single man under the supervison of a speed boss, gang boss, etc., runs day after day at the given speed and feed, and gets work out in the time that the instruction card calls for, and when a premium is kept for him in the office for having done the work in the required time, you begin to have a moral suasion on that workman which is very powerful. At first he won't take the premium if it is contrary to the laws of his union, but as time goes on and it piles up and amounts to a big item, he will be apt to step into the office and ask for his premium, and before long your man will be a thorough convert to the new system. Now, after one man has been persuaded, by means of the four functional foremen, etc., that he will earn more money under the new system than under the laws of the union, you can then take the next man, and so convert one after another right through your shop, and as time goes on public opinion will swing around more and more rapidly your way.

I have a profound respect for the workmen of the United States; they are in the main sensible men — not all of them, of course, but they are just as sensible as are those on the side of the management. There are some fools among them; so there are among the men who manage industrial plants. They are in many respects misguided men, and they require a great deal of information that they have not got. So do most managers.

All that most workmen need to make them do what is right is a series of proper object lessons. When they are convinced that a system is offered them which will yield them larger returns than the union provides for, they will promptly acquiesce. The necessary object lessons can best be given by centering the efforts of the management upon one spot. The mistake that ninety-nine men out of a hundred make is that they have attempted to influence a large body of men at once instead of taking one man at a time.

Another important factor is the question of time. If any one expects large results in six months or a year in a very large works he is looking for the impossible. If any one expects to convert union men to a higher rate of production, coupled with high wages, in six months or a year, he is expecting next to an impossibility. But if he is patient enough to wait for two or three years, he can go among almost any set of workmen in the country and get results.

Some method of disciplining the men is unfortunately a necessary element of all systems of management. It is important that a consistent, carefully considered plan should be adopted for this as for all other details of the art. No system of discipline is at all complete which is not sufficiently broad to cover the great variety in the character and disposition of the various men to be found in a shop. There is a large class of men who require really no discipline in the ordinary acceptance of the term; men who are so sensitive, conscientious and desirous of doing just what is right that a suggestion, a few words of explanation, or at most a brotherly admonition is all that they require. In all cases, therefore, one should begin with every new man by talking to him in the most friendly way, and this should be repeated several times over until it is evident that mild treatment does not produce the desired effect.

Certain men are both thick-skinned and coarse-grained, and these individuals are apt to mistake a mild manner and a kindly way of saying things for timidity or weakness. With such men the severity both of words and manner should be gradually increased until either the desired result has been attained or the possibilities of the English language have been exhausted.

Up to this point all systems of discipline should be alike. There will be found in all shops, however, a certain number of men with whom talk, either mild or severe, will have little or no effect, unless it produces the conviction that something more tangible and disagreeable will come next. The question is what this something shall be.

Discharging the men is, of course, effective as far as that individual is concerned, and this is in all cases the last step; but it is desirable to have several remedies between talking and discharging more severe than the one and less drastic than the other.

Usually one or more of the following expedients are adopted for this purpose:

First. Lowering the man's wages.

Second. Laying him off for a longer or shorter period of time.

Third. Fining him.

Fourth. Giving him a series of "bad marks," and when these sum up to more than a given number per week or month, applying one or the other of the first three remedies.

The general objections to the first and second expedients is that for a large number of offenses they are too severe, so that the disciplinarian hesitates to apply them. The men find this out, and some of them will take advantage of this and keep much of the time close to the limit. In laying a man off, also, the employer is apt to suffer as much in many cases as the man, through having machinery lying idle or work delayed. The fourth remedy is also objectionable because some men will deliberately take close to their maximum of "bad marks."

In the writer's experience, the fining system, if justly and properly applied, is more effective and much to be preferred to either of the others. He has applied this system of discipline in various works with uniform success over a long period of years, and so far as he knows, none of those who have tried it under his directions have abandoned it.

The success of the fining system depends upon two elements:

First. The impartiality, good judgment and justice with which it is applied.

Second. Every cent of the fines imposed should in some form be returned to the workmen. If any part of the fines is retained by the company, it is next to impossible to keep the workmen from believing that at least a part of the motive in fining them is to make money out of them; and this thought works so much harm as to more than overbalance the good effects of the system. If, however, all of the fines are in some way promptly returned to the men, they recognize it as purely a system of discipline, and it is so direct, effective and uniformly just that the best men soon appreciate its value and approve of it quite as much as the company.

In many cases the writer has first formed a mutual beneficial association among the employés, to which all of the men as well as the company contribute. An accident insurance association is much safer and less liable to be abused than a general sickness or life insurance association; so that, when practicable, an association of this sort should be formed and managed by the men. All of the fines can then be turned over each week to this association and so find their way directly back to the men.

Like all other elements, the fining system should not be plunged into head first. It should be worked up to gradually and with judgment, choosing at first only the most flagrant cases for fining and those offenses which affect the welfare of some of the other workmen. It will not be properly and most effectively applied until small offenses as well as great receive their appropriate fine. The writer has fined men from one cent to as high as sixty dollars per fine. It is most important that the fines should be applied absolutely impartially to all employés, high and low. The writer has invariably fined himself just as he would the men under him for all offenses committed.

The fine is best applied in the form of a request to contribute a certain amount to the mutual beneficial association, with the understanding that unless this request is complied with the man will be discharged.

In certain cases the fining system may not produce the desired result, so that coupled with it as an additional means of disciplining the men should be the first and second expedients of "lowering wages" and "laying the men off for a longer or shorter time."

The writer does not at all depreciate the value of the many semi-philanthropic and paternal aids and improvements, such as comfortable lavatories, eating rooms, lecture halls, and free lectures, night schools, kindergartens, baseball and athletic grounds, village improvement societies, and mutual beneficial associations, unless done for advertising purposes. This kind of so-called welfare work all tends to improve and elevate the workmen and make life better worth living. Viewed from the managers' standpoint they are valuable aids in making more intelligent and better workmen, and in promoting a kindly feeling among the men for their employers. They are, however, of distinctly secondary importance, and should never be allowed to engross the attention of the superintendent to the detriment of the more important and fundamental elements of management. They should come in all establishments, but they should come only after the great problem of work and wages has been permanently settled to the satisfaction of both parties. The solution of this problem will take more than the entire time of the management in the average case for several years.

Mr. Patterson, of the National Cash Register Company, of Dayton, Ohio, has presented to the world a grand object lesson of the combination of many philanthropic schemes with, in many respects, a practical and efficient management. He stands out a pioneer in this work and an example of a kind-hearted and truly successful man. Yet I feel that the recent strike in his works demonstrates all the more forcibly my contention that the establishment of the semi-philanthropic schemes should follow instead of preceding the solution of the wages question; unless, as is very rarely the case, there are brains, energy and money enough available in a company to establish both elements at the same time.

Unfortunately there is no school of management. There is no single establishment where a relatively large part of the details of management can be seen, which represent the best of their kinds. The finest developments are for the most part isolated, and in many cases almost buried with the mass of rubbish which surrounds them.

Among the many improvements for which the originators will probably never receive the credit which they deserve the following may be mentioned.

The remarkable system for analyzing all of the work upon new machines as the drawings arrived from the drafting-room and of directing the movement and grouping of the various parts as they progressed through the shop, which was developed and used for several years by Mr. Wm. H. Thorne, of Wm. Sellers & Co., of Philadelphia, while the company was under the general management of Mr. J. Sellers Bancroft. Unfortunately the full benefit of this method was never realized owing to the lack of the other functional elements which should have accompanied it.

And then the employment bureau which forms such an important element of the Western Electric Company in Chicago; the complete and effective system for managing the messenger boys introduced by Mr. Almon Emrie while superintendent of the Ingersoll Sargent Drill Company, of Easton, Pa.; the mnemonic system of order numbers invented by Mr. Oberlin Smith and amplified by Mr. Henry R. Towne, of The Yale & Towne Company, of Stamford, Conn.; and the system of inspection introduced by Mr. Chas. D. Rogers in the works of the American Screw Company, at Providence, R. I.

and the many good points in the apprentice system developed by Mr. Vauclain, of the Baldwin Locomotive Works, of Philadelphia.

The card system of shop returns invented and introduced as a complete system by Captain Henry Metcalfe, U. S. A., in the government shops of the Frankford Arsenal represents another such distinct advance in the art of management. The writer appreciates the difficulty of this undertaking as he was at the same time engaged in the slow evolution of a similar system in the Midvale Steel Works, which, however, was the result of a gradual development instead of a complete, well thought out invention as was that of Captain Metcalfe.

The writer is indebted to most of these gentlemen and to many others, but most of all to the Midvale Steel Company, for elements of the system which he has described.

The rapid and successful application of the general principles involved in any system will depend largely upon the adoption of those details which have been found in actual service to be most useful. There are many such elements which the writer feels should be described in minute detail. It would, however, be improper to burden this record with matters of such comparatively small importance.

INDEX

A *Piece Rate System*, 58.
Ability, rising through especial, 17.
Accident insurance associations, 119, 120, 198.
American Machinist cited, 38.
American Screw Works, 73.
American Society of Mechanical Engineers, 5, 37, 58, 80.
Analysis of orders for machines, 111, 112; of inquiries for new work, 111, 114.
Apprentice system of Mr. Vauclain, 202.
Assembling sheet for time study, 160, 161.
Average man, work of, compared with first-class man, 24, 28, 50.

Balance clerk, duties of, 113, 114.
Barth, Carl G., law of cycle of operations discovered by, 173; developed a slide rule, 180.
Belts, the tightening of, 125, 126.
Bench work, time study for, 111–113.
Bethlehem Steel Co., 46; case of, used in illustration of shop management, 46–56, 73; functional foremanship in, 105, 106; concentration of departments in, 110.
Bicycle balls, inspection of, 85–90.
Bonus, men do better when it is offered, 175; time, 176.
Bosses, gang, duties of, in military type of organization, 96–98; eight under functional management, 99, 100; executive functional, four types of, 100; gang, 100, 101; speed, 101; inspectors, 101; repair, 101, 102; of the planning room, four types of, 102; improvement due to introduction of, 108; and over-foremen, in system of functional foremanship, 108, 109.
Boycott, the, 191.

Change in management. *See* Management, change in.
Chemical manufactories, case of rival used as illustration, 18.
Clerks, order of work and route, duties of, 102; instruction card, duties of, 102, 103; time and cost, duties of, 103.
Contract system, 35.
Coöperation, quotation on, from *A Piece Rate System*, 37; no scope for personal ambition in, 37; remoteness of the reward, 37.
Cost, of items manufactured, entered in planning room, 115; of production, lowered by separating brain work from manual labor, 121.
Cycles of elementary operations, 172; mathematical law of, 173, 174.

Day work, 20; task work applied to 71.
Deceit involved in soldiering, 35, 41.
Details must be carefully standardized, 65.
Differential rate system of piece work, 58, 76; compared with task work with a bonus, 76–80; applied to inspection of bicycle balls, 85–90; applied to large engineering establishment, 92–94.
Disciplinarian, shop, duties of, 103, 104, 119.
Disciplining of men, 195–199.
Dividends, relation between the payment of, and shop management, 19, 20.
Dodge, James M., 80, 143.
Dollar, the, 6.
Drifting, objection made to the use of the word, 41.

Economy in industrial engineering, 6, 7.

203

Employers and men, personal relations between, 21, 22, 182–188.
Employment bureau, 118, 119.
Emrie, Almon, his system for managing messenger boys, 201.
Engineer as an Economist, the, 5.
Engineering, analogy between modern methods of shop management and modern, 66–68.
Exception principle, example of, 109; coming more and more into use, 126.
Executive functional bosses, duties of, 100–102.
Expense exhibits, 115.

Fining system, 197–199.
First-class man, his work compared with average man's, 24, 25, 50; wages of, 25, 27; conditions of development, 28; treatment of at Bethlehem Steel Co., 55.
Foremanship, functional. *See* Functional foremanship.
Foremen, their duties under military type of organization, 94; functional, 98, 99, 108, 109; the selecting and training of, 138–140; best to begin by training in the full number of, 140; difficulty of selecting in advance those who are likely to prove successful as, 140, 141; different types of men should be chosen as, 141–143; inspector first to be chosen, 144.
Formulæ in time study, 159, 162, 163, 165.
Four principles of good shop management, 63, 64, 69, 70, 71, 75.
Functional bosses, executive, duties of, 99, 102; of the planning room, duties of, 102–104.
Functional foremanship, advantages of, 104, 105; how to realize full possibilities of, 105, 106; in limited use, 106; managers apologize for, 106; introduced into Midvale Steel Co., 107; best way to introduce, 107, 108; and overforemen, 108, 109; analogy of, to management of large school, 109; selection and training of foremen, 138–140; a difficulty in introducing, 145; objected to, 146. *See* Foremen.

Functional management, what it consists in, 99, 100.

Gang bosses, duties of, in military type of organization, 96–98; duties of, in functional management, 100, 101; improvement due to introduction of, 108.
Gantt, H. L., 70, 77, 180.

Halsey, F. A., 38; quoted, 42.
Hand work, time study for, 111–113.
High pay for success, 64.
High wages and low labor cost the foundation of the best shop management, 22, 23, 25, 27, 46; principles to be followed to obtain, 63.

Improvement of system on plant, 120.
Information bureau, 116.
Inspectors, duties of, 101; improvement due to introduction of, 108; first to be chosen, 144.
Instruction card, for lathe work, 171; description of, 180–182.
Instruction card, clerks, duties of, 102, 103.
Insurance associations, accident, 119, 120, 198.

Labor cost low, the foundation of the best shop management, 22; conditions of high and low, 23.
Labor unions, 186–194; the ideal, 56, 57.
Large daily task, 63.
Lathe work, instruction card for, 171. Limiting of amount of work by unions, 188, 189.
Loafing, 30.
Loss in case of failure, 64.

Machine tools, methods employed in solving the time problem for, 178, 179.
Machines, analysis of orders for, 111, 112; time study for operations done by, 111, 113.
Machinist, in system of functional foremanship, 146.
Maintenance of system and plant, 116–118.
Man, well-rounded, qualities which go to make up, 96.

INDEX 205

Management, Shop, unevenness in development of its elements, 17-19; lack of apparent relation between, and the payment of dividends, 17, 19, 20; rise of men of especial ability, 17; master spirit in, 18; should be looked upon as an art, 18, 60, 63; elements of the successful, 19; in this country behind modern management, 20; art of, defined, 21; relation between employer and men, 21 ff.; high wages and low labor cost the foundation of the best, 22, 63; ignorance in regard to the amount of time required for work, 24, 30, 34; indifference to proper systems, 30; indifference toward the men, 30; contract system, 35; failure of coöperative experiments, 37, 38; Towne-Halsey system of, compared with task system, 42-45; accurate time study the basis of good, 46, 58; example of, at Bethlehem Steel Co., 46-56; old and modern methods compared, 59; difficulties of radical changes in, 60, 64; managers too overwhelmed by work to give thought to, 61, 62; a good organization of more importance than a good plant, 62; four principles which should be followed to unite high wages with low labor cost, 63, 64, 69, 70, 71, 75; a large daily task desirable, 63, 69; standard conditions, 64, 71; high pay for success, 64, 70; loss in case of failure, 64, 71; necessity and economy of a planning department, 64-67; analogy between modern engineering and modern methods of, 66-68; freedom from strikes under scientific, 68; task system, 69, 76, 80; differential rate system, 76-94; shops under-officered, 94.

Change in, functional management, 98-100; should not be made without foresight of what is involved, 128-130; object of, 130, 131; men must be brought to see what is meant by, 131, 132; instruction of men as regards, 132, 133; men must rise from one plane of efficiency to another, 133, 134; should be made gradually, 134, 135; change in, the first step of should be the selection of competent reorganizer, 135, 136; where beginnings should be made in, 136-138; the selecting and training of functional foremen, 138; inspectors first to be chosen, 144; the task idea, 144, 145; a difficulty in, 145.

Master spirit, rise of, from humble position, 17; good management of his particular department, 18.
Messenger boys, Mr. Almon's system for managing, 201.
Messenger system, 118.
Metal tools, improvement in, 8, 9.
Metcalfe, Captain Henry, his card system of shop returns, 202.
Methods, desirability of standardizing, 123, 124.
Midvale Steel Works, under the old system, 44; functional foremanship introduced into, 107; repair force in, 118; study of time problem carried on in, 179; no strikes in, 183; the policy of, 183.
Military plan of management, 92, 98, 99.
Minimum wage, 190.
Mnemonic symbol system, 115, 116.
Mnemonic system of Messrs. Smith and Towne, 201.
Mutual accident insurance associations, 119, 120, 198.

Natural laziness, 30.
Non-producers, and producers, relative numbers of, 121, 122; what is meant by, 122; the diminishing of the number of, 147.
Note sheet for time study, 151-158.

Order of Work and Route clerk, duties of, 102.
Orders for machines, analysis of, 111, 112.
Organization, the building of an efficient, slow and costly, 62; good, of more importance than a good plant, 62, 63. See Management.
Over-foremen, in system of functional foremanship, 108, 109. See Foremen.

Patterson, Mr., of the National Cash Register Co., 200.
Pay department, 115.
Piece Rate System, A, quoted, 37, 81; cited, 58; its main object overlooked, 58; quoted, 177, 183–186.
Piece work, feeling of antagonism under, 35; adoption of, at Bethlehem Steel Co., 50, 53; task work applied to, 73.
Planning department, 64; expense of, 65; necessity and economy of, 67; four functional bosses of, 102–104; where best placed, 109, 110; general management should belong to, 110; leading functions of, 112–120; objections sometimes made to its doing the thinking for the men, 146.
Plant and system, maintenance of, 116–118; improvement of, 120.
Post Office Delivery, 118.
Premium Plan, 43.
Producers, and non-producers, relative numbers of, 121, 122; what is meant by, 122.
Promoting of men, 142, 143.
Purdue University, 5.

Reorganization. *See* Management, change in.
Repair bosses, duties of, 101, 102; improvement due to introduction of, 108.
Repetition in work, 28.
Reports, 117, 126, 127.
Rogers, Charles D., 73, 201.
Rush order department, 120.

Sales department, inquiries for new work received in, analysis of, 114.
Science of Industrial Management, 9, 11.
Selecting and training of men, 138–143.
Sewing-machine, 8.
Shop disciplinarian, duties of, 103, 104, 119.
Shop Management, great value of the monograph, 7, 9.
Sinclair, George M., 180.
Slide rules, 113, 180.
Smart and Honest, 40.
Smith, Oberlin, his mnemonic system of order numbers, 201.
Soldiering, 30–34; under the Towne-Halsey plan, 40; enforced by fellow workmen, 32, 34, 67.
Speed, of a first-class and an average man, 175; need of a book on, 177.
Speed bosses, duties of, 101; improvement due to introduction of, 108.
Speed element in Towne-Halsey and task system compared, 44, 45.
Standard conditions, 64.
Standardizing, desirability of, 123–126.
Standards, 116, 175.
Stop watch, 155.
Strikes, freedom from, under scientific management, 68; none in Midvale Steel Co., 183.
Study of unit times. *See* Time study.
Subdivision of job into unit operations, 168–172.
Symonds Rolling Machine Co., 83.
System and plant, maintenance of, 116–118; improvement of, 120.

Task idea, 144, 145.
Task system compared with Towne-Halsey system, 42.
Task work, 69, 85; with bonus, 70; applied to day work, 71–73; applied to piece work 73; compared with differential piece work, 76–80.
Taylor-White process of treating tool steel, 124.
Taylor, Dr. F. W., his valuable contribution to the art of industrial engineering, 5, 7; *Shop Management*, 7, 9; *The Art of Cutting Metals*, 8; *A Piece Rate System*, 58.
The Art of Cutting Metals, 8.
Thompson, Sanford E., 91; his study of unit times, 150; implements developed by, 150–154.
Thorne, Wm. H., his method of analyzing work upon new machines, 201.
Tickler, use of, 116–118.
Time and cost clerk, duties of, 103.
Time card, and workmen, 127, 128.
Time study, 24, 30, 34, 45; basis of good management, 46, 58, 65;

INDEX

under Towne-Halsey plan, 38, 45; advocated, 46; study of at Bethlehem Steel Co., 48, 52–56; comparison of older methods with modern plan, 59; quickest time, 59; for hand work, 111–113; for operations done by machines, 111, 113; advantages of, 148; difficulties of, 149; made by Mr. Thompson, 150; implements of, developed by Mr. Thompson, 150–154; note sheet, 151–158; watch book, 152, 153; stop watch, 155; of several men at once, 158; formulæ in, 159, 162, 163, 165; assembling sheet, 160, 161; table for shoveling earth in average contract work, 164; every detail necessary in, 165, 166; practical trials of results desirable in, 166; should lead to accurate prediction of time, 167, 168, 174; subdivision of job into units, 168–172; classes of work which can be submitted to, 176, 177; need of literature on the subject, 177; for machine tools, methods employed in, 178, 179; in Midvale Steel Co., 179–182; pay, etc., best determined by, 187.

Tools, desirability of standardizing, 123–126; machine, methods employed in solving the problem for, 178, 179.

Towne, Henry R., 5; *The Engineer as an Economist*, 5; mnemonic system of order numbers amplified by, 201.

Towne-Halsey system of management, described, 38–42, 59; and task system compared, 42; writer approves the plan of, 39, 61.

Training and selecting of men, 138–143.

Transportation, time study for, 111–113.

Trusts, component companies of, built up through especial ability of one or two men, 17.

Typewriting-machine, 8.

Union men, how to deal with, 191–194.
Unions, labor, 186–194.
Unit times, study of. *See* Time Study.

Vauclain, Mr., of the Baldwin Locomotive Works, his apprentice system, 202.
Vise work, time study for, 111–113.

Wadleigh, A. B., 54.
Wage, minimum, 190.
Wages, for first-class men, 25–27; should be regulated to fit special work, 28.
Ward, Artemus, quoted, 70.
Watch book, 152, 153.
Welfare work, 199, 200.
White, J. Maunsel, part discoverer of the Taylor-White process of treating tool steel, 124.
Workman, and employer, interests should be mutual, 20; and employer, relations between, 21, 182–188; average and first-class, 24; should be given highest class of work for which he is fitted, 28; 29; should be called upon to do his best, 28, 29; should be paid according to his work, 29; loafing and systematic soldiering, 30–34; objection to piece work, 34; under contract system, 35; in military type of organization, 99; in functional management, 99, 100; and use of time card, 127, 128; must be brought to see what change in organization means, 131, 132; instruction of, as regards reorganization, 132, 133; must rise from one plane of efficiency to another, 133, 134; looks upon change as antagonistic to his interests, 137; different types of men should be chosen, 141–143; his mistake in limiting amount of work, 188, 189; needs proper object lessons, 195; the disciplining of, 195–199 *See* Union men.

The Principles of
Scientific Management

The Principles of
Scientific Management

BY

FREDERICK WINSLOW TAYLOR, M.E., Sc.D.
PAST PRESIDENT OF THE AMERICAN SOCIETY OF
MECHANICAL ENGINEERS

INTRODUCTION

PRESIDENT ROOSEVELT, in his address to the Governors at the White House, prophetically remarked that "The conservation of our national resources is only preliminary to the larger question of national efficiency."

The whole country at once recognized the importance of conserving our material resources and a large movement has been started which will be effective in accomplishing this object. As yet, however, we have but vaguely appreciated the importance of "the larger question of increasing our national efficiency."

We can see our forests vanishing, our water-powers going to waste, our soil being carried by floods into the sea; and the end of our coal and our iron is in sight. But our larger wastes of human effort, which go on every day through such of our acts as are blundering, ill-directed, or inefficient, and which Mr. Roosevelt refers to as a lack of "national efficiency," are less visible, less tangible, and are but vaguely appreciated.

We can see and feel the waste of material things. Awkward, inefficient, or ill-directed movements of men, however, leave nothing visible or tangible behind them. Their appreciation calls for an act of memory, an effort of the imagination. And for this reason, even though our daily loss from this source is greater than from our waste of material things, the one has stirred us deeply, while the other has moved us but little.

As yet there has been no public agitation for "greater national efficiency," no meetings have been called to consider how this is to be brought about. And still there are signs that the need for greater efficiency is widely felt.

The search for better, for more competent men, from the presidents of our great companies down to our household servants, was never more vigorous than it is now. And more than ever before is the demand for competent men in excess of the supply.

What we are all looking for, however, is the ready-made, competent man; the man whom some one else has trained. It is only when we fully realize that our duty, as well as our opportunity, lies in systematically cooperating to train and to make this competent man, instead of in hunting for a man whom some one else has trained, that we shall be on the road to national efficiency.

In the past the prevailing idea has been well expressed in the saying that "Captains of industry are born, not made"; and the theory has been that if one could get the right man, methods could be safely left to him. In the future it will be appreciated that our leaders must be trained right as well as born right, and that no great man can (with the old system of personal management) hope to compete with a number of ordinary men who have been properly organized so as efficiently to cooperate.

In the past the man has been first; in the future the system must be first. This in no sense, however, implies that great men are not needed. On the contrary, the first object of any good system must be that of developing first-class men; and under systematic management the best man rises to the top more certainly and more rapidly than ever before.

This paper has been written:

First. To point out, through a series of simple illustrations, the great loss which the whole country is suffering through inefficiency in almost all of our daily acts.

Second. To try to convince the reader that the remedy for this inefficiency lies in systematic management, rather than in searching for some unusual or extraordinary man.

Third. To prove that the best management is a true science, resting upon clearly defined laws, rules, and principles, as a foundation. And further to show that the fundamental principles of scientific management are applicable to all kinds of human activities, from our simplest individual acts to the work of our great corporations, which call for the most elaborate cooperation. And, briefly, through a series of illustrations, to convince the reader that whenever these principles are correctly applied, results must follow which are truly astounding.

This paper was originally prepared for presentation to The American Society of Mechanical Engineers. The illustrations chosen are such as, it is believed, will especially appeal to engineers and to managers of industrial and manufacturing establishments, and also quite as much to all of the men who are working in these establishments. It is hoped, however, that it will be clear to other readers that the same principles can be applied with equal force to all social activities: to the management of our homes; the management of our farms; the management of the business of our tradesmen, large and small; of our churches, our philanthropic institutions, our universities, and our governmental departments.

The Principles of Scientific Management

CHAPTER I

Fundamentals of Scientific Management

THE principal object of management should be to secure the maximum prosperity for the employer, coupled with the maximum prosperity for each employé.

The words "maximum prosperity" are used, in their broad sense, to mean not only large dividends for the company or owner, but the development of every branch of the business to its highest state of excellence, so that the prosperity may be permanent.

In the same way maximum prosperity for each employé means not only higher wages than are usually received by men of his class, but, of more importance still, it also means the development of each man to his state of maximum efficiency, so that he may be able to do, generally speaking, the highest grade of work for which his natural abilities fit him, and it further means giving him, when possible, this class of work to do.

It would seem to be so self-evident that maximum prosperity for the employer, coupled with maximum prosperity for the employé, ought to be the two leading objects of management, that even to state this fact should be unnecessary. And yet there is no question that, throughout the industrial world, a large part of the organization of employers, as well as employés, is for war rather than for peace, and that perhaps the majority on either side do not believe that it is possible so to arrange their mutual relations that their interests become identical.

The majority of these men believe that the fundamental interests of employés and employers are necessarily antagonistic. Scientific management, on the contrary, has for its very foundation the firm conviction that the true interests of the two are one and the same; that prosperity for the employer cannot exist through a long term of years unless it is accompanied by prosperity for the employé, and *vice versa;* and that it is possible to give the workman what he most wants — high wages — and the employer what he wants — a low labor cost — for his manufactures.

It is hoped that some at least of those who do not sympathize with each of these objects may be led to modify their views; that some employers, whose attitude toward their workmen has been that of trying to get the largest amount of work out of them for the smallest possible wages, may be led to see that a more liberal policy toward their men will pay them better; and that some of those workmen who begrudge a fair and even a large profit to their employers, and who feel that all of the fruits of their labor should belong to them, and that those for whom they work and the capital invested in the business are entitled to little or nothing, may be led to modify these views.

No one can be found who will deny that in the case of any single individual the greatest prosperity can exist only when that individual has reached his highest state of efficiency; that is, when he is turning out his largest daily output.

The truth of this fact is also perfectly clear in the case of two men working together. To illustrate: if you and your workman have become so skilful that you and he together are making two pairs of shoes in a day, while your competitor and his workman are making only one pair, it is clear that after selling your two pairs of shoes you can pay your workman much higher wages than your competitor who produces only one pair of shoes is able to pay his man, and that there will still be enough money left over for you to have a larger profit than your competitor.

In the case of a more complicated manufacturing establishment, it should also be perfectly clear that the greatest permanent prosperity for the workman, coupled with the greatest prosperity for the employer, can be brought about only when the work of the establishment is done with the smallest combined expenditure of human effort, plus nature's resources, plus the cost for the use of capital in the shape of machines, buildings, etc. Or, to state the same thing in a different way: that the greatest prosperity can exist only as the result of the greatest possible productivity of the men and machines of the establishment — that is, when each man and each machine are turning out the largest possible output; because unless your men and your machines are daily turning out more work than others around you, it is clear that competition will prevent your paying higher wages to your workmen than are paid to those of your competitor. And what is true as to the possibility of paying high wages in the case of two companies competing close beside one another is also true as to whole districts of the country and even as to nations which are in competition. In a word, that maximum prosperity can exist only as the result of maximum productivity. Later in this paper illustrations will be given of several companies which are earning large dividends and at the same time paying from 30 per cent. to 100 per cent. higher wages to their men than are paid to similar men immediately around them, and with whose employers they are in competition. These illustrations will cover different types of work, from the most elementary to the most complicated.

If the above reasoning is correct, it follows that the most important object of both the workmen and the management should be the training and development of each individual in the establishment, so that he can do (at his fastest pace and with the maximum of efficiency) the highest class of work for which his natural abilities fit him.

These principles appear to be so self-evident that many men may think it almost childish to state them. Let us, however, turn to the facts, as they actually exist in this country and in England. The English and American peoples are the greatest sportsmen in the world. Whenever an American workman plays baseball, or an English workman plays cricket, it is safe to say that he strains every nerve to secure victory for his side. He does his very best to make the largest possible number of runs. The universal sentiment is so strong that any man who fails to give out all there is in him in sport is branded as a "quitter," and treated with contempt by those who are around him.

When the same workman returns to work on the following day, instead of using every effort to turn out the largest possible amount of work, in a majority of the cases this man deliberately plans to do as little as he safely can — to turn out far less work than he is well able to do — in many instances to do not more than one-third to one-half of a proper day's work. And in fact if he were to do his best to turn out his largest possible day's work, he would be abused by his fellow-workers for so doing, even more than if he had proved himself a "quitter" in sport. Underworking, that is, deliberately working slowly so as to avoid doing a full day's work, "soldiering," as it is called in this country, "hanging it out," as it is called in England, "ca canae," as it is called in Scotland, is almost universal in industrial establishments, and prevails also to a large extent in the building trades; and the writer asserts without fear of contradiction that this constitutes the greatest evil with which the working-people of both England and America are now afflicted. It will be shown later in this paper that doing away with slow working and "soldiering" in all its forms and so arranging the relations between employer and employé that each workman will work to his very best advantage and at his best speed, accompanied by the intimate cooperation with the management and the help (which the workman should receive) from the management, would result on the average in nearly doubling the output of each man and each machine. What other reforms, among those which are being discussed by these two nations, could do as much toward promoting prosperity, toward the diminution of poverty, and the alleviation of suffering? America and England have been recently agitated over such subjects as the tariff, the control of the large corporations on the one hand, and of hereditary power on the other hand, and over various more or less socialistic proposals for taxation, etc. On these subjects both peoples have been profoundly stirred, and yet hardly a voice has been raised to call attention to this vastly greater and more important subject of "soldiering," which directly and powerfully affects the wages, the prosperity, and the life of almost every working-man, and also quite as much the prosperity of every industrial establishment in the nation.

FUNDAMENTALS OF SCIENTIFIC MANAGEMENT 15

The elimination of "soldiering" and of the several causes of slow working would so lower the cost of production that both our home and foreign markets would be greatly enlarged, and we could compete on more than even terms with our rivals. It would remove one of the fundamental causes for dull times, for lack of employment, and for poverty, and therefore would have a more permanent and far-reaching effect upon these misfortunes than any of the curative remedies that are now being used to soften their consequences. It would insure higher wages and make shorter working hours and better working and home conditions possible.

Why is it, then, in the face of the self-evident fact that maximum prosperity can exist only as the result of the determined effort of each workman to turn out each day his largest possible day's work, that the great majority of our men are deliberately doing just the opposite, and that even when the men have the best of intentions their work is in most cases far from efficient?

There are three causes for this condition, which may be briefly summarized as:

First. The fallacy, which has from time immemorial been almost universal among workmen, that a material increase in the output of each man or each machine in the trade would result in the end in throwing a large number of men out of work.

Second. The defective systems of management which are in common use, and which make it necessary for each workman to soldier, or work slowly, in order that he may protect his own best interests.

Third. The inefficient rule-of-thumb methods, which are still almost universal in all trades, and in practising which our workmen waste a large part of their effort.

This paper will attempt to show the enormous gains which would result from the substitution by our workmen of scientific for rule-of-thumb methods.

To explain a little more fully these three causes:

First. The great majority of workmen still believe that if they were to work at their best speed they would be doing a great injustice to the whole trade by throwing a lot of men out of work, and yet the history of the development of each trade shows that each improvement, whether it be the invention of a new machine or the introduction of a better method, which results in increasing the productive capacity of the men in the trade and cheapening the costs, instead of throwing men out of work make in the end work for more men.

The cheapening of any article in common use almost immediately results in a largely increased demand for that article. Take the case of shoes, for instance. The introduction of machinery for doing every element of the work which was formerly done by hand has resulted in making shoes at a fraction of their former labor cost, and in selling them so cheap that now almost every man, woman, and child in the working-classes buys one or two pairs of shoes per year, and wears shoes all the time, whereas formerly each workman bought perhaps one pair of shoes every five years, and went barefoot most of the time, wearing shoes only as a luxury or as a matter of the sternest necessity. In spite of the enormously increased output of shoes per workman, which has come with shoe machinery, the demand for shoes has so increased that there are relatively more men working in the shoe industry now than ever before.

The workmen in almost every trade have before them an object lesson of this kind, and yet, because they are ignorant of the history of their own trade even, they still firmly believe, as their fathers did before them, that it is against their best interests for each man to turn out each day as much work as possible.

Under this fallacious idea a large proportion of the workmen of both countries each day deliberately work slowly so as to curtail the output. Almost every labor union has made, or is contemplating making, rules which have for their object curtailing the output of their members, and those men who have the greatest influence with the working-people, the labor leaders as well as many people with philanthropic feelings who are helping them, are daily spreading this fallacy and at the same time telling them that they are overworked.

A great deal has been and is being constantly said about "sweat-shop" work and conditions. The writer has great sympathy with those who are overworked, but on the whole a greater sympathy for those who are *under paid*. For every individual, however, who is overworked, there are a hundred who intentionally underwork — greatly underwork — every day of their lives, and who for this reason deliberately aid in establishing those conditions which in the end inevitably result in low wages. And yet hardly a single voice is being raised in an endeavor to correct this evil.

As engineers and managers, we are more intimately acquainted with these facts than any other class in the community, and are therefore best fitted to lead in a movement to combat this fallacious idea by educating not only the workmen but the whole of the country as to the true facts. And yet we are practically doing nothing in this direction, and are leaving this field entirely in the hands of the labor agitators (many of whom are misinformed and misguided), and of sentimentalists who are ignorant as to actual working conditions.

Second. As to the second cause for soldiering — the relations which exist between employers and employés under almost all of the systems of management which are in common use — it is impossible in a few words to make it clear to one not familiar with this problem why it is that the *ignorance of employers* as to the proper time in which work of various kinds should be done makes it for the interest of the workman to "soldier."

The writer therefore quotes herewith from a paper read before The American Society of Mechanical Engineers, in June, 1903, entitled "Shop Management," which it is hoped will explain fully this cause for soldiering:

"This loafing or soldiering proceeds from two causes. First, from the natural instinct and tendency of men to take it easy, which may be called natural soldiering. Second, from more intricate second thought and reasoning caused by their relations with other men, which may be called systematic soldiering.

"There is no question that the tendency of the average man (in all walks of life) is toward working at a slow, easy gait, and that it is only after a good deal of thought and observation on his part or as a result of example, conscience, or external pressure that he takes a more rapid pace.

"There are, of course, men of unusual energy, vitality, and ambition who naturally choose the fastest gait, who set up their own standards, and who work hard, even though it may be against their best interests. But these few uncommon men only serve by forming a contrast to emphasize the tendency of the average.

"This common tendency to 'take it easy' is greatly increased by bringing a number of men together on similar work and at a uniform standard rate of pay by the day.

"Under this plan the better men gradually but surely slow down their gait to that of the poorest and least efficient. When a naturally energetic man works for a few days beside a lazy one, the logic of the situation is unanswerable.

'Why should I work hard when that lazy fellow gets the same pay that I do and does only half as much work?'

"A careful time study of men working under these conditions will disclose facts which are ludicrous as well as pitiable.

"To illustrate: The writer has timed a naturally energetic workman who, while going and coming from work, would walk at a speed of from three to four miles per hour, and not infrequently trot home after a day's work. On arriving at his work he would immediately slow down to a speed of about one mile an hour. When, for example, wheeling a loaded wheelbarrow, he would go at a good fast pace even up hill in order to be as short a time as possible under load, and immediately on the return walk slow down to a mile an hour, improving every opportunity for delay short of actually sitting down. In order to be sure not to do more than his lazy neighbor, he would actually tire himself in his effort to go slow.

"These men were working under a foreman of good reputation and highly thought of by his employer, who, when his attention was called to this state of things, answered: 'Well, I can keep them from sitting down, but the devil can't make them get a move on while they are at work.'

"The natural laziness of men is serious, but by far the greatest evil from which both workmen and employers are suffering is the *systematic soldiering* which is almost universal under all of the ordinary

FUNDAMENTALS OF SCIENTIFIC MANAGEMENT 21

schemes of management and which results from a careful study on the part of the workmen of what will promote their best interests.

"The writer was much interested recently in hearing one small but experienced golf caddy boy of twelve explaining to a green caddy, who had shown special energy and interest, the necessity of going slow and lagging behind his man when he came up to the ball, showing him that since they were paid by the hour, the faster they went the less money they got, and finally telling him that if he went too fast the other boys would give him a licking.

"This represents a type of *systematic soldiering* which is not, however, very serious, since it is done with the knowledge of the employer, who can quite easily break it up if he wishes.

"The greater part of the *systematic soldiering*, however, is done by the men with the deliberate object of keeping their employers ignorant of how fast work can be done.

"So universal is soldiering for this purpose that hardly a competent workman can be found in a large establishment, whether he works by the day or on piece work, contract work, or under any of the ordinary systems, who does not devote a considerable part of his time to studying just how slow he can work and still convince his employer that he is going at a good pace.

"The causes for this are, briefly, that practically all employers determine upon a maximum sum which they feel it is right for each of their classes of employees to earn per day, whether their men work by the day or piece.

"Each workman soon finds out about what this figure is for his particular case, and he also realizes that when his employer is convinced that a man is capable of doing more work than he has done, he will find sooner or later some way of compelling him to do it with little or no increase of pay.

"Employers derive their knowledge of how much of a given class of work can be done in a day from either their own experience, which has frequently grown hazy with age, from casual and unsystematic observation of their men, or at best from records which are kept, showing the quickest time in which each job has been done. In many cases the employer will feel almost certain that a given job can be done faster than it has been, but he rarely cares to take the drastic measures necessary to force men to do it in the quickest time, unless he has an actual record proving conclusively how fast the work can be done.

"It evidently becomes for each man's interest, then, to see that no job is done faster than it has been in the past. The younger and less experienced men are taught this by their elders, and all possible persuasion and social pressure is brought to bear upon the greedy and selfish men to keep them from making new records which result in temporarily increasing their wages, while all those who come after them are made to work harder for the same old pay.

"Under the best day work of the ordinary type,

when accurate records are kept of the amount of work done by each man and of his efficiency, and when each man's wages are raised as he improves, and those who fail to rise to a certain standard are discharged and a fresh supply of carefully selected men are given work in their places, both the natural loafing and systematic soldiering can be largely broken up. This can only be done, however, when the men are thoroughly convinced that there is no intention of establishing piece work even in the remote future, and it is next to impossible to make men believe this when the work is of such a nature that they believe piece work to be practicable. In most cases their fear of making a record which will be used as a basis for piece work will cause them to soldier as much as they dare.

"It is, however, under piece work that the art of systematic soldiering is thoroughly developed; after a workman has had the price per piece of the work he is doing lowered two or three times as a result of his having worked harder and increased his output, he is likely entirely to lose sight of his employer's side of the case and become imbued with a grim determination to have no more cuts if soldiering can prevent it. Unfortunately for the character of the workman, soldiering involves a deliberate attempt to mislead and deceive his employer, and thus upright and straightforward workmen are compelled to become more or less hypocritical. The employer is soon looked upon as an antagonist, if not an enemy, and the mutual confidence which should exist between a leader and his men, the enthusiasm, the feeling that they are all working for the same end and will share in the results is entirely lacking.

"The feeling of antagonism under the ordinary piece-work system becomes in many cases so marked on the part of the men that any proposition made by their employers, however reasonable, is looked upon with suspicion, and soldiering becomes such a fixed habit that men will frequently take pains to restrict the product of machines which they are running when even a large increase in output would involve no more work on their part."

Third. As to the third cause for slow work, considerable space will later in this paper be devoted to illustrating the great gain, both to employers and employés, which results from the substitution of scientific for rule-of-thumb methods in even the smallest details of the work of every trade. The enormous saving of time and therefore increase in the output which it is possible to effect through eliminating unnecessary motions and substituting fast for slow and inefficient motions for the men working in any of our trades can be fully realized only after one has personally seen the improvement which results from a thorough motion and time study, made by a competent man.

To explain briefly: owing to the fact that the workmen in all of our trades have been taught the details of their work by observation of those immediately around them, there are many different ways in common use for doing the same thing, perhaps forty, fifty, or a hundred ways of doing each act in each trade, and for the same reason there is a great variety in the implements used for each class of work. Now, among the various methods and implements used in each element of each trade there is always one method and one implement which is quicker and better than any of the rest. And this one best method and best implement can only be discovered or developed through a scientific study and analysis of all of the methods and implements in use, together with accurate, minute, motion and time study. This involves the gradual substitution of science for rule of thumb throughout the mechanic arts.

This paper will show that the underlying philosophy of all of the old systems of management in common use makes it imperative that each workman shall be left with the final responsibility for doing his job practically as he thinks best, with comparatively little help and advice from the management. And it will also show that because of this isolation of workmen, it is in most cases impossible for the men working under these systems to do their work in accordance with the rules and laws of a science or art, even where one exists.

The writer asserts as a general principle (and he proposes to give illustrations tending to prove the fact later in this paper) that in almost all of the mechanic arts the science which underlies each act of each workman is so great and amounts to so much that the workman who is best suited to actually doing the work is incapable of fully understanding this science, without the guidance and help of those who are working with him or over him, either through lack of education or through insufficient mental capacity. In order that the work may be done in accordance with scientific laws, it is necessary that there shall be a far more equal division of the responsibility between the management and the workmen than exists under any of the ordinary types of management. Those in the management whose duty it is to develop this science should also guide and help the workman in working under it, and should assume a much larger share of the responsibility for results than under usual conditions is assumed by the management.

The body of this paper will make it clear that, to work according to scientific laws, the management must take over and perform much of the work which is now left to the men; almost every act of the workman should be preceded by one or more preparatory acts of the management which enable him to do his work better and quicker than he otherwise could. And each man should daily be taught by and receive the most friendly help from those who are over him, instead of being, at the one extreme, driven or coerced by his bosses, and at the other left to his own unaided devices.

This close, intimate, personal cooperation between the management and the men is of the essence of modern scientific or task management.

It will be shown by a series of practical illustrations that, through this friendly cooperation, namely, through sharing equally in every day's burden, all of the great obstacles (above described) to obtaining the maximum output for each man and each machine in the establishment are swept away. The 30 per cent. to 100 per cent. increase in wages which the workmen are able to earn beyond what they receive under the old type of management, coupled with the daily intimate shoulder to shoulder contact with the management, entirely removes all cause for soldiering. And in a few years, under this system, the workmen have before them the object lesson of seeing that a great increase in the output per man results in giving employment to more men, instead of throwing men out of work, thus completely eradicating the fallacy that a larger output for each man will throw other men out of work.

It is the writer's judgment, then, that while much can be done and should be done by writing and talking toward educating not only workmen, but all classes in the community, as to the importance of obtaining the maximum output of each man and each machine, it is only through the adoption of modern scientific management that this great problem can be finally solved. Probably most of the readers of this paper will say that all of this is mere theory. On the contrary, the theory, or philosophy, of scientific management is just beginning to be understood, whereas the management itself has been a gradual evolution, extending over a period of nearly thirty years. And during this time the employés of one company after another, including a large range and diversity of industries, have gradually changed from the ordinary to the scientific type of management. At least 50,000 workmen in the United States are now employed under this system; and they are receiving from 30 per cent. to 100 per cent. higher wages daily than are paid to men of similar caliber with whom they are surrounded, while the companies employing them are more prosperous than ever before. In these companies the output, per man and per machine, has on an average been doubled. During all these years there has never been a single strike among the men working under this system. In place of the suspicious watchfulness and the more or less open warfare which characterizes the ordinary types of management, there is universally friendly cooperation between the management and the men.

Several papers have been written, describing the expedients which have been adopted and the details which have been developed under scientific management and the steps to be taken in changing from the ordinary to the scientific type. But unfortunately most of the readers of these papers have mistaken the mechanism for the true essence. Scientific management fundamentally consists of certain broad general principles, a certain philosophy, which can be applied in many ways, and a description of what any one man or men may believe to be the best mechanism for applying these general principles should in no way be confused with the principles themselves.

It is not here claimed that any single panacea exists for all of the troubles of the working-people or of employers. As long as some people are born lazy or inefficient, and others are born greedy and brutal, as long as vice and crime are with us, just so long will a certain amount of poverty, misery, and unhappiness be with us also. No system of management, no single expedient within the control of any man or any set of men can insure continuous prosperity to either workmen or employers. Prosperity depends upon so many factors entirely beyond the control of any one set of men, any state, or even any one country, that certain periods will inevitably come when both sides must suffer, more or less. It is claimed, however, that under scientific management the intermediate periods will be far more prosperous, far happier, and more free from discord and dissension. And also, that the periods will be fewer, shorter and the suffering less. And this will be particularly true in any one town, any one section of the country, or any one state which first substitutes the principles of scientific management for the rule of thumb.

That these principles are certain to come into general use practically throughout the civilized world, sooner or later, the writer is profoundly convinced, and the sooner they come the better for all the people.

CHAPTER II

THE PRINCIPLES OF SCIENTIFIC MANAGEMENT

THE writer has found that there are three questions uppermost in the minds of men when they become interested in scientific management.

First. Wherein do the principles of scientific management differ essentially from those of ordinary management?

Second. Why are better results attained under scientific management than under the other types?

Third. Is not the most important problem that of getting the right man at the head of the company? And if you have the right man cannot the choice of the type of management be safely left to him?

One of the principal objects of the following pages will be to give a satisfactory answer to these questions.

THE FINEST TYPE OF ORDINARY MANAGEMENT

Before starting to illustrate the principles of scientific management, or "task management" as it is briefly called, it seems desirable to outline what the writer believes will be recognized as the best type of management which is in common use. This is done so that the great difference between the best of the ordinary management and scientific management may be fully appreciated.

In an industrial establishment which employs say from 500 to 1000 workmen, there will be found in many cases at least twenty to thirty different trades. The workmen in each of these trades have had their knowledge handed down to them by word of mouth, through the many years in which their trade has been developed from the primitive condition, in which our far-distant ancestors each one practised the rudiments of many different trades, to the present state of great and growing subdivision of labor, in which each man specializes upon some comparatively small class of work.

The ingenuity of each generation has developed quicker and better methods for doing every element of the work in every trade. Thus the methods which are now in use may in a broad sense be said to be an evolution representing the survival of the fittest and best of the ideas which have been developed since the starting of each trade. However, while this is true in a broad sense, only those who are intimately acquainted with each of these trades are fully aware of the fact that in hardly any element of any trade is there uniformity in the methods which are used. Instead of having only one way which is generally accepted as a standard, there are in daily use, say, fifty or a hundred different ways of doing each element of the work. And a little thought will make it clear that this must inevitably be the case, since our methods have been handed down from man to man by word of mouth, or have, in most cases, been almost unconsciously learned through personal observation. Practically in no instances have they been codified or systematically analyzed or described. The ingenuity and experience of each generation — of each decade, even, have without doubt handed over better methods to the next. This mass of rule-of-thumb or traditional knowledge may be said to be the principal asset or possession of every tradesman. Now, in the best of the ordinary types of management, the managers recognize frankly the fact that the 500 or 1000 workmen, included in the twenty to thirty trades, who are under them, possess this mass of traditional knowledge, a large part of which is not in the possession of the management. The management, of course, includes foremen and superintendents, who themselves have been in most cases first-class workers at their trades. And yet these foremen and superintendents know, better than any one else, that their own knowledge and personal skill falls far short of the combined knowledge and dexterity of all the workmen under them. The most experienced managers therefore frankly place before their workmen the problem of doing the work in the best and most economical way. They recognize the task before them as that of inducing each workman to use his best endeavors, his hardest work, all his traditional knowledge, his skill, his ingenuity, and his good-will — in a word, his "initiative," so as to yield the largest possible return to his employer. The problem before the management, then, may be briefly said to be that of obtaining the best *initiative* of every workman. And the writer uses the word "initiative" in its broadest sense, to cover all of the good qualities sought for from the men.

On the other hand, no intelligent manager would hope to obtain in any full measure the initiative of his workmen unless he felt that he was giving them something more than they usually receive from their employers. Only those among the readers of this paper who have been managers or who have worked themselves at a trade realize how far the average workman falls short of giving his employer his full initiative. It is well within the mark to state that in nineteen out of twenty industrial establishments the workmen believe it to be directly against their interests to give their employers their best initiative, and that instead of working hard to do the largest possible amount of work and the best quality of work for their employers, they deliberately work as slowly as they dare while they at the same time try to make those over them believe that they are working fast.[1]

The writer repeats, therefore, that in order to have any hope of obtaining the initiative of his workmen the manager must give some *special incentive* to his men beyond that which is given to the average of the trade. This incentive can be given in several different ways, as, for example,

[1] The writer has tried to make the reason for this unfortunate state of things clear in a paper entitled "Shop Management," read before the American Society of Mechanical Engineers."

the hope of rapid promotion or advancement; higher wages, either in the form of generous piecework prices or of a premium or bonus of some kind for good and rapid work; shorter hours of labor; better surroundings and working conditions than are ordinarily given, etc., and, above all, this special incentive should be accompanied by that personal consideration for, and friendly contact with, his workmen which comes only from a genuine and kindly interest in the welfare of those under him. It is only by giving a special inducement or "incentive" of this kind that the employer can hope even approximately to get the "initiative" of his workmen. Under the ordinary type of management the necessity for offering the workman a special inducement has come to be so generally recognized that a large proportion of those most interested in the subject look upon the adoption of some one of the modern schemes for paying men (such as piece work, the premium plan, or the bonus plan, for instance) as practically the whole system of management. Under scientific management, however, the particular pay system which is adopted is merely one of the subordinate elements.

Broadly speaking, then, the best type of management in ordinary use may be defined as management in which the workmen give their best *initiative* and in return receive some *special incentive* from their employers. This type of management will be referred to as the management of *"initiative and incentive"* in contradistinction to scientific management, or task management, with which it is to be compared.

The writer hopes that the management of "initiative and incentive" will be recognized as representing the best type in ordinary use, and in fact he believes that it will be hard to persuade the average manager that anything better exists in the whole field than this type. The task which the writer has before him, then, is the difficult one of trying to prove in a thoroughly convincing way that there is another type of management which is not only better but overwhelmingly better than the management of "initiative and incentive."

The universal prejudice in favor of the management of "initiative and incentive" is so strong that no mere theoretical advantages which can be pointed out will be likely to convince the average manager that any other system is better. It will be upon a series of practical illustrations of the actual working of the two systems that the writer will depend in his efforts to prove that scientific management is so greatly superior to other types. Certain elementary principles, a certain philosophy, will however be recognized as the essence of that which is being illustrated in all of the practical examples which will be given. And the broad principles in which the scientific system differs from the ordinary or "rule-of-thumb" system are so simple in their nature that it seems desirable to describe them before starting with the illustrations.

Under the old type of management success depends

almost entirely upon getting the "initiative" of the workmen, and it is indeed a rare case in which this initiative is really attained. Under scientific management the "initiative" of the workmen (that is, their hard work, their good-will, and their ingenuity) is obtained with absolute uniformity and to a greater extent than is possible under the old system; and in addition to this improvement on the part of the men, the managers assume new burdens, new duties, and responsibilities never dreamed of in the past. The managers assume, for instance, the burden of gathering together all of the traditional knowledge which in the past has been possessed by the workmen and then of classifying, tabulating, and reducing this knowledge to rules, laws, and formulæ which are immensely helpful to the workmen in doing their daily work. In addition to developing a *science* in this way, the management take on three other types of duties which involve new and heavy burdens for themselves.

These new duties are grouped under four heads:

First. They develop a science for each element of a man's work, which replaces the old rule-of-thumb method.

Second. They scientifically select and then train, teach, and develop the workman, whereas in the past he chose his own work and trained himself as best he could.

Third. They heartily cooperate with the men so as to insure all of the work being done in accordance with the principles of the science which has been developed.

Fourth. There is an almost equal division of the work and the responsibility between the management and the workmen. The management take over all work for which they are better fitted than the workmen, while in the past almost all of the work and the greater part of the responsibility were thrown upon the men.

It is this combination of the initiative of the workmen, coupled with the new types of work done by the management, that makes scientific management so much more efficient than the old plan.

Three of these elements exist in many cases, under the management of "initiative and incentive," in a small and rudimentary way, but they are, under this management, of minor importance, whereas under scientific management they form the very essence of the whole system.

The fourth of these elements, "an almost equal division of the responsibility between the management and the workmen," requires further explanation. The philosophy of the management of "initiative and incentive" makes it necessary for each workman to bear almost the entire responsibility for the general plan as well as for each detail of his work, and in many cases for his implements as well. In addition to this he must do all of the actual physical labor. The development of a science, on the other hand, involves the establishment of many rules, laws, and formulæ which replace the judgment of the individual workman and which can be effectively used only after having been systematically recorded, indexed, etc. The practical use of scientific data also calls for a room in which to keep the books, records,[1] etc., and a desk for the planner to work at. Thus all of the planning which under the old system was done by the workman, as a result of his personal experience, must of necessity under the new system be done by the management in accordance with the laws of the science; because even if the workman was well suited to the development and use of scientific data, it would be physically impossible for him to work at his machine and at a desk at the same time. It is also clear that in most cases one type of man is needed to plan ahead and an entirely different type to execute the work.

The man in the planning room, whose specialty under scientific management is planning ahead, invariably finds that the work can be done better and more economically by a subdivision of the labor; each act of each mechanic, for example, should be preceded by various preparatory acts done by other men. And all of this involves, as we have said, "an almost equal division of the responsibility and the work between the management and the workman."

To summarize: Under the management of "initiative and incentive" practically the whole problem is "up to the workman," while under scientific management fully one-half of the problem is "up to the management."

[1] For example, the records containing the data used under scientific management in an ordinary machine-shop fill thousands of pages.

Perhaps the most prominent single element in modern scientific management is the task idea. The work of every workman is fully planned out by the management at least one day in advance, and each man receives in most cases complete written instructions, describing in detail the task which he is to accomplish, as well as the means to be used in doing the work. And the work planned in advance in this way constitutes a task which is to be solved, as explained above, not by the workman alone, but in almost all cases by the joint effort of the workman and the management. This task specifies not only what is to be done but how it is to be done and the exact time allowed for doing it. And whenever the workman succeeds in doing his task right, and within the time limit specified, he receives an addition of from 30 per cent. to 100 per cent. to his ordinary wages. These tasks are carefully planned, so that both good and careful work are called for in their performance, but it should be distinctly understood that in no case is the workman called upon to work at a pace which would be injurious to his health. The task is always so regulated that the man who is well suited to his job will thrive while working at this rate during a long term of years and grow happier and more prosperous, instead of being overworked. Scientific management consists very largely in preparing for and carrying out these tasks.

The writer is fully aware that to perhaps most of the readers of this paper the four elements which differentiate the new management from the old will at first appear to be merely high-sounding phrases; and he would again repeat that he has no idea of convincing the reader of their value merely through announcing their existence. His hope of carrying conviction rests upon demonstrating the tremendous force and effect of these four elements through a series of practical illustrations. It will be shown, first, that they can be applied absolutely to all classes of work, from the most elementary to the most intricate; and second, that when they are applied, the results must of necessity be overwhelmingly greater than those which it is possible to attain under the management of initiative and incentive.

The first illustration is that of handling pig iron, and this work is chosen because it is typical of perhaps the crudest and most elementary form of labor which is performed by man. This work is done by men with no other implements than their hands. The pig-iron handler stoops down, picks up a pig weighing about 92 pounds, walks for a few feet or yards and then drops it on to the ground or upon a pile. This work is so crude and elementary in its nature that the writer firmly believes that it would be possible to train an intelligent gorilla so as to become a more efficient pig-iron handler than any man can be. Yet it will be shown that the science of handling pig iron is so great and amounts to so much that it is impossible for the man who is best suited to this type of work to understand the principles of this science, or even to work in accordance with these principles without the aid of a man better educated than he is. And the further illustrations to be given will make it clear that in almost all of the mechanic arts the science which underlies each workman's act is so great and amounts to so much that the workman who is best suited actually to do the work is incapable (either through lack of education or through insufficient mental capacity) of understanding this science. This is announced as a general principle, the truth of which will become apparent as one illustration after another is given. After showing these four elements in the handling of pig iron, several illustrations will be given of their application to different kinds of work in the field of the mechanic arts, at intervals in a rising scale, beginning with the simplest and ending with the more intricate forms of labor.

One of the first pieces of work undertaken by us, when the writer started to introduce scientific management into the Bethlehem Steel Company, was to handle pig iron on task work. The opening of the Spanish War found some 80,000 tons of pig iron placed in small piles in an open field adjoining the works. Prices for pig iron had been so low that it could not be sold at a profit, and it therefore had been stored. With the opening of the Spanish War the price of pig iron rose, and this large accumulation of iron was sold. This gave us a good opportunity to show the workmen, as well as the owners and managers of the works, on a fairly large scale the advantages of task work over the old-fashioned day work and piece work, in doing a very elementary class of work.

The Bethlehem Steel Company had five blast furnaces, the product of which had been handled by a pig-iron gang for many years. This gang, at this time, consisted of about 75 men. They were good, average pig-iron handlers, were under an excellent foreman who himself had been a pig-iron handler, and the work was done, on the whole, about as fast and as cheaply as it was anywhere else at that time.

A railroad switch was run out into the field, right along the edge of the piles of pig iron. An inclined plank was placed against the side of a car, and each man picked up from his pile a pig of iron weighing about 92 pounds, walked up the inclined plank and dropped it on the end of the car.

We found that this gang were loading on the average about $12\frac{1}{2}$ long tons per man per day. We were surprised to find, after studying the matter, that a first-class pig-iron handler ought to handle between 47[1] and 48 long tons per day, instead of $12\frac{1}{2}$ tons. This task seemed to us so very large that we were obliged to go over our work several times before we were absolutely sure that we were right. Once we were sure, however, that 47 tons was a proper day's work for a first-class pig-iron handler, the task which faced us as managers under the modern scientific plan was clearly before us. It was our duty to see that the 80,000 tons of pig

[1] See foot-note at foot of page 60.

iron was loaded on to the cars at the rate of 47 tons per man per day, in place of $12\frac{1}{2}$ tons, at which rate the work was then being done. And it was further our duty to see that this work was done without bringing on a strike among the men, without any quarrel with the men, and to see that the men were happier and better contented when loading at the new rate of 47 tons than they were when loading at the old rate of $12\frac{1}{2}$ tons.

Our first step was the scientific selection of the workman. In dealing with workmen under this type of management, it is an inflexible rule to talk to and deal with only one man at a time, since each workman has his own special abilities and limitations, and since we are not dealing with men in masses, but are trying to develop each individual man to his highest state of efficiency and prosperity. Our first step was to find the proper workman to begin with. We therefore carefully watched and studied these 75 men for three or four days, at the end of which time we had picked out four men who appeared to be physically able to handle pig iron at the rate of 47 tons per day. A careful study was then made of each of these men. We looked up their history as far back as practicable and thorough inquiries were made as to the character, habits, and the ambition of each of them. Finally we selected one from among the four as the most likely man to start with. He was a little Pennsylvania Dutchman who had been observed to trot back home for a mile or so after his work in the evening, about as fresh as he was when he came trotting down to work in the morning. We found that upon wages of $1.15 a day he had succeeded in buying a small plot of ground, and that he was engaged in putting up the walls of a little house for himself in the morning before starting to work and at night after leaving. He also had the reputation of being exceedingly "close," that is, of placing a very high value on a dollar. As one man whom we talked to about him said, "A penny looks about the size of a cart-wheel to him." This man we will call Schmidt.

The task before us, then, narrowed itself down to getting Schmidt to handle 47 tons of pig iron per day and making him glad to do it. This was done as follows. Schmidt was called out from among the gang of pig-iron handlers and talked to somewhat in this way:

"Schmidt, are you a high-priced man?"

"Vell, I don't know vat you mean."

"Oh yes, you do. What I want to know is whether you are a high-priced man or not."

"Vell, I don't know vat you mean."

"Oh, come now, you answer my questions. What I want to find out is whether you are a high-priced man or one of these cheap fellows here. What I want to find out is whether you want to earn $1.85 a day or whether you are satisfied with $1.15, just the same as all those cheap fellows are getting."

"Did I vant $1.85 a day? Vas dot a high-priced man? Vell, yes, I vas a high-priced man."

"Oh, you're aggravating me. Of course you want

$1.85 a day — every one wants it! You know perfectly well that that has very little to do with your being a high-priced man. For goodness' sake answer my questions, and don't waste any more of my time. Now come over here. You see that pile of pig iron?"

"Yes."

"You see that car?"

"Yes."

"Well, if you are a high-priced man, you will load that pig iron on that car to-morrow for $1.85. Now do wake up and answer my question. Tell me whether you are a high-priced man or not."

"Vell — did I got $1.85 for loading dot pig iron on dot car to-morrow?"

"Yes, of course you do, and you get $1.85 for loading a pile like that every day right through the year. That is what a high-priced man does, and you know it just as well as I do."

"Vell, dot's all right. I could load dot pig iron on the car to-morrow for $1.85, and I get it every day, don't I?"

"Certainly you do — certainly you do."

"Vell, den, I vas a high-priced man."

"Now, hold on, hold on. You know just as well as I do that a high-priced man has to do exactly as he's told from morning till night. You have seen this man here before, haven't you?"

"No, I never saw him."

"Well, if you are a high-priced man, you will do exactly as this man tells you to-morrow, from morning till night. When he tells you to pick up a pig and walk, you pick it up and you walk, and when he tells you to sit down and rest, you sit down. You do that right straight through the day. And what's more, no back talk. Now a high-priced man does just what he's told to do, and no back talk. Do you understand that? When this man tells you to walk, you walk; when he tells you to sit down, you sit down, and you don't talk back at him. Now you come on to work here to-morrow morning and I'll know before night whether you are really a high-priced man or not."

This seems to be rather rough talk. And indeed it would be if applied to an educated mechanic, or even an intelligent laborer. With a man of the mentally sluggish type of Schmidt it is appropriate and not unkind, since it is effective in fixing his attention on the high wages which he wants and away from what, if it were called to his attention, he probably would consider impossibly hard work.

What would Schmidt's answer be if he were talked to in a manner which is usual under the management of "initiative and incentive"? say, as follows:

"Now, Schmidt, you are a first-class pig-iron handler and know your business well. You have been handling at the rate of $12\frac{1}{2}$ tons per day. I have given considerable study to handling pig iron, and feel sure that you could do a much larger day's work than you have been doing. Now don't you think that if you really tried you could handle 47 tons of pig iron per day, instead of $12\frac{1}{2}$ tons?"

What do you think Schmidt's answer would be to this?

Schmidt started to work, and all day long, and at regular intervals, was told by the man who stood over him with a watch, "Now pick up a pig and walk. Now sit down and rest. Now walk — now rest," etc. He worked when he was told to work, and rested when he was told to rest, and at half-past five in the afternoon had his $47\frac{1}{2}$ tons loaded on the car. And he practically never failed to work at this pace and do the task that was set him during the three years that the writer was at Bethlehem. And throughout this time he averaged a little more than $1.85 per day, whereas before he had never received over $1.15 per day, which was the ruling rate of wages at that time in Bethlehem. That is, he received 60 per cent. higher wages than were paid to other men who were not working on task work. One man after another was picked out and trained to handle pig iron at the rate of $47\frac{1}{2}$ tons per day until all of the pig iron was handled at this rate, and the men were receiving 60 per cent. more wages than other workmen around them.

The writer has given above a brief description of three of the four elements which constitute the essence of scientific management: first, the careful selection of the workman, and, second and third, the method of first inducing and then training and helping the workman to work according to the scientific method. Nothing has as yet been said about the science of handling pig iron. The writer trusts, however, that before leaving this illustration the reader will be thoroughly convinced that there is a science of handling pig iron, and further that this science amounts to so much that the man who is suited to handle pig iron cannot possibly understand it, nor even work in accordance with the laws of this science, without the help of those who are over him.

The writer came into the machine-shop of the Midvale Steel Company in 1878, after having served an apprenticeship as a pattern-maker and as a machinist. This was close to the end of the long period of depression following the panic of 1873, and business was so poor that it was impossible for many mechanics to get work at their trades. For this reason he was obliged to start as a day laborer instead of working as a mechanic. Fortunately for him, soon after he came into the shop the clerk of the shop was found stealing. There was no one else available, and so, having more education than the other laborers (since he had been prepared for college) he was given the position of clerk. Shortly after this he was given work as a machinist in running one of the lathes, and, as he turned out rather more work than other machinists were doing on similar lathes, after several months was made gang-boss over the lathes.

Almost all of the work of this shop had been done on piece work for several years. As was usual then, and in fact as is still usual in most of the shops in this country, the shop was really run by the workmen, and not by the bosses. The workmen together had carefully planned just how fast each job should be done, and they had set a pace for each machine throughout the shop, which was limited to about one-third of a good day's work. Every new workman who came into the shop was told at once by the other men exactly how much of each kind of work he was to do, and unless he obeyed these instructions he was sure before long to be driven out of the place by the men.

As soon as the writer was made gang-boss, one after another of the men came to him and talked somewhat as follows:

"Now, Fred, we're very glad to see that you've been made gang-boss. You know the game all right, and we're sure that you're not likely to be a piece-work hog. You come along with us, and everything will be all right, but if you try breaking any of these rates you can be mighty sure that we'll throw you over the fence."

The writer told them plainly that he was now working on the side of the management, and that he proposed to do whatever he could to get a fair day's work out of the lathes. This immediately started a war; in most cases a friendly war, because the men who were under him were his personal friends, but none the less a war, which as time went on grew more and more bitter. The writer used every expedient to make them do a fair day's work, such as discharging or lowering the wages of the more stubborn men who refused to make any improvement, and such as lowering the piece-work price, hiring green men, and personally teaching them how to do the work, with the promise from them that when they had learned how, they would then do a fair day's work. While the men constantly brought such pressure to bear (both inside and outside the works) upon all those who started to increase their output that they were finally compelled to do about as the rest did, or else quit. No one who has not had this experience can have an idea of the bitterness which is gradually developed in such a struggle. In a war of this kind the workmen have one expedient which is usually effective. They use their ingenuity to contrive various ways in which the machines which they are running are broken or damaged — apparently by accident, or in the regular course of work — and this they always lay at the door of the foreman, who has forced them to drive the machine so hard that it is overstrained and is being ruined. And there are few foremen indeed who are able to stand up against the combined pressure of all of the men in the shop. In this case the problem was complicated by the fact that the shop ran both day and night.

The writer had two advantages, however, which are not possessed by the ordinary foreman, and these came, curiously enough, from the fact that he was not the son of a working man.

First, owing to the fact that he happened not to be of working parents, the owners of the company believed that he had the interest of the works more at heart than the other workmen, and they therefore had more confidence in his word than they did in that of the machinists who were under him. So that, when the machinists reported to the Superintendent that the machines were being smashed up because an incompetent foreman was overstraining them, the Superintendent accepted the word of the writer when he said that these men were deliberately breaking their machines as a part of the piece-work war which was going on, and he also allowed the writer to make the only effective answer to this Vandalism on the part of the men, namely: "There will be no more accidents to the machines in this shop. If any part of a machine is broken the man in charge of it must pay at least a part of the cost of its repair, and the fines collected in this way will all be handed over to the mutual beneficial association to help care for sick workmen." This soon stopped the wilful breaking of machines.

Second. If the writer had been one of the workmen, and had lived where they lived, they would have brought such social pressure to bear upon him that it would have been impossible to have stood out against them. He would have been called "scab" and other foul names every time he appeared on the street, his wife would have been abused, and his children would have been stoned. Once or twice he was begged by some of his friends among the workmen not to walk home, about two and a half miles along the lonely path by the side of the railway. He was told that if he continued to do this it would be at the risk of his life. In all such cases, however, a display of timidity is apt to increase rather than diminish the risk, so the writer told these men to say to the other men in the shop that he proposed to walk home every night right up that railway track; that he never had carried and never would carry any weapon of any kind, and that they could shoot and be d——.

After about three years of this kind of struggling, the output of the machines had been materially increased, in many cases doubled, and as a result the writer had been promoted from one gang-bossship to another until he became foreman of the shop. For any right-minded man, however, this success is in no sense a recompense for the bitter relations which he is forced to maintain with all of those around him. Life which is one continuous struggle with other men is hardly worth living. His workman friends came to him continually and asked him, in a personal, friendly way, whether he would advise them, for their own best interest, to turn out more work. And, as a truthful man, he had to tell them that if he were in their place he would fight against turning out any more work, just as they were doing, because under the piecework system they would be allowed to earn no more wages than they had been earning, and yet they would be made to work harder.

Soon after being made foreman, therefore, he decided to make a determined effort to in some way change the system of management, so that the interests of the workmen and the management should become the same, instead of antagonistic. This resulted, some three years later, in the starting of the type of management which is described in papers presented to the American Society of Mechanical Engineers entitled "A Piece-Rate System" and "Shop Management."

In preparation for this system the writer realized that the greatest obstacle to harmonious cooperation between the workmen and the management lay in the ignorance of the management as to what really constitutes a proper day's work for a workman. He fully realized that, although he was foreman of the shop, the combined knowledge and skill of the workmen who were under him was certainly ten times as great as his own. He therefore obtained the permission of Mr. William Sellers, who was at that time the President of the Midvale Steel Company, to spend some money in a careful, scientific study of the time required to do various kinds of work.

Mr. Sellers allowed this more as a reward for having, to a certain extent, "made good" as foreman of the shop in getting more work out of the men, than for any other reason. He stated, however, that he did not believe that any scientific study of this sort would give results of much value.

Among several investigations which were undertaken at this time, one was an attempt to find some rule, or law, which would enable a foreman to know in advance how much of any kind of heavy laboring work a man who was well suited to his job ought to do in a day; that is, to study the tiring effect of heavy labor upon a first-class man. Our first step was to employ a young college graduate to look up all that had been written on the subject in English, German, and French. Two classes of experiments had been made: one by physiologists who were studying the endurance of the human animal, and the other by engineers who wished to determine what fraction of a horse-power a man-power was. These experiments had been made largely upon men who were lifting loads by means of turning the crank of a winch from which weights were suspended, and others who were engaged in walking, running, and lifting weights in various ways. However, the records of these investigations were so meager that no law of any value could be deduced from them. We therefore started a series of experiments of our own.

Two first-class laborers were selected, men who had proved themselves to be physically powerful and who were also good steady workers. These men were paid double wages during the experiments, and were told that they must work to the best of their ability at all times, and that we should make certain tests with them from time to time to find whether they were "soldiering" or not, and that the moment either one of them started to try to deceive us he would be discharged. They worked to the best of their ability throughout the time that they were being observed.

Now it must be clearly understood that in these

experiments we were not trying to find the maximum work that a man could do on a short spurt or for a few days, but that our endeavor was to learn what really constituted a full day's work for a first-class man; the best day's work that a man could properly do, year in and year out, and still thrive under. These men were given all kinds of tasks, which were carried out each day under the close observation of the young college man who was conducting the experiments, and who at the same time noted with a stop-watch the proper time for all of the motions that were made by the men. Every element in any way connected with the work which we believed could have a bearing on the result was carefully studied and recorded. What we hoped ultimately to determine was what fraction of a horse-power a man was able to exert, that is, how many foot-pounds of work a man could do in a day.

After completing this series of experiments, therefore, each man's work for each day was translated into foot-pounds of energy, and to our surprise we found that there was no constant or uniform relation between the foot-pounds of energy which the man exerted during a day and the tiring effect of his work. On some kinds of work the man would be tired out when doing perhaps not more than one-eighth of a horse-power, while in others he would be tired to no greater extent by doing half a horse-power of work. We failed, therefore, to find any law which was an accurate guide to the maximum day's work for a first-class workman.

56 THE PRINCIPLES OF SCIENTIFIC MANAGEMENT

A large amount of very valuable data had been obtained, which enabled us to know, for many kinds of labor, what was a proper day's work. It did not seem wise, however, at this time to spend any more money in trying to find the exact law which we were after. Some years later, when more money was available for this purpose, a second series of experiments was made, similar to the first, but somewhat more thorough. This, however, resulted as the first experiments, in obtaining valuable information but not in the development of a law. Again, some years later, a third series of experiments was made, and this time no trouble was spared in our endeavor to make the work thorough. Every minute element which could in any way affect the problem was carefully noted and studied, and two college men devoted about three months to the experiments. After this data was again translated into foot-pounds of energy exerted for each man each day, it became perfectly clear that there is no direct relation between the horse-power which a man exerts (that is, his foot-pounds of energy per day) and the tiring effect of the work on the man. The writer, however, was quite as firmly convinced as ever that some definite, clear-cut law existed as to what constitutes a full day's work for a first-class laborer, and our data had been so carefully collected and recorded that he felt sure that the necessary information was included somewhere in the records. The problem of developing this law from the accumulated facts was therefore handed over to Mr. Carl G. Barth, who is a better mathematician than any of the rest of us, and we decided to investigate the problem in a new way, by graphically representing each element of the work through plotting curves, which should give us, as it were, a bird's-eye view of every element. In a comparatively short time Mr. Barth had discovered the law governing the tiring effect of heavy labor on a first-class man. And it is so simple in its nature that it is truly remarkable that it should not have been discovered and clearly understood years before. The law which was developed is as follows:

The law is confined to that class of work in which the limit of a man's capacity is reached because he is tired out. It is the law of heavy laboring, corresponding to the work of the cart horse, rather than that of the trotter. Practically all such work consists of a heavy pull or a push on the man's arms, that is, the man's strength is exerted by either lifting or pushing something which he grasps in his hands. And the law is that for each given pull or push on the man's arms it is possible for the workman to be under load for only a definite percentage of the day. For example, when pig iron is being handled (each pig weighing 92 pounds), a first-class workman can only be under load 43 per cent. of the day. He must be entirely free from load during 57 per cent. of the day. And as the load becomes lighter, the percentage of the day under which the man can remain under load increases. So that, if the workman is handling a half-pig weighing 46 pounds, he can then be under load 58 per cent. of the day, and only has to rest during 42 per cent. As the weight grows lighter the man can remain under load during a larger and larger percentage of the day, until finally a load is reached which he can carry in his hands all day long without being tired out. When that point has been arrived at this law ceases to be useful as a guide to a laborer's endurance, and some other law must be found which indicates the man's capacity for work.

When a laborer is carrying a piece of pig iron weighing 92 pounds in his hands, it tires him about as much to stand still under the load as it does to walk with it, since his arm muscles are under the same severe tension whether he is moving or not. A man, however, who stands still under a load is exerting no horse-power whatever, and this accounts for the fact that no constant relation could be traced in various kinds of heavy laboring work between the foot-pounds of energy exerted and the tiring effect of the work on the man. It will also be clear that in all work of this kind it is necessary for the arms of the workman to be completely free from load (that is, for the workman to rest) at frequent intervals. Throughout the time that the man is under a heavy load the tissues of his arm muscles are in process of degeneration, and frequent periods of rest are required in order that the blood may have a chance to restore these tissues to their normal condition.

To return now to our pig-iron handlers at the

Bethlehem Steel Company. If Schmidt had been allowed to attack the pile of 47 tons of pig iron without the guidance or direction of a man who understood the art, or science, of handling pig iron, in his desire to earn his high wages he would probably have tired himself out by 11 or 12 o'clock in the day. He would have kept so steadily at work that his muscles would not have had the proper periods of rest absolutely needed for recuperation, and he would have been completely exhausted early in the day. By having a man, however, who understood this law, stand over him and direct his work, day after day, until he acquired the habit of resting at proper intervals, he was able to work at an even gait all day long without unduly tiring himself.

Now one of the very first requirements for a man who is fit to handle pig iron as a regular occupation is that he shall be so stupid and so phlegmatic that he more nearly resembles in his mental make-up the ox than any other type. The man who is mentally alert and intelligent is for this very reason entirely unsuited to what would, for him, be the grinding monotony of work of this character. Therefore the workman who is best suited to handling pig iron is unable to understand the real science of doing this class of work. He is so stupid that the word "percentage" has no meaning to him, and he must consequently be trained by a man more intelligent than himself into the habit of working in accordance with the laws of this science before he can be successful.

The writer trusts that it is now clear that even in the case of the most elementary form of labor that is known, there is a science, and that when the man best suited to this class of work has been carefully selected, when the science of doing the work has been developed, and when the carefully selected man has been trained to work in accordance with this science, the results obtained must of necessity be overwhelmingly greater than those which are possible under the plan of "initiative and incentive."

Let us, however, again turn to the case of these pig-iron handlers, and see whether, under the ordinary type of management, it would not have been possible to obtain practically the same results.

The writer has put the problem before many good managers, and asked them whether, under premium work, piece work, or any of the ordinary plans of management, they would be likely even to approximate 47 tons [1] per man per day, and not a

[1] Many people have questioned the accuracy of the statement that first-class workmen can load 47½ tons of pig iron from the ground on to a car in a day. For those who are skeptical, therefore, the following data relating to this work are given:

First. That our experiments indicated the existence of the following law: that a first-class laborer, suited to such work as handling pig iron, could be under load only 42 per cent. of the day and must be free from load 58 per cent. of the day.

Second. That a man in loading pig iron from piles placed on the ground in an open field on to a car which stood on a track adjoining these piles, ought to handle (and that they did handle regularly) 47½ long tons (2240 pounds per ton) per day.

That the price paid for loading this pig iron was $3\tfrac{9}{10}$ cents per ton, and that the men working at it averaged $1.85 per day, whereas, in the past, they had been paid only $1.15 per day.

In addition to these facts, the following are given:

man has suggested that an output of over 18 to 25 tons could be attained by any of the ordinary expedients. It will be remembered that the Bethlehem men were loading only $12\frac{1}{2}$ tons per man.

To go into the matter in more detail, however: As to the scientific selection of the men, it is a fact that in this gang of 75 pig-iron handlers only about one man in eight was physically capable of handling $47\frac{1}{2}$ tons per day. With the very best of intentions, the other seven out of eight men were physically unable to work at this pace. Now the one man in eight who was able to do this work was in no sense superior to the other men who were working on the

$47\frac{1}{2}$ long tons equal 106,400 pounds of pig iron per day.

At 92 pounds per pig, equals 1156 pigs per day.

42 per cent. of a day under load equals 600 minutes; multiplied by 0.42 equals 252 minutes under load.

252 minutes divided by 1156 pigs equals 0.22 minutes per pig under load.

A pig-iron handler walks on the level at the rate of one foot in 0.006 minutes. The average distance of the piles of pig iron from the car was 36 feet. It is a fact, however, that many of the pig-iron handlers ran with their pig as soon as they reached the inclined plank. Many of them also would run down the plank after loading the car. So that when the actual loading went on, many of them moved at a faster rate than is indicated by the above figures. Practically the men were made to take a rest, generally by sitting down, after loading ten to twenty pigs. This rest was in addition to the time which it took them to walk back from the car to the pile. It is likely that many of those who are skeptical about the possibility of loading this amount of pig iron do not realize that while these men were walking back they were entirely free from load, and that therefore their muscles had, during that time, the opportunity for recuperation. It will be noted that with an average distance of 36 feet of the pig iron from the car, these men walked about eight miles under load each day and eight miles free from load.

If any one who is interested in these figures will multiply them and divide them, one into the other, in various ways, he will find that all of the facts stated check up exactly.

gang. He merely happened to be a man of the type of the ox, — no rare specimen of humanity, difficult to find and therefore very highly prized. On the contrary, he was a man so stupid that he was unfitted to do most kinds of laboring work, even. The selection of the man, then, does not involve finding some extraordinary individual, but merely picking out from among very ordinary men the few who are especially suited to this type of work. Although in this particular gang only one man in eight was suited to doing the work, we had not the slightest difficulty in getting all the men who were needed — some of them from inside of the works and others from the neighboring country — who were exactly suited to the job.

Under the management of "initiative and incentive" the attitude of the management is that of "putting the work up to the workmen." What likelihood would there be, then, under the old type of management, of these men properly selecting themselves for pig-iron handling? Would they be likely to get rid of seven men out of eight from their own gang and retain only the eighth man? No! And no expedient could be devised which would make these men properly select themselves. Even if they fully realized the necessity of doing so in order to obtain high wages (and they are not sufficiently intelligent properly to grasp this necessity), the fact that their friends or their brothers who were working right alongside of them would temporarily be thrown out of a job because they were not suited to this kind of work would entirely prevent them from properly selecting themselves, that is, from removing the seven out of eight men on the gang who were unsuited to pig-iron handling.

As to the possibility, under the old type of management, of inducing these pig-iron handlers (after they had been properly selected) to work in accordance with the science of doing heavy laboring, namely, having proper scientifically determined periods of rest in close sequence to periods of work. As has been indicated before, the essential idea of the ordinary types of management is that each workman has become more skilled in his own trade than it is possible for any one in the management to be, and that, therefore, the details of how the work shall best be done must be left to him. The idea, then, of taking one man after another and training him under a competent teacher into new working habits until he continually and habitually works in accordance with scientific laws, which have been developed by some one else, is directly antagonistic to the old idea that each workman can best regulate his own way of doing the work. And besides this, the man suited to handling pig iron is too stupid properly to train himself. Thus it will be seen that with the ordinary types of management the development of scientific knowledge to replace rule of thumb, the scientific selection of the men, and inducing the men to work in accordance with these scientific principles are entirely out of the question. And this because the philosophy of the old management puts the entire responsibility upon the workmen, while the philosophy of the new places a great part of it upon the management.

With most readers great sympathy will be aroused because seven out of eight of these pig-iron handlers were thrown out of a job. This sympathy is entirely wasted, because almost all of them were immediately given other jobs with the Bethlehem Steel Company. And indeed it should be understood that the removal of these men from pig-iron handling, for which they were unfit, was really a kindness to themselves, because it was the first step toward finding them work for which they were peculiarly fitted, and at which, after receiving proper training, they could permanently and legitimately earn higher wages.

Although the reader may be convinced that there is a certain science back of the handling of pig iron, still it is more than likely that he is still skeptical as to the existence of a science for doing other kinds of laboring. One of the important objects of this paper is to convince its readers that every single act of every workman can be reduced to a science. With the hope of fully convincing the reader of this fact, therefore, the writer proposes to give several more simple illustrations from among the thousands which are at hand.

For example, the average man would question whether there is much of any science in the work of shoveling. Yet there is but little doubt, if any intelligent reader of this paper were deliberately to set out to find what may be called the foundation of the science of shoveling, that with perhaps 15 to 20 hours of thought and analysis he would be almost sure to have arrived at the essence of this science. On the other hand, so completely are the rule-of-thumb ideas still dominant that the writer has never met a single shovel contractor to whom it had ever even occurred that there was such a thing as the science of shoveling. This science is so elementary as to be almost self-evident.

For a first-class shoveler there is a given shovel load at which he will do his biggest day's work. What is this shovel load? Will a first-class man do more work per day with a shovel load of 5 pounds, 10 pounds, 15 pounds, 20, 25, 30, or 40 pounds? Now this is a question which can be answered only through carefully made experiments. By first selecting two or three first-class shovelers, and paying them extra wages for doing trustworthy work, and then gradually varying the shovel load and having all the conditions accompanying the work carefully observed for several weeks by men who were used to experimenting, it was found that a first-class man would do his biggest day's work with a shovel load of about 21 pounds. For instance, that this man would shovel a larger tonnage per day with a 21-pound load than with a 24-pound load or than with an 18-pound load on his shovel. It is, of course, evident that no shoveler can always take a load of exactly 21 pounds on his shovel, but nevertheless, although his load may vary 3 or 4 pounds one way or the other, either below or above the 21 pounds, he will do his biggest day's work when his average for the day is about 21 pounds.

The writer does not wish it to be understood that this is the whole of the art or science of shoveling. There are many other elements, which together go to make up this science. But he wishes to indicate the important effect which this one piece of scientific knowledge has upon the work of shoveling.

At the works of the Bethlehem Steel Company, for example, as a result of this law, instead of allowing each shoveler to select and own his own shovel, it became necessary to provide some 8 to 10 different kinds of shovels, etc., each one appropriate to handling a given type of material; not only so as to enable the men to handle an average load of 21 pounds, but also to adapt the shovel to several other requirements which become perfectly evident when this work is studied as a science. A large shovel tool room was built, in which were stored not only shovels but carefully designed and standardized labor implements of all kinds, such as picks, crowbars, etc. This made it possible to issue to each workman a shovel which would hold a load of 21 pounds of whatever class of material they were to handle: a small shovel for ore, say, or a large one for ashes. Iron ore is one of the heavy materials which are handled in a works of this kind, and rice coal, owing to the fact that it is so slippery on the shovel, is one of the lightest materials. And it was found on studying the rule-of-thumb plan at the Bethlehem Steel Company, where each shoveler owned his own shovel, that he would frequently go from shoveling ore, with a load of about 30 pounds per shovel, to handling rice coal, with a load on the same shovel of less than 4 pounds. In the one case, he was so overloaded that it was impossible for him to do a full day's work, and in the other case he was so ridiculously underloaded that it was manifestly impossible to even approximate a day's work.

Briefly to illustrate some of the other elements which go to make up the science of shoveling, thousands of stop-watch observations were made to study just how quickly a laborer, provided in each case with the proper type of shovel, can push his shovel into the pile of materials and then draw it out properly loaded. These observations were made first when pushing the shovel into the body of the pile. Next when shoveling on a dirt bottom, that is, at the outside edge of the pile, and next with a wooden bottom, and finally with an iron bottom. Again a similar accurate time study was made of the time required to swing the shovel backward and then throw the load for a given horizontal distance, accompanied by a given height. This time study was made for various combinations of distance and height. With data of this sort before him, coupled with the law of endurance described in the case of the pig-iron handlers, it is evident that the man who is directing shovelers can first teach them the exact methods which should be employed to use their strength to the very best advantage, and can then assign them daily tasks which are so just that the workman can each day be sure of earning the large bonus which is paid whenever he successfully performs this task.

There were about 600 shovelers and laborers of this general class in the yard of the Bethlehem Steel Company at this time. These men were scattered in their work over a yard which was, roughly, about two miles long and half a mile wide. In order that each workman should be given his proper implement and his proper instructions for doing each new job, it was necessary to establish a detailed system for directing men in their work, in place of the old plan of handling them in large groups, or gangs, under a few yard foremen. As each workman came into the works in the morning, he took out of his own special pigeonhole, with his number on the outside, two pieces of paper, one of which stated just what implements he was to get from the tool room and where he was to start to work, and the second of which gave the history of his previous day's work; that is, a statement of the work which he had done, how much he had earned the day before, etc. Many of these men were foreigners and unable to read and write, but they all knew at a glance the essence of this report, because yellow paper showed the man that he had failed to do his full task the day before, and informed him that he had not earned as much as $1.85 a day, and that none but high-priced men would be allowed to stay permanently with this

THE PRINCIPLES OF SCIENTIFIC MANAGEMENT 69

gang. The hope was further expressed that he would earn his full wages on the following day. So that whenever the men received white slips they knew that everything was all right, and whenever they received yellow slips they realized that they must do better or they would be shifted to some other class of work.

Dealing with every workman as a separate individual in this way involved the building of a labor office for the superintendent and clerks who were in charge of this section of the work. In this office every laborer's work was planned out well in advance, and the workmen were all moved from place to place by the clerks with elaborate diagrams or maps of the yard before them, very much as chessmen are moved on a chess-board, a telephone and messenger system having been installed for this purpose. In this way a large amount of the time lost through having too many men in one place and too few in another, and through waiting between jobs, was entirely eliminated. Under the old system the workmen were kept day after day in comparatively large gangs, each under a single foreman, and the gang was apt to remain of pretty nearly the same size whether there was much or little of the particular kind of work on hand which this foreman had under his charge, since each gang had to be kept large enough to handle whatever work in its special line was likely to come along.

When one ceases to deal with men in large gangs or groups, and proceeds to study each workman as an individual, if the workman fails to do his task, some competent teacher should be sent to show him exactly how his work can best be done, to guide, help, and encourage him, and, at the same time, to study his possibilities as a workman. So that, under the plan which individualizes each workman, instead of brutally discharging the man or lowering his wages for failing to make good at once, he is given the time and the help required to make him proficient at his present job, or he is shifted to another class of work for which he is either mentally or physically better suited.

All of this requires the kindly cooperation of the management, and involves a much more elaborate organization and system than the old-fashioned herding of men in large gangs. This organization consisted, in this case, of one set of men, who were engaged in the development of the science of laboring through time study, such as has been described above; another set of men, mostly skilled laborers themselves, who were teachers, and who helped and guided the men in their work; another set of toolroom men who provided them with the proper implements and kept them in perfect order, and another set of clerks who planned the work well in advance, moved the men with the least loss of time from one place to another, and properly recorded each man's earnings, etc. And this furnishes an elementary illustration of what has been referred to as cooperation between the management and the workmen.

The question which naturally presents itself is whether an elaborate organization of this sort can be made to pay for itself; whether such an organization is not top-heavy. This question will best be answered by a statement of the results of the third year of working under this plan.

	Old Plan	New Plan Task Work
The number of yard laborers was reduced from between	400 & 600 down to about	140
Average number of tons per man per day	16	59
Average earnings per man per day	$1.15	$1.88
Average cost of handling a ton of 2240 lbs.	$0.072	$0.033

And in computing the low cost of $0.033 per ton, the office and tool-room expenses, and the wages of all labor superintendents, foremen, clerks, time-study men, etc., are included.

During this year the total saving of the new plan over the old amounted to $36,417.69, and during the six months following, when all of the work of the yard was on task work, the saving was at the rate of between $75,000 and $80,000 per year.

Perhaps the most important of all the results attained was the effect on the workmen themselves. A careful inquiry into the condition of these men developed the fact that out of the 140 workmen only two were said to be drinking men. This does not, of course, imply that many of them did not take an occasional drink. The fact is that a steady drinker would find it almost impossible to keep up with the pace which was set, so that they were practically all sober. Many, if not most of them, were saving money, and they all lived better than they had before. These men constituted the finest body of picked laborers that the writer has ever seen together, and they looked upon the men who were over them, their bosses and their teachers, as their very best friends; not as nigger drivers, forcing them to work extra hard for ordinary wages, but as friends who were teaching them and helping them to earn much higher wages than they had ever earned before. It would have been absolutely impossible for any one to have stirred up strife between these men and their employers. And this presents a very simple though effective illustration of what is meant by the words "prosperity for the employé, coupled with prosperity for the employer," the two principal objects of management. It is evident also that this result has been brought about by the application of the four fundamental principles of scientific management.

As another illustration of the value of a scientific study of the motives which influence workmen in their daily work, the loss of ambition and initiative will be cited, which takes place in workmen when they are herded into gangs instead of being treated as separate individuals. A careful analysis had demonstrated the fact that when workmen are herded together in gangs, each man in the gang becomes far less efficient than when his personal ambition is stimulated; that when men work in gangs, their individual efficiency falls almost invariably down to or below the level of the worst man in the gang; and that they are all pulled down instead of being elevated by being herded together. For this reason a general order had been issued in the Bethlehem Steel Works that not more than four men were to be allowed to work in a labor gang without a special permit, signed by the General Superintendent of the works, this special permit to extend for one week only. It was arranged that as far as possible each laborer should be given a separate individual task. As there were about 5000 men at work in the establishment, the General Superintendent had so much to do that there was but little time left for signing these special permits.

After gang work had been by this means broken up, an unusually fine set of ore shovelers had been developed, through careful selection and individual, scientific training. Each of these men was given a separate car to unload each day, and his wages depended upon his own personal work. The man who unloaded the largest amount of ore was paid the highest wages, and an unusual opportunity came for demonstrating the importance of individualizing each workman. Much of this ore came from the Lake Superior region, and the same ore was delivered both in Pittsburg and in Bethlehem in exactly similar cars. There was a shortage of ore handlers in Pittsburg, and hearing of the fine gang of laborers that had been developed at Bethlehem, one of the Pittsburg steel works sent an agent to hire the

Bethlehem men. The Pittsburg men offered $4\frac{9}{10}$ cents a ton for unloading exactly the same ore, with the same shovels, from the same cars, that were unloaded in Bethlehem for $3\frac{2}{10}$ cents a ton. After carefully considering this situation, it was decided that it would be unwise to pay more than $3\frac{2}{10}$ cents per ton for unloading the Bethlehem cars, because, at this rate, the Bethlehem laborers were earning a little over $1.85 per man per day, and this price was 60 per cent. more than the ruling rate of wages around Bethlehem.

A long series of experiments, coupled with close observation, had demonstrated the fact that when workmen of this caliber are given a carefully measured task, which calls for a big day's work on their part, and that when in return for this extra effort they are paid wages up to 60 per cent. beyond the wages usually paid, that this increase in wages tends to make them not only more thrifty but better men in every way; that they live rather better, begin to save money, become more sober, and work more steadily. When, on the other hand, they receive much more than a 60 per cent. increase in wages, many of them will work irregularly and tend to become more or less shiftless, extravagant, and dissipated. Our experiments showed, in other words, that it does not do for most men to get rich too fast.

After deciding, for this reason, not to raise the wages of our ore handlers, these men were brought into the office one at a time, and talked to somewhat as follows·

"Now, Patrick, you have proved to us that you are a high-priced man. You have been earning every day a little more than $1.85, and you are just the sort of man that we want to have in our ore-shoveling gang. A man has come here from Pittsburg, who is offering $4\frac{9}{10}$ cents per ton for handling ore while we can pay only $3\frac{2}{10}$ cents per ton. I think, therefore, that you had better apply to this man for a job. Of course, you know we are very sorry to have you leave us, but you have proved yourself a high-priced man, and we are very glad to see you get this chance of earning more money. Just remember, however, that at any time in the future, when you get out of a job, you can always come right back to us. There will always be a job for a high-priced man like you in our gang here."

Almost all of the ore handlers took this advice, and went to Pittsburg, but in about six weeks most of them were again back in Bethlehem unloading ore at the old rate of $3\frac{2}{10}$ cents a ton. The writer had the following talk with one of these men after he had returned:

"Patrick, what are you doing back here? I thought we had gotten rid of you."

"Well, sir, I'll tell you how it was. When we got out there Jimmy and I were put on to a car with eight other men. We started to shovel the ore out just the same as we do here. After about half an hour I saw a little devil alongside of me doing pretty near nothing, so I said to him, 'Why don't you go to work? Unless we get the ore out of this car we won't get any money on pay-day.' He turned to me and said, 'Who in —— are you?' 'Well,' I said, 'that's none of your business'; and the little devil stood up to me and said, 'You'll be minding your own business, or I'll throw you off this car!' 'Well, I could have spit on him and drowned him, but the rest of the men put down their shovels and looked as if they were going to back him up; so I went round to Jimmy and said (so that the whole gang could hear it), 'Now, Jimmy, you and I will throw a shovelful whenever this little devil throws one, and not another shovelful.' So we watched him, and only shoveled when he shoveled. — When pay-day came around, though, we had less money than we got here at Bethlehem. After that Jimmy and I went in to the boss, and asked him for a car to ourselves, the same as we got at Bethlehem, but he told us to mind our own business. And when another pay-day came around we had less money than we got here at Bethlehem, so Jimmy and I got the gang together and brought them all back here to work again."

When working each man for himself, these men were able to earn higher wages at $3\frac{2}{10}$ cents a ton than they could earn when they were paid $4\frac{9}{10}$ cents a ton on gang work; and this again shows the great gain which results from working according to even the most elementary of scientific principles. But it also shows that in the application of the most elementary principles it is necessary for the management to do their share of the work in cooperating with the workmen. The Pittsburg managers knew just how the results had been attained at Bethlehem, but they were unwilling to go to the small trouble and expense required to plan ahead and assign a separate car to each shoveler, and then keep an individual record of each man's work, and pay him just what he had earned.

Bricklaying is one of the oldest of our trades. For hundreds of years there has been little or no improvement made in the implements and materials used in this trade, nor in fact in the method of laying bricks. In spite of the millions of men who have practised this trade, no great improvement has been evolved for many generations. Here, then, at least, one would expect to find but little gain possible through scientific analysis and study. Mr. Frank B. Gilbreth, a member of our Society, who had himself studied bricklaying in his youth, became interested in the principles of scientific management, and decided to apply them to the art of bricklaying. He made an intensely interesting analysis and study of each movement of the bricklayer, and one after another eliminated all unnecessary movements and substituted fast for slow motions. He experimented with every minute element which in any way affects the speed and the tiring of the bricklayer.

He developed the exact position which each of the feet of the bricklayer should occupy with relation to the wall, the mortar box, and the pile of bricks, and so made it unnecessary for him to take a step or two toward the pile of bricks and back again each time a brick is laid.

He studied the best height for the mortar box and brick pile, and then designed a scaffold, with a table on it, upon which all of the materials are placed, so as to keep the bricks, the mortar, the man, and the wall in their proper relative positions. These scaffolds are adjusted, as the wall grows in height, for all of the bricklayers by a laborer especially detailed for this purpose, and by this means the bricklayer is saved the exertion of stooping down to the level of his feet for each brick and each trowelful of mortar and then straightening up again. Think of the waste of effort that has gone on through all these years, with each bricklayer lowering his body, weighing, say, 150 pounds, down two feet and raising it up again every time a brick (weighing about 5 pounds) is laid in the wall! And this each bricklayer did about one thousand times a day.

As a result of further study, after the bricks are unloaded from the cars, and before bringing them to the bricklayer, they are carefully sorted by a laborer, and placed with their best edge up on a simple wooden frame, constructed so as to enable him to take hold of each brick in the quickest time and in the most advantageous position. In this way the bricklayer avoids either having to turn the brick over or end for end to examine it before laying it, and he saves, also, the time taken in deciding which is the best edge and end to place on the outside of the wall. In most cases, also, he saves the time taken in disentangling the brick from a disorderly pile on the scaffold. This "pack" of bricks (as Mr. Gilbreth calls his loaded wooden frames) is placed by the helper in its proper position on the adjustable scaffold close to the mortar box.

We have all been used to seeing bricklayers tap each brick after it is placed on its bed of mortar several times with the end of the handle of the trowel so as to secure the right thickness for the joint. Mr. Gilbreth found that by tempering the mortar just right, the bricks could be readily bedded to the proper depth by a downward pressure of the hand with which they are laid. He insisted that his mortar mixers should give special attention to tempering the mortar, and so save the time consumed in tapping the brick.

Through all of this minute study of the motions to be made by the bricklayer in laying bricks under standard conditions, Mr. Gilbreth has reduced his movements from eighteen motions per brick to five, and even in one case to as low as two motions per brick. He has given all of the details of this analysis to the profession in the chapter headed "Motion Study," of his book entitled "Bricklaying System," published by Myron C. Clerk Publishing Company, New York and Chicago; E. F. N. Spon, of London.

An analysis of the expedients used by Mr. Gilbreth in reducing the motions of his bricklayers from eighteen to five shows that this improvement has been made in three different ways:

First. He has entirely dispensed with certain movements which the bricklayers in the past believed were necessary, but which a careful study and trial on his part have shown to be useless.

Second. He has introduced simple apparatus, such as his adjustable scaffold and his packets for holding the bricks, by means of which, with a very small amount of cooperation from a cheap laborer, he entirely eliminates a lot of tiresome and time-consuming motions which are necessary for the bricklayer who lacks the scaffold and the packet.

Third. He teaches his bricklayers to make simple motions with both hands at the same time, where before they completed a motion with the right hand and followed it later with one from the left hand.

For example, Mr. Gilbreth teaches his bricklayer to pick up a brick in the left hand at the same instant that he takes a trowelful of mortar with the right hand. This work with two hands at the same time is, of course, made possible by substituting a deep mortar box for the old mortar board (on which the mortar spread out so thin that a step or two had to be taken to reach it) and then placing the mortar box and the brick pile close together, and at the proper height on his new scaffold.

These three kinds of improvements are typical of the ways in which needless motions can be entirely eliminated and quicker types of movements substituted for slow movements when scientific motion study, as Mr. Gilbreth calls his analysis, time study, as the writer has called similar work, are applied in any trade.

Most practical men would (knowing the opposition of almost all tradesmen to making any change in their methods and habits), however, be skeptical as to the possibility of actually achieving any large results from a study of this sort. Mr. Gilbreth reports that a few months ago, in a large brick building which he erected, he demonstrated on a commercial scale the great gain which is possible from practically applying his scientific study. With union bricklayers, in laying a factory wall, twelve inches thick, with two kinds of brick, faced and ruled joints on both sides of the wall, he averaged, after his selected workmen had become skilful in his new methods, 350 bricks per man *per hour;* whereas the average speed of doing this work with the old methods was, in that section of the country, 120 bricks per man per hour. His bricklayers were taught his new method of bricklaying by their foreman. Those who failed to profit by their teaching were dropped, and each man, as he became proficient under the new method, received a substantial (not a small) increase in his wages. With a view to individualizing his workmen and stimulating each man to do his best, Mr. Gilbreth also developed an ingenious method for measuring and recording the number of bricks laid by each man, and for telling each workman at frequent intervals how many bricks he had succeeded in laying.

It is only when this work is compared with the

conditions which prevail under the tyranny of some of our misguided bricklayers' unions that the great waste of human effort which is going on will be realized. In one foreign city the bricklayers' union have restricted their men to *275 bricks per day* on work of this character when working for the city, and *375* per day when working for private owners. The members of this union are probably sincere in their belief that this restriction of output is a benefit to their trade. It should be plain to all men, however, that this deliberate loafing is almost criminal, in that it inevitably results in making every workman's family pay higher rent for their housing, and also in the end drives work and trade away from their city, instead of bringing it to it.

Why is it, in a trade which has been continually practised since before the Christian era, and with implements practically the same as they now are, that this simplification of the bricklayer's movements, this great gain, has not been made before?

It is highly likely that many times during all of these years individual bricklayers have recognized the possibility of eliminating each of these unnecessary motions. But even if, in the past, he did invent each one of Mr. Gilbreth's improvements, no bricklayer could alone increase his speed through their adoption because it will be remembered that in all cases several bricklayers work together in a row and that the walls all around a building must grow at the same rate of speed. No one bricklayer, then, can work much faster than the one next to him. Nor has any one workman the authority to make other men cooperate with him to do faster work. It is only through *enforced* standardization of methods, *enforced* adoption of the best implements and working conditions, and *enforced* cooperation that this faster work can be assured. And the duty of enforcing the adoption of standards and of enforcing this cooperation rests with the *management* alone. The *management* must supply continually one or more teachers to show each new man the new and simpler motions, and the slower men must be constantly watched and helped until they have risen to their proper speed. All of those who, after proper teaching, either will not or cannot work in accordance with the new methods and at the higher speed must be discharged by the *management*. The *management* must also recognize the broad fact that workmen will not submit to this more rigid standardization and will not work extra hard, unless they receive extra pay for doing it.

All of this involves an individual study of and treatment for each man, while in the past they have been handled in large groups.

The *management* must also see that those who prepare the bricks and the mortar and adjust the scaffold, etc., for the bricklayers, cooperate with them by doing their work just right and always on time; and they must also inform each bricklayer at frequent intervals as to the progress he is making, so that he may not unintentionally fall off in his pace. Thus it will be seen that it is the assumption by the management of new duties and new kinds of work never done by employers in the past that makes this great improvement possible, and that, without this new help from the management, the workman even with full knowledge of the new methods and with the best of intentions could not attain these startling results.

Mr. Gilbreth's method of bricklaying furnishes a simple illustration of true and effective cooperation. Not the type of cooperation in which a mass of workmen on one side together cooperate with the management; but that in which several men in the management (each one in his own particular way) help each workman individually, on the one hand, by studying his needs and his shortcomings and teaching him better and quicker methods, and, on the other hand, by seeing that all other workmen with whom he comes in contact help and cooperate with him by doing their part of the work right and fast.

The writer has gone thus fully into Mr. Gilbreth's method in order that it may be perfectly clear that this increase in output and that this harmony could not have been attained under the management of "initiative and incentive" (that is, by putting the problem up to the workman and leaving him to solve it alone) which has been the philosophy of the past. And that his success has been due to the use of the four elements which constitute the essence of scientific management.

First. The development (by the management, not the workman) of the science of bricklaying, with rigid rules for each motion of every man, and the perfection and standardization of all implements and working conditions.

Second. The careful selection and subsequent training of the bricklayers into first-class men, and the elimination of all men who refuse to or are unable to adopt the best methods.

Third. Bringing the first-class bricklayer and the science of bricklaying together, through the constant help and watchfulness of the management, and through paying each man a large daily bonus for working fast and doing what he is told to do.

Fourth. An almost equal division of the work and responsibility between the workman and the management. All day long the management work almost side by side with the men, helping, encouraging, and smoothing the way for them, while in the past they stood one side, gave the men but little help, and threw on to them almost the entire responsibility as to methods, implements, speed, and harmonious cooperation.

Of these four elements, the first (the development of the science of bricklaying) is the most interesting and spectacular. Each of the three others is, however, quite as necessary for success.

It must not be forgotten that back of all this, and directing it, there must be the optimistic, determined, and hard-working leader who can wait patiently as well as work.

In most cases (particularly when the work to be done is intricate in its nature) the "development of the science" is the most important of the four great elements of the new management. There are instances, however, in which the "scientific selection of the workman" counts for more than anything else. A case of this type is well illustrated in the very simple though unusual work of inspecting bicycle balls.

When the bicycle craze was at its height some years ago several million small balls made of hardened steel were used annually in bicycle bearings. And among the twenty or more operations used in making steel balls, perhaps the most important was that of inspecting them after final polishing so as to remove all fire-cracked or otherwise imperfect balls before boxing.

The writer was given the task of systematizing the largest bicycle ball factory in this country. This company had been running for from eight to ten years on ordinary day work before he undertook its reorganization, so that the one hundred and twenty or more girls who were inspecting the balls were "old hands" and skilled at their jobs.

It is impossible even in the most elementary work to change rapidly from the old independence of individual day work to scientific cooperation.

In most cases, however, there exist certain imperfections in working conditions which can at once be improved with benefit to all concerned.

In this instance it was found that the inspectors

(girls) were working ten and one-half hours per day (with a Saturday half holiday.)

Their work consisted briefly in placing a row of small polished steel balls on the back of the left hand, in the crease between two of the fingers pressed together, and while they were rolled over and over, they were minutely examined in a strong light, and with the aid of a magnet held in the right hand, the defective balls were picked out and thrown into especial boxes. Four kinds of defects were looked for — dented, soft, scratched, and fire-cracked — and they were mostly so minute as to be invisible to an eye not especially trained to this work. It required the closest attention and concentration, so that the nervous tension of the inspectors was considerable, in spite of the fact that they were comfortably seated and were not physically tired.

A most casual study made it evident that a very considerable part of the ten and one-half hours during which the girls were supposed to work was really spent in idleness because the working period was too long.

It is a matter of ordinary common sense to plan working hours so that the workers can really "work while they work" and "play while they play," and not mix the two.

Before the arrival of Mr. Sanford E. Thompson, who undertook a scientific study of the whole process, we decided, therefore, to shorten the working hours.

The old foreman who had been over the inspecting

room for years was instructed to interview one after another of the better inspectors and the more influential girls and persuade them that they could do just as much work in ten hours each day as they had been doing in ten and one-half hours. Each girl was told that the proposition was to shorten the day's work to ten hours and pay them the same day's pay they were receiving for the ten and one-half hours.

In about two weeks the foreman reported that all of the girls he had talked to agreed that they could do their present work just as well in ten hours as in ten and one-half and that they approved of the change.

The writer had not been especially noted for his tact so he decided that it would be wise for him to display a little more of this quality by having the girls vote on the new proposition. This decision was hardly justified, however, for when the vote was taken the girls were unanimous that $10\frac{1}{2}$ hours was good enough for them and they wanted no innovation of any kind.

This settled the matter for the time being. A few months later tact was thrown to the winds and the working hours were arbitrarily shortened in successive steps to 10 hours, $9\frac{1}{2}$, 9, and $8\frac{1}{2}$ (the pay per day remaining the same); and with each shortening of the working day the output increased instead of diminishing.

The change from the old to the scientific method in this department was made under the direction of Mr. Sanford E. Thompson, perhaps the most experienced man in motion and time study in this country, under the general superintendence of Mr. H. L. Gautt.

In the Physiological departments of our universities experiments are regularly conducted to determine what is known as the "personal coefficient" of the man tested. This is done by suddenly bringing some object, the letter A or B for instance, within the range of vision of the subject, who, the instant he recognizes the letter has to do some definite thing, such as to press a particular electric button. The time which elapses from the instant the letter comes in view until the subject presses the button is accurately recorded by a delicate scientific instrument.

This test shows conclusively that there is a great difference in the "personal coefficient" of different men. Some individuals are born with unusually quick powers of perception accompanied by quick responsive action. With some the message is almost instantly transmitted from the eye to the brain, and the brain equally quickly responds by sending the proper message to the hand.

Men of this type are said to have a low "personal coefficient," while those of slow perception and slow action have a *high* "personal coefficient."

Mr. Thompson soon recognized that the quality most needed for bicycle ball inspectors was a low "personal coefficient." Of course the ordinary qualities of endurance and industry were also called for.

For the ultimate good of the girls as well as the

company, however, it became necessary to exclude all girls who lacked a low "personal coefficient." And unfortunately this involved laying off many of the most intelligent, hardest working, and most trustworthy girls merely because they did not possess the quality of quick perception followed by quick action.

While the gradual selection of girls was going on other changes were also being made.

One of the dangers to be guarded against, when the pay of the man or woman is made in any way to depend on the quantity of the work done, is that in the effort to increase the quantity the quality is apt to deteriorate.

It is necessary in almost all cases, therefore, to take definite steps to insure against any falling off in quality before moving in any way towards an increase in quantity.

In the work of these particular girls quality was the very essence. They were engaged in picking out all defective balls.

The first step, therefore, was to make it impossible for them to slight their work without being found out. This was accomplished through what is known as over-inspection. Each one of four of the most trustworthy girls was given each day a lot of balls to inspect which had been examined the day before by one of the regular inspectors; the number identifying the lot to be over-inspected having been changed by the foreman so that none of the over-inspectors knew whose work they were examining. In addition to this one of the lots inspected by the four over-inspectors was examined on the following day by the chief inspector, selected on account of her especial accuracy and integrity.

An effective expedient was adopted for checking the honesty and accuracy of the over-inspection. Every two or three days a lot of balls was especially prepared by the foreman, who counted out a definite number of perfect balls, and added a recorded number of defective balls of each kind. Neither the inspectors nor the over-inspectors had any means of distinguishing this prepared lot from the regular commercial lots. And in this way all temptation to slight their work or make false returns was removed.

After insuring in this way against deterioration in quality, effective means were at once adopted to increase the output. Improved day work was substituted for the old slipshod method. An accurate daily record was kept both as to the quantity and quality of the work done in order to guard against any personal prejudice on the part of the foreman and to insure absolute impartiality and justice for each inspector. In a comparatively short time this record enabled the foreman to stir the ambition of all the inspectors by increasing the wages of those who turned out a large quantity and good quality, while at the same time lowering the pay of those who did indifferent work and discharging others who proved to be incorrigibly slow or careless. A careful examination was then made of the way in which each girl spent her time and an accurate time study was undertaken, through the use of a stop-watch and record blanks, to determine how fast each kind of inspection should be done, and to establish the exact conditions under which each girl could do her quickest and best work, while at the same time guarding against giving her a task so severe that there was danger from over fatigue or exhaustion. This investigation showed that the girls spent a considerable part of their time either in partial idleness, talking and half working, or in actually doing nothing.

Even when the hours of labor had been shortened from $10\frac{1}{2}$ to $8\frac{1}{2}$ hours, a close observation of the girls showed that after about an hour and one-half of consecutive work they began to get nervous. They evidently needed a rest. It is wise to stop short of the point at which overstrain begins, so we arranged for them to have a ten minutes period for recreation at the end of each hour and one quarter. During these recess periods (two of ten minutes each in the morning and two in the afternoon) they were obliged to stop work and were encouraged to leave their seats and get a complete change of occupation by walking around and talking, etc.

In one respect no doubt some people will say that these girls were brutally treated. They were seated so far apart that they could not conveniently talk while at work.

Shortening their hours of labor, however, and providing so far as we knew the most favorable working conditions made it possible for them to really work steadily instead of pretending to do so. And it is only after this stage in the reorganization is reached, when the girls have been properly selected and on the one hand such precautions have been taken as to guard against the possibility of overdriving them, while, on the other hand, the temptation to slight their work has been removed and the most favorable working conditions have been established, that the final step should be taken which insures them what they most want, namely, *high wages*, and the employers what they most want, namely, the maximum output and best quality of work, — which means *a low labor cost.*

This step is to give each girl each day a carefully measured task which demands a full day's work from a competent operative, and also to give her a large premium or bonus whenever she accomplishes this task.

This was done in this case through establishing what is known as differential rate piece work.[1] Under this system the pay of each girl was increased in proportion to the quantity of her output and also still more in proportion to the accuracy of her work.

As will be shown later, the differential rate (the lots inspected by the over-inspectors forming the basis for the differential) resulted in a large gain in the quantity of work done and at the same time in a marked improvement in the quality.

Before they finally worked to the best advantage

[1] See paper read before the American Society of Mechanical Engineers, by Fred. W. Taylor, Vol. XVI, p. 856, entitled "Piece Rate System."

it was found to be necessary to measure the output of each girl as often as once every hour, and to send a teacher to each individual who was found to be falling behind to find what was wrong, to straighten her out, and to encourage and help her to catch up.

There is a general principle back of this which should be appreciated by all of those who are especially interested in the management of men.

A reward, if it is to be effective in stimulating men to do their best work, must come soon after the work has been done. But few men are able to look forward for more than a week or perhaps at most a month, and work hard for a reward which they are to receive at the end of this time.

The average workman must be able to measure what he has accomplished and clearly see his reward at the end of each day if he is to do his best. And more elementary characters, such as the young girls inspecting bicycle balls, or *children*, for instance, should have proper encouragement either in the shape of personal attention from those over them or an actual reward in sight as often as once an hour.

This is one of the principal reasons why cooperation or "profit-sharing" either through selling stock to the employés or through dividends on wages received at the end of the year, etc., have been at the best only mildly effective in stimulating men to work hard. The nice time which they are sure to have to-day if they take things easily and go slowly proves more attractive than steady hard work with a possible reward to be shared with others six months later. A second reason for the inefficiency of profit-sharing schemes had been that no form of cooperation has yet been devised in which each individual is allowed free scope for his personal ambition. Personal ambition always has been and will remain a more powerful incentive to exertion than a desire for the general welfare. The few misplaced drones, who do the loafing and share equally in the profits, with the rest, under cooperation are sure to drag the better men down toward their level.

Other and formidable difficulties in the path of cooperative schemes are, the equitable division of the profits, and the fact that, while workmen are always ready to share the profits, they are neither able nor willing to share the losses. Further than this, in many cases, it is neither right nor just that they should share either the profits or the losses, since these may be due in great part to causes entirely beyond their influence or control, and to which they do not contribute.

To come back to the girls inspecting bicycle balls, however, the final outcome of all the changes was that *thirty-five girls did the work formerly done by one hundred and twenty.* And that the *accuracy of the work at the higher speed was two-thirds greater than at the former slow speed.*

The good that came to the girls was,

First. That they averaged from 80 to 100 per cent. higher wages than they formerly received.

Second. Their hours of labor were shortened from $10\frac{1}{2}$ to $8\frac{1}{2}$ per day, with a Saturday half holiday. And they were given four recreation periods properly distributed through the day, which made overworking impossible for a healthy girl.

Third. Each girl was made to feel that she was the object of especial care and interest on the part of the management, and that if anything went wrong with her she could always have a helper and teacher in the management to lean upon.

Fourth. All young women should be given two consecutive days of rest (with pay) each month, to be taken whenever they may choose. It is my impression that these girls were given this privilege, although I am not quite certain on this point.

The benefits which came to the company from these changes were:

First. A substantial improvement in the quality of the product.

Second. A material reduction in the cost of inspection, in spite of the extra expense involved in clerk work, teachers, time study, over-inspectors, and in paying higher wages.

Third. That the most friendly relations existed between the management and the employés, which rendered labor troubles of any kind or a strike impossible.

These good results were brought about by many changes which substituted favorable for unfavorable working conditions. It should be appreciated, however, that the one element which did more than all of the others was, the careful selection of girls with quick perception to replace those whose perceptions were slow — (the substitution of girls with a low personal coefficient for those whose personal coefficient was high) — the scientific selection of the workers.

The illustrations have thus far been purposely confined to the more elementary types of work, so that a very strong doubt must still remain as to whether this kind of cooperation is desirable in the case of more intelligent mechanics, that is, in the case of men who are more capable of generalization, and who would therefore be more likely, of their own volition, to choose the more scientific and better methods. The following illustrations will be given for the purpose of demonstrating the fact that in the higher classes of work the scientific laws which are developed are so intricate that the high-priced mechanic needs (even more than the cheap laborer) the cooperation of men better educated than himself in finding the laws, and then in selecting, developing, and training him to work in accordance with these laws. These illustrations should make perfectly clear our original proposition that in practically all of the mechanic arts the science which underlies each workman's act is so great and amounts to so much that the workman who is best suited to actually doing the work is incapable, either through lack of education or through insufficient mental capacity, of understanding this science.

A doubt, for instance, will remain in the minds perhaps of most readers (in the case of an establishment which manufactures the same machine, year in and year out, in large quantities, and in which, therefore, each mechanic repeats the same limited series of operations over and over again), whether the ingenuity of each workman and the help which he from time to time receives from his foreman will not develop such superior methods and such a personal dexterity that no scientific study which could be made would result in a material increase in efficiency.

A number of years ago a company employing about three hundred men, which had been manufacturing the same machine for ten to fifteen years, sent for us to report as to whether any gain could be made through the introduction of scientific management. Their shops had been run for many years under a good superintendent and with excellent foremen and workmen, on piece work. The whole establishment was, without doubt, in better physical condition than the average machine-shop in this country. The superintendent was distinctly displeased when told that through the adoption of task management the output, with the same number of men and machines, could be more than doubled. He said that he believed that any such statement was mere boasting, absolutely false, and instead of inspiring him with confidence, he was disgusted that any one should make such an impudent claim. He, however, readily assented to the proposition that he should select any one of the machines whose output he considered as representing the average of the shop, and that we should then demonstrate on this machine that through scientific methods its ouptut could be more than doubled.

The machine selected by him fairly represented the work of the shop. It had been run for ten or twelve years past by a first-class mechanic who was more than equal in his ability to the average workmen in the establishment. In a shop of this sort, in which similar machines are made over and over again, the work is necessarily greatly subdivided, so that no one man works upon more than a comparatively small number of parts during the year. A careful record was therefore made, in the presence of both parties, of the time actually taken in finishing each of the parts which this man worked upon. The total time required by him to finish each piece, as well as the exact speeds and feeds which he took, were noted, and a record was kept of the time which he took in setting the work in the machine and removing it. After obtaining in this way a statement of what represented a fair average of the work done in the shop, we applied to this one machine the principles of scientific management.

By means of four quite elaborate slide-rules, which have been especially made for the purpose of determining the all-round capacity of metal-cutting machines, a careful analysis was made of every element of this machine in its relation to the work in hand. Its pulling power at its various speeds, its feeding capacity, and its proper speeds were determined by means of the slide-rules, and changes were then made in the countershaft and driving pulleys so as to run it at its proper speed. Tools, made of high-speed steel, and of the proper shapes, were properly dressed, treated, and ground. (It should be understood, however, that in this case the high-speed steel which had heretofore been in general use in the shop was also used in our demonstration.) A large special slide-rule was then made, by means of which the exact speeds and feeds were indicated at which each kind of work could be done in the shortest possible time in this particular lathe. After preparing in this way so that the workman should work according to the new method, one after another, pieces of work were finished in the lathe, corresponding to the work which had been done in our preliminary trials, and the gain in time made through running the machine according to scientific principles ranged from two and one-half times the speed in the slowest instance to nine times the speed in the highest.

The change from rule-of-thumb management to scientific management involves, however, not only a study of what is the proper speed for doing the work and a remodeling of the tools and the implements in the shop, but also a complete change in the mental attitude of all the men in the shop toward their work and toward their employers. The physical improvements in the machines necessary to insure large gains, and the motion study followed by minute study with a stop-watch of the time in which each workman should do his work, can be made comparatively quickly. But the change in the mental attitude and in the habits of the three hundred or more workmen can be brought about only slowly and through a long series of object-lessons, which finally demonstrates to each man the great advantage which he will gain by heartily cooperating in his every-day work with the men in the management. Within three years, however, in this shop, the output had been more than doubled per man and per machine. The men had been carefully selected and in almost all cases promoted from a lower to a higher order of work, and so instructed by their teachers (the functional foremen) that they were able to earn higher wages than ever before. The average increase in the daily earnings of each man was about 35 per cent., while, at the same time, the sum total of the wages paid for doing a given amount of work was lower than before. This increase in the speed of doing the work, of course, involved a substitution of the quickest hand methods for the old independent rule-of-thumb methods, and an elaborate analysis of the hand work done by each man. (By hand work is meant such work as depends upon the manual dexterity and speed of a workman, and which is independent of the work done by the machine.) The time saved by scientific hand work was in many cases greater even than that saved in machine-work.

It seems important to fully explain the reason why, with the aid of a slide-rule, and after having studied the art of cutting metals, it was possible for the scientifically equipped man, who had never before seen these particular jobs, and who had never worked on this machine, to do work from two and one-half to nine times as fast as it had been done before by a good mechanic who had spent his whole time for some ten to twelve years in doing this very work upon this particular machine. In a word, this was possible because the art of cutting metals involves a true science of no small magnitude, a science, in fact, so intricate that it is impossible for any machinist who is suited to running a lathe year in and year out either to understand it or to work according to its laws without the help of men who have made this their specialty. Men who are unfamiliar with machine-shop work are prone to look upon the manufacture of each piece as a special problem, independent of any other kind of machine-work. They are apt to think, for instance, that the problems connected with making the parts of an engine require the especial study, one may say almost the life study, of a set of engine-making mechanics, and that these problems are entirely different from those which would be met with in machining lathe or planer parts. In fact, however, a study of those elements which are peculiar either to engine parts or to lathe parts is trifling, compared with the great study of the art, or science, of cutting metals, upon a knowledge of which rests the ability to do really fast machine-work of all kinds.

The real problem is how to remove chips fast from a casting or a forging, and how to make the piece smooth and true in the shortest time, and it matters but little whether the piece being worked upon is part, say, of a marine engine, a printing-press, or an automobile. For this reason, the man with the slide-rule, familiar with the science of cutting metals, who had never before seen this particular work, was able completely to distance the skilled mechanic who had made the parts of this machine his specialty for years.

It is true that whenever intelligent and educated men find that the responsibility for making progress in any of the mechanic arts rests with them, instead of upon the workmen who are actually laboring at the trade, that they almost invariably start on the road which leads to the development of a science where, in the past, has existed mere traditional or rule-of-thumb knowledge. When men, whose education has given them the habit of generalizing and everywhere looking for laws, find themselves confronted with a multitude of problems, such as exist in every trade and which have a general similarity one to another, it is inevitable that they should try to gather these problems into certain logical groups, and then search for some general laws or rules to guide them in their solution. As has been pointed out, however, the underlying principles of the management of "initiative and incentive," that is, the underlying philosophy of this management, necessarily leaves the solution of all of these problems in the hands of each individual workman, while the philosophy of scientific management places their solution in the hands of the management. The workman's whole time is each day taken in actually doing the work with his hands, so that, even if he had the necessary education and habits of generalizing in his thought, he lacks the time and the opportunity for developing these laws, because the study of even a simple law involving say time study requires the cooperation of two men, the one doing the work while the other times him with a stop-watch. And even if the workman were to develop laws where before existed only rule-of-thumb knowledge, his personal interest would lead him almost inevitably to keep his discoveries secret, so that he could, by means of this special knowledge, personally do more work than other men and so obtain higher wages.

Under scientific management, on the other hand, it becomes the duty and also the pleasure of those who are engaged in the management not only to develop laws to replace rule of thumb, but also to teach impartially all of the workmen who are under them the quickest ways of working. The useful results obtained from these laws are always so great that any company can well afford to pay for the time and the experiments needed to develop them. Thus under scientific management exact scientific knowledge and methods are everywhere, sooner or later, sure to replace rule of thumb, whereas under the old type of management working in accordance with scientific laws is an impossibility.

The development of the art or science of cutting metals is an apt illustration of this fact. In the fall of 1880, about the time that the writer started to make the experiments above referred to, to determine what constitutes a proper day's work for a laborer, he also obtained the permission of Mr. William Sellers, the President of the Midvale Steel Company, to make a series of experiments to determine what angles and shapes of tools were the best for cutting steel, and also to try to determine the proper cutting speed for steel. At the time that these experiments were started it was his belief that they would not last longer than six months, and, in fact, if it had been known that a longer period than this would be required, the permission to spend a considerable sum of money in making them would not have been forthcoming.

A 66-inch diameter vertical boring-mill was the first machine used in making these experiments, and large locomotive tires, made out of hard steel of uniform quality, were day after day cut up into chips in gradually learning how to make, shape, and use the cutting tools so that they would do faster work. At the end of six months sufficient practical information had been obtained to far more than repay the cost of materials and wages which had been expended in experimenting. And yet the comparatively small number of experiments which had been made served principally to make it clear that the actual knowledge attained was but a small fraction of that which still remained to be developed, and which was badly needed by us, in our daily attempt to direct and help the machinists in their tasks.

Experiments in this field were carried on, with

occasional interruption, through a period of about 26 years, in the course of which ten different experimental machines were especially fitted up to do this work. Between 30,000 and 50,000 experiments were carefully recorded, and many other experiments were made, of which no record was kept. In studying these laws more than 800,000 pounds of steel and iron was cut up into chips with the experimental tools, and it is estimated that from $150,000 to $200,000 was spent in the investigation.

Work of this character is intensely interesting to any one who has any love for scientific research. For the purpose of this paper, however, it should be fully appreciated that the motive power which kept these experiments going through many years, and which supplied the money and the opportunity for their accomplishment, was not an abstract search after scientific knowledge, but was the very practical fact that we lacked the exact information which was needed every day, in order to help our machinists to do their work in the best way and in the quickest time.

All of these experiments were made to enable us to answer correctly the two questions which face every machinist each time that he does a piece of work in a metal-cutting machine, such as a lathe, planer, drill press, or milling machine. These two questions are:

In order to do the work in the quickest time,
At what cutting speed shall I run my machine? and
What feed shall I use?
They sound so simple that they would appear

to call for merely the trained judgment of any good mechanic. In fact, however, after working 26 years, it has been found that the answer in every case involves the solution of an intricate mathematical problem, in which the effect of twelve independent variables must be determined.

Each of the twelve following variables has an important effect upon the answer. The figures which are given with each of the variables represent the effect of this element upon the cutting speed. For example, after the first variable (A) we quote, "The proportion is as 1 in the case of semi-hardened steel or chilled iron to 100 in the case of a very soft, low-carbon steel." The meaning of this quotation is that soft steel can be cut 100 times as fast as the hard steel or chilled iron. The ratios which are given, then, after each of these elements, indicate the wide range of judgment which practically every machinist has been called upon to exercise in the past in determining the best speed at which to run the machine and the best feed to use.

(A) The quality of the metal which is to be cut; *i.e.*, its hardness or other qualities which affect the cutting speed. The proportion is as 1 in the case of semi-hardened steel or chilled iron to 100 in the case of very soft, low-carbon steel.

(B) The chemical composition of the steel from which the tool is made, and the heat treatment of the tool. The proportion is as 1 in tools made from tempered carbon steel to 7 in the best high-speed tools.

(C) The thickness of the shaving, or, the thickness of the spiral strip or band of metal which is to be removed by the tool. The proportion is as 1 with thickness of shaving $\frac{3}{16}$ of an inch to $3\frac{1}{2}$ with thickness of shaving $\frac{1}{64}$ of an inch.

(D) The shape or contour of the cutting edge of the tool. The proportion is as 1 in a thread tool to 6 in a broad-nosed cutting tool.

(E) Whether a copious stream of water or other cooling medium is used on the tool. The proportion is as 1 for tool running dry to 1.41 for tool cooled by a copious stream of water.

(F) The depth of the cut. The proportion is as 1 with $\frac{1}{2}$-inch depth of cut to 1.36 with $\frac{1}{8}$-inch depth of cut.

(G) The duration of the cut, *i.e.*, the time which a tool must last under pressure of the shaving without being reground. The proportion is as 1 when tool is to be ground every $1\frac{1}{2}$ hours to 1.20 when tool is to be ground every 20 minutes.

(H) The lip and clearance angles of the tool. The proportion is as 1 with lip angle of 68 degrees to 1.023 with lip angle of 61 degrees.

(J) The elasticity of the work and of the tool on account of producing chatter. The proportion is as 1 with tool chattering to 1.15 with tool running smoothly.

(K) The diameter of the casting or forging which is being cut.

(L) The pressure of the chip or shaving upon the cutting surface of the tool.

(M) The pulling power and the speed and feed changes of the machine.

It may seem preposterous to many people that it should have required a period of 26 years to investigate the effect of these twelve variables upon the cutting speed of metals. To those, however, who have had personal experience as experimenters, it will be appreciated that the great difficulty of the problem lies in the fact that it contains so many variable elements. And in fact the great length of time consumed in making each single experiment was caused by the difficulty of holding eleven variables constant and uniform throughout the experiment, while the effect of the twelfth variable was being investigated. Holding the eleven variables constant was far more difficult than the investigation of the twelfth element.

As, one after another, the effect upon the cutting speed of each of these variables was investigated, in order that practical use could be made of this knowledge, it was necessary to find a mathematical formula which expressed in concise form the laws which had been obtained. As examples of the twelve formulæ which were developed, the three following are given:

$$P = 45{,}000\, D^{14/15} F^{3/4}$$

$$V = \frac{90}{T^{1/8}}$$

$$V = \frac{11.9}{F^{0.665} \left(\dfrac{48}{3}D\right)^{0.2373 + \frac{2.4}{18 + 24D}}}$$

After these laws had been investigated and the various formulæ which mathematically expressed them had been determined, there still remained the difficult task of how to solve one of these complicated mathematical problems quickly enough to make this knowledge available for every-day use. If a good mathematician who had these formulæ before him were to attempt to get the proper answer (*i.e.*, to get the correct cutting speed and feed by working in the ordinary way) it would take him from two to six hours, say, to solve a single problem; far longer to solve the mathematical problem than would be taken in most cases by the workmen in doing the whole job in his machine. Thus a task of considerable magnitude which faced us was that of finding a quick solution of this problem, and as we made progress in its solution, the whole problem was from time to time presented by the writer to one after another of the noted mathematicians in this country. They were offered any reasonable fee for a rapid, practical method to be used in its solution. Some of these men merely glanced at it; others, for the sake of being courteous, kept it before them for some two or three weeks. They all gave us practically the same answer: that in many cases it was possible to solve mathematical problems which contained four variables, and in some cases problems with five or six variables, but that it was manifestly impossible to solve a problem containing twelve variables in any other way than by the slow process of "trial and error."

A quick solution was, however, so much of a necessity in our every-day work of running machine-shops, that in spite of the small encouragement received from the mathematicians, we continued at irregular periods, through a term of fifteen years, to give a large amount of time searching for a simple solution. Four or five men at various periods gave practically their whole time to this work, and finally, while we were at the Bethlehem Steel Company, the slide-rule was developed which is illustrated on Folder No. 11 of the paper "On the Art of Cutting Metals," and is described in detail in the paper presented by Mr. Carl G. Barth to the American Society of Mechanical Engineers, entitled "Slide-rules for the Machine-shop, as a part of the Taylor System of Management" (Vol. XXV of The Transactions of the American Society of Mechanical Engineers). By means of this slide-rule, one of these intricate problems can be solved in less than a half minute by any good mechanic, whether he understands anything about mathematics or not, thus making available for every-day, practical use the years of experimenting on the art of cutting metals.

This is a good illustration of the fact that some way can always be found of making practical, every-day use of complicated scientific data, which appears to be beyond the experience and the range of the technical training of ordinary practical men. These slide-rules have been for years in constant daily use by machinists having no knowledge of mathematics.

A glance at the intricate mathematical formulæ (see page 109) which represent the laws of cutting metals should clearly show the reason why it is impossible for any machinist, without the aid of these laws, and who depends upon his personal experience, correctly to guess at the answer to the two questions,

What speed shall I use?

What feed shall I use?

even though he may repeat the same piece of work many times.

To return to the case of the machinist who had been working for ten to twelve years in machining the same pieces over and over again, there was but a remote chance in any of the various kinds of work which this man did that he should hit upon the one best method of doing each piece of work out of the hundreds of possible methods which lay before him. In considering this typical case, it must also be remembered that the metal-cutting machines throughout our machine-shops have practically all been speeded by their makers by guesswork, and without the knowledge obtained through a study of the art of cutting metals. In the machine-shops systematized by us we have found that there is not one machine in a hundred which is speeded by its makers at anywhere near the correct cutting speed. So that, in order to compete with the science of cutting metals, the machinist, before he could use proper speeds, would first have to put new pulleys on the countershaft of his machine, and also make in most cases changes in the shapes and treatment of his tools, etc. Many of these changes are matters entirely beyond his control, even if he knows what ought to be done.

If the reason is clear to the reader why the rule-of-thumb knowledge obtained by the machinist who is engaged on *repeat work* cannot possibly compete with the true science of cutting metals, it should be even more apparent why the high-class mechanic, who is called upon to do a *great variety* of work from day to day, is even less able to compete with this science. The high-class mechanic who does a different kind of work each day, in order to do each job in the quickest time, would need, in addition to a thorough knowledge of the art of cutting metals, a vast knowledge and experience in the quickest way of doing each kind of hand work. And the reader, by calling to mind the gain which was made by Mr. Gilbreth through his motion and time study in laying bricks, will appreciate the great possibilities for quicker methods of doing all kinds of hand work which lie before every tradesman after he has the help which comes from a scientific motion and time study of his work.

For nearly thirty years past, time-study men connected with the management of machine-shops have been devoting their whole time to a scientific motion study, followed by accurate time study, with a stop-watch, of all of the elements connected with the machinist's work. When, therefore. the teachers, who form one section of the management, and who are cooperating with the working men, are in possession both of the science of cutting metals and of the equally elaborate motion-study and time-study science connected with this work, it is not difficult to appreciate why even the highest class mechanic is unable to do his best work without constant daily assistance from his teachers. And if this fact has been made clear to the reader, one of the important objects in writing this paper will have been realized.

It is hoped that the illustrations which have been given make it apparent why scientific management must inevitably in all cases produce overwhelmingly greater results, both for the company and its employés, than can be obtained with the management of "initiative and incentive." And it should also be clear that these results have been attained, not through a marked superiority in the mechanism of one type of management over the mechanism of another, but rather through the substitution of one set of underlying principles for a totally different set of principles, — by the substitution of one philosophy for another philosophy in industrial management.

To repeat then throughout all of these illustrations, it will be seen that the useful results have hinged mainly upon (1) the substitution of a science for the individual judgment of the workman; (2) the scientific selection and development of the workman, after each man has been studied, taught, and trained, and one may say experimented with, instead of allowing the workmen to select themselves and develop in a haphazard way; and (3) the intimate cooperation of the management with the workmen, so that they together do the work in accordance with the scientific laws which have been developed, instead of leaving the solution of each problem in the hands of the individual workman. In applying these new principles, in place of the old individual effort of each workman, both sides share almost equally in the daily performance of each task, the management doing that part of the work for which they are best fitted, and the workmen the balance.

It is for the illustration of this philosophy that this paper has been written, but some of the elements involved in its general principles should be further discussed.

The development of a science sounds like a formidable undertaking, and in fact anything like a thorough study of a science such as that of cutting metals necessarily involves many years of work. The science of cutting metals, however, represents in its complication, and in the time required to develop it, almost an extreme case in the mechanic arts. Yet even in this very intricate science, within a few months after starting, enough knowledge had been obtained to much more than pay for the work of experimenting. This holds true in the case of practically all scientific development in the mechanic arts. The first laws developed for cutting metals were crude, and contained only a partial knowledge of the truth, yet this imperfect knowledge was vastly better than the utter lack of exact information or the very imperfect rule of thumb which existed before, and it enabled the workmen, with the help of the management, to do far quicker and better work.

For example, a very short time was needed to discover one or two types of tools which, though imperfect as compared with the shapes developed years afterward, were superior to all other shapes and kinds in common use. These tools were adopted as standard and made possible an immediate increase in the speed of every machinist who used them. These types were superseded in a comparatively short time by still other tools which remained standard until they in their turn made way for later improvements.[1]

The science which exists in most of the mechanic arts is, however, far simpler than the science of cutting metals. In almost all cases, in fact, the laws or rules which are developed are so simple that the average man would hardly dignify them with

[1] Time and again the experimenter in the mechanic arts will find himself face to face with the problem as to whether he had better make immediate practical use of the knowledge which he has attained, or wait until some positive finality in his conclusions has been reached. He recognizes clearly the fact that he has already made some definite progress, but sees the possibility (even the probability) of still further improvement. Each particular case must of course be independently considered, but the general conclusion we have reached is that in most instances it is wise to put one's conclusions as soon as possible to the rigid test of practical use. The one indispensable condition for such a test, however, is that the experimenter shall have full opportunity, coupled with sufficient authority, to insure a thorough and impartial trial. And this, owing to the almost universal prejudice in favor of the old, and to the suspicion of the new, is difficult to get.

the name of a science. In most trades, the science is developed through a comparatively simple analysis and time study of the movements required by the workmen to do some small part of his work, and this study is usually made by a man equipped merely with a stop-watch and a properly ruled notebook. Hundreds of these "time-study men" are now engaged in developing elementary scientific knowledge where before existed only rule of thumb. Even the motion study of Mr. Gilbreth in bricklaying (described on pages 77 to 84) involves a much more elaborate investigation than that which occurs in most cases. The general steps to be taken in developing a simple law of this class are as follows:

First. Find, say, 10 or 15 different men (preferably in as many separate establishments and different parts of the country) who are especially skilful in doing the particular work to be analyzed.

Second. Study the exact series of elementary operations or motions which each of these men uses in doing the work which is being investigated, as well as the implements each man uses.

Third. Study with a stop-watch the time required to make each of these elementary movements and then select the quickest way of doing each element of the work.

Fourth. Eliminate all false movements, slow movements, and useless movements.

Fifth. After doing away with all unnecessary movements, collect into one series the quickest and best movements as well as the best implements.

This one new method, involving that series of motions which can be made quickest and best, is then substituted in place of the ten or fifteen inferior series which were formerly in use. This best method becomes standard, and remains standard, to be taught first to the teachers (or functional foremen) and by them to every workman in the establishment until it is superseded by a quicker and better series of movements. In this simple way one element after another of the science is developed.

In the same way each type of implement used in a trade is studied. Under the philosophy of the management of "initiative and incentive" each workman is called upon to use his own best judgment, so as to do the work in the quickest time, and from this results in all cases a large variety in the shapes and types of implements which are used for any specific purpose. Scientific management requires, first, a careful investigation of each of the many modifications of the same implement, developed under rule of thumb; and second, after a time study has been made of the speed attainable with each of these implements, that the good points of several of them shall be united in a single standard implement, which will enable the workman to work faster and with greater ease than he could before. This one implement, then, is adopted as standard in place of the many different kinds before in use, and it remains standard for all workmen to use until superseded by an implement which has been shown, through motion and time study, to be still better.

With this explanation it will be seen that the development of a science to replace rule of thumb is in most cases by no means a formidable undertaking, and that it can be accomplished by ordinary, every-day men without any elaborate scientific training; but that, on the other hand, the successful use of even the simplest improvement of this kind calls for records, system, and cooperation where in the past existed only individual effort.

There is another type of scientific investigation which has been referred to several times in this paper, and which should receive special attention, namely, the accurate study of the motives which influence men. At first it may appear that this is a matter for individual observation and judgment, and is not a proper subject for exact scientific experiments. It is true that the laws which result from experiments of this class, owing to the fact that the very complex organism—the human being — is being experimented with, are subject to a larger number of exceptions than is the case with laws relating to material things. And yet laws of this kind, which apply to a large majority of men, unquestionably exist, and when clearly defined are of great value as a guide in dealing with men. In developing these laws, accurate, carefully planned and executed experiments, extending through a term of years, have been made, similar in a general way to the experiments upon various other elements which have been referred to in this paper.

Perhaps the most important law belonging to this class, in its relation to scientific management, is the effect which the task idea has upon the efficiency of the workman. This, in fact, has become such an important element of the mechanism of scientific management, that by a great number of people scientific management has come to be known as "task management."

There is absolutely nothing new in the task idea. Each one of us will remember that in his own case this idea was applied with good results in his schoolboy days. No efficient teacher would think of giving a class of students an indefinite lesson to learn. Each day a definite, clear-cut task is set by the teacher before each scholar, stating that he must learn just so much of the subject; and it is only by this means that proper, systematic progress can be made by the students. The average boy would go very slowly if, instead of being given a task, he were told to do as much as he could. All of us are grown-up children, and it is equally true that the average workman will work with the greatest satisfaction, both to himself and to his employer, when he is given each day a definite task which he is to perform in a given time, and which constitutes a proper day's work for a good workman. This furnishes the workman with a clear-cut standard, by which he can throughout the day measure his own progress, and the accomplishment of which affords him the greatest satisfaction.

The writer has described in other papers a series of experiments made upon workmen, which have resulted in demonstrating the fact that it is impossible, through any long period of time, to get workmen to work much harder than the average men around them, unless they are assured a large and a permanent increase in their pay. This series of experiments, however, also proved that plenty of workmen can be found who are willing to work at their best speed, provided they are given this liberal increase in wages. The workman must, however, be fully assured that this increase beyond the average is to be permanent. Our experiments have shown that the exact percentage of increase required to make a workman work at his highest speed depends upon the kind of work which the man is doing.

It is absolutely necessary, then, when workmen are daily given a task which calls for a high rate of speed on their part, that they should also be insured the necessary high rate of pay whenever they are successful. This involves not only fixing for each man his daily task, but also paying him a large bonus, or premium, each time that he succeeds in doing his task in the given time. It is difficult to appreciate in full measure the help which the proper use of these two elements is to the workman in elevating him to the highest standard of efficiency and speed in his trade, and then keeping him there, unless one has seen first the old plan and afterward the new tried upon the same man. And in fact until one has seen similar accurate experiments made upon various grades of workmen engaged in doing widely different types of work. The remarkable and almost uniformly good results from the *correct* application of the task and the bonus must be seen to be appreciated.

These two elements, the task and the bonus (which, as has been pointed out in previous papers, can be applied in several ways), constitute two of the most important elements of the mechanism of scientific management. They are especially important from the fact that they are, as it were, a climax, demanding before they can be used almost all of the other elements of the mechanism; such as a planning department, accurate time study, standardization of methods and implements, a routing system, the training of functional foremen or teachers, and in many cases instruction cards, slide-rules, etc. (Referred to later in rather more detail on page 129.)

The necessity for systematically teaching workmen how to work to the best advantage has been several times referred to. It seems desirable, therefore, to explain in rather more detail how this teaching is done. In the case of a machine-shop which is managed under the modern system, detailed written instructions as to the best way of doing each piece of work are prepared in advance, by men in the planning department. These instructions represent the combined work of several men in the planning room, each of whom has his own specialty, or function. One of them, for instance, is a specialist on the proper speeds and cutting tools to be used. He uses the slide-rules which have been above described as an aid, to guide him in obtaining proper speeds, etc. Another man analyzes the best and quickest motions to be made by the workman in setting the work up in the machine and removing it, etc. Still a third, through the time-study records which have been accumulated, makes out a timetable giving the proper speed for doing each element of the work. The directions of all of these men, however, are written on a single instruction card, or sheet.

These men of necessity spend most of their time in the planning department, because they must be close to the records and data which they continually use in their work, and because this work requires the use of a desk and freedom from interruption. Human nature is such, however, that many of the workmen, if left to themselves, would pay but little attention to their written instructions. It is necessary, therefore, to provide teachers (called functional foremen) to see that the workmen both understand and carry out these written instructions.

Under functional management, the old-fashioned single foreman is superseded by eight different men, each one of whom has his own special duties, and these men, acting as the agents for the planning department (see paragraph 234 to 245 of the paper entitled "Shop Management"), are the expert teachers, who are at all times in the shop, helping and directing the workmen. Being each one chosen for his knowledge and personal skill in his specialty, they are able not only to tell the workman what he should do, but in case of necessity they do the work themselves in the presence of the workman, so as to show him not only the best but also the quickest methods.

One of these teachers (called the inspector) sees to it that he understands the drawings and instructions for doing the work. He teaches him how to do work of the right quality; how to make it fine and exact where it should be fine, and rough and quick where accuracy is not required, — the one being just as important for success as the other. The second teacher (the gang boss) shows him how to set up the job in his machine, and teaches him to make all of his personal motions in the quickest and best way. The third (the speed boss) sees that the machine is run at the best speed and that the proper tool is used in the particular way which will enable the machine to finish its product in the shortest possible time. In addition to the assistance given by these teachers, the workman receives orders and help from four other men; from the "repair boss" as to the adjustment, cleanliness, and general care of his machine, belting, etc.; from the "time clerk," as to everything relating to his pay and to proper written reports and returns; from the "route clerk," as to the order in which he does his work and as to the movement of the work from one part of the shop to another; and, in case a workman gets into any trouble with any of his various bosses, the "disciplinarian" interviews him.

It must be understood, of course, that all workmen engaged on the same kind of work do not require the same amount of individual teaching and attention from the functional foremen. The men who are new at a given operation naturally require far more teaching and watching than those who have been a long time at the same kind of jobs.

Now, when through all of this teaching and this minute instruction the work is apparently made so smooth and easy for the workman, the first impression is that this all tends to make him a mere automaton, a wooden man. As the workmen frequently say when they first come under this system, "Why, I am not allowed to think or move without some one interfering or doing it for me!" The same criticism and objection, however, can be raised against all other modern subdivision of labor. It does not follow, for example, that the modern surgeon is any more narrow or wooden a man than the early settler of this country. The frontiersman, however, had to be not only a surgeon, but also an architect, housebuilder, lumberman, farmer, soldier, and doctor, and he had to settle his law cases with a gun. You would hardly say that the life of the modern surgeon is any more narrowing, or that he is more of a wooden man than the frontiersman. The many problems to be met and solved by the surgeon are just as intricate and difficult and as developing and broadening in their way as were those of the frontiersman.

And it should be remembered that the training of the surgeon has been almost identical in type with the teaching and training which is given to the workman under scientific management. The surgeon, all through his early years, is under the closest supervision of more experienced men, who show him in the minutest way how each element of his work is best done. They provide him with the finest implements, each one of which has been the subject of special study and development, and then insist upon his using each of these implements in the very best way. All of this teaching, however, in no way narrows him. On the contrary he is quickly given the very best knowledge of his predecessors; and, provided (as he is, right from the start) with standard implements and methods which represent the best knowledge of the world up to date, he is able to use his own originality and ingenuity to make *real additions to the world's knowledge, instead of reinventing things which are old.* In a similar way the workman who is cooperating with his many teachers under scientific management has an opportunity to develop which is at least as good as and generally better than that which he had when the whole problem was "up to him" and he did his work entirely unaided.

If it were true that the workman would develop into a larger and finer man without all of this teaching, and without the help of the laws which have been formulated for doing his particular job, then it would follow that the young man who now comes to college to have the help of a teacher in mathematics, physics, chemistry, Latin, Greek, etc., would do better to study these things unaided and by himself. The only difference in the two cases is that students come to their teachers, while from the nature of the work done by the mechanic under scientific management, the teachers must go to him. What really happens is that, with the aid of the science which is invariably developed, and through the instructions from his teachers, each workman of a given intellectual capacity is enabled to do a much higher, more interesting, and finally more developing and more profitable kind of work than he was before able to do. The laborer who before was unable to do anything beyond, perhaps, shoveling and wheeling dirt from place to place, or carrying the work from one part of the shop to another, is in many cases taught to do the more elementary machinist's work, accompanied by the agreeable surroundings and the interesting variety and higher wages which go with the machinist's trade. The cheap machinist or helper, who before was able to run perhaps merely a drill press, is taught to do the more intricate and higher priced lathe and planer work, while the highly skilled and more intelligent machinists become functional foremen and teachers. And so on, right up the line.

It may seem that with scientific management there is not the same incentive for the workman to use his ingenuity in devising new and better methods of doing the work, as well as in improving his implements, that there is with the old type of management. It is true that with scientific management the workman is not allowed to use whatever implements and methods he sees fit in the daily practise of his work. Every encouragement, however, should be given him to suggest improvements, both in methods and in implements. And whenever a workman proposes an improvement, it should be the policy of the management to make a careful analysis of the new method, and if necessary conduct a series of experiments to determine accurately the relative merit of the new suggestion and of the old standard. And whenever the new method is found to be markedly superior to the old, it should be adopted as the standard for the whole establishment. The workman should be given the full credit for the improvement, and should be paid a cash premium as a reward for his ingenuity. In this way the true initiative of the workmen is better attained under scientific management than under the old individual plan.

The history of the development of scientific management up to date, however, calls for a word of warning. The mechanism of management must not be mistaken for its essence, or underlying philosophy. Precisely the same mechanism will in one case produce disastrous results and in another the most beneficent. The same mechanism which will produce the finest results when made to serve the underlying principles of scientific management, will lead to failure and disaster if accompanied by the wrong spirit in those who are using it. Hundreds of people have already mistaken the mechanism of this system for its essence. Messrs. Gantt, Barth, and the writer have presented papers to the American Society of Mechanical Engineers on the subject of scientific management. In these papers the mechanism which is used has been described at some length. As elements of this mechanism may be cited:

Time study, with the implements and methods for properly making it.

Functional or divided foremanship and its superiority to the old-fashioned single foreman.

The standardization of all tools and implements used in the trades, and also of the acts or movements of workmen for each class of work.

The desirability of a planning room or department.

The "exception principle" in management.

The use of slide-rules and similar time-saving implements.

Instruction cards for the workman.

The task idea in management, accompanied by a large bonus for the successful performance of the task.

The "differential rate."

Mnemonic systems for classifying manufactured products as well as implements used in manufacturing.

A routing system.

Modern cost system, etc., etc.

These are, however, merely the elements or details of the mechanism of management. Scientific management, in its essence, consists of a certain philosophy, which results, as before stated, in a combination of the four great underlying principles of management:[1]

When, however, the elements of this mechanism, such as time study, functional foremanship, etc., are used without being accompanied by the true philosophy of management, the results are in many cases disastrous. And, unfortunately, even when men who are thoroughly in sympathy with the principles of scientific management undertake to change too rapidly from the old type to the new, without heeding the warnings of those who have had years of experience in making this change, they frequently meet with serious troubles, and sometimes with strikes, followed by failure.

The writer, in his paper on "Shop Management," has called especial attention to the risks which managers run in attempting to change rapidly from the old to the new management. In many cases, however, this warning has not been heeded. The physical changes which are needed, the actual time study which has to be made, the standardization of all implements connected with the work,

[1] *First.* The development of a true science. *Second.* The scientific selection of the workman. *Third.* His scientific education and development. *Fourth.* Intimate friendly cooperation between the management and the men.

the necessity for individually studying each machine and placing it in perfect order, all take time, but the faster these elements of the work are studied and improved, the better for the undertaking. On the other hand, the really great problem involved in a change from the management of "initiative and incentive" to scientific management consists in a complete revolution in the mental attitude and the habits of all of those engaged in the management, as well of the workmen. And this change can be brought about only gradually and through the presentation of many object-lessons to the workman, which, together with the teaching which he receives, thoroughly convince him of the superiority of the new over the old way of doing the work. This change in the mental attitude of the workman imperatively demands time. It is impossible to hurry it beyond a certain speed. The writer has over and over again warned those who contemplated making this change that it was a matter, even in a simple establishment, of from two to three years, and that in some cases it requires from four to five years.

The first few changes which affect the workmen should be made exceedingly slowly, and only one workman at a time should be dealt with at the start. Until this single man has been thoroughly convinced that a great gain has come to him from the new method, no further change should be made. Then one man after another should be tactfully changed over. After passing the point at which from one-fourth to one-third of the men in the employ of the company have been changed from the old to the new, very rapid progress can be made, because at about this time there is, generally, a complete revolution in the public opinion of the whole establishment and practically all of the workmen who are working under the old system become desirous to share in the benefits which they see have been received by those working under the new plan.

Inasmuch as the writer has personally retired from the business of introducing this system of management (that is, from all work done in return for any money compensation), he does not hesitate again to emphasize the fact that those companies are indeed fortunate who can secure the services of experts who have had the necessary practical experience in introducing scientific management, and who have made a special study of its principles. It is not enough that a man should have been a manager in an establishment which is run under the new principles. The man who undertakes to direct the steps to be taken in changing from the old to the new (particularly in any establishment doing elaborate work) must have had personal experience in overcoming the especial difficulties which are always met with, and which are peculiar to this period of transition. It is for this reason that the writer expects to devote the rest of his life chiefly to trying to help those who wish to take up this work as their profession, and to advising the managers and owners of companies in general as to the steps which they should take in making this change.

As a warning to those who contemplate adopting scientific management, the following instance is given. Several men who lacked the extended experience which is required to change without danger of strikes, or without interference with the success of the business, from the management of "initiative and incentive" to scientific management, attempted rapidly to increase the output in quite an elaborate establishment, employing between three thousand and four thousand men. Those who undertook to make this change were men of unusual ability, and were at the same time enthusiasts and I think had the interests of the workmen truly at heart. They were, however, warned by the writer, before starting, that they must go exceedingly slowly, and that the work of making the change in this establishment could not be done in less than from three to five years. This warning they entirely disregarded. They evidently believed that by using much of the mechanism of scientific management, in combination with the principles of the management of "initiative and incentive," instead of with the principles of scientific management, that they could do, in a year or two, what had been proved in the past to require at least double this time. The knowledge obtained from accurate time study, for example, is a powerful implement, and can be used, in one case to promote harmony between the workmen and the management, by gradually educating, training, and leading the workmen into new and better methods of doing the work, or, in the other case, it may be used more or less as a club to drive the workmen into doing a larger day's work for approximately the same pay that they received in the past. Unfortunately the men who had charge of this work did not take the time and the trouble required to train functional foremen, or teachers, who were fitted gradually to lead and educate the workmen. They attempted, through the old-style foreman, armed with his new weapon (accurate time study), to drive the workmen, against their wishes, and without much increase in pay, to work much harder, instead of gradually teaching and leading them toward new methods, and convincing them through object-lessons that task management means for them somewhat harder work, but also far greater prosperity. The result of all this disregard of fundamental principles was a series of strikes, followed by the downfall of the men who attempted to make the change, and by a return to conditions throughout the establishment far worse than those which existed before the effort was made.

This instance is cited as an object-lesson of the futility of using the mechanism of the new management while leaving out its essence, and also of trying to shorten a necessarily long operation in entire disregard of past experience. It should be emphasized that the men who undertook this work were both able and earnest, and that failure was not due to lack of ability on their part, but to their undertaking to do the impossible. These particular men will not again make a similar mistake, and it is hoped that their experience may act as a warning to others.

In this connection, however, it is proper to again state that during the thirty years that we have been engaged in introducing scientific management there has not been a single strike from those who were working in accordance with its principles, even during the critical period when the change was being made from the old to the new. If proper methods are used by men who have had experience in this work, there is absolutely no danger from strikes or other troubles.

The writer would again insist that in no case should the managers of an establishment, the work of which is elaborate, undertake to change from the old to the new type unless the directors of the company fully understand and believe in the fundamental principles of scientific management and unless they appreciate all that is involved in making this change, particularly the time required, and unless they want scientific management greatly.

Doubtless some of those who are especially interested in working men will complain because under scientific management the workman, when he is shown how to do twice as much work as he formerly did, is not paid twice his former wages, while others who are more interested in the dividends than the workmen will complain that under this system the men receive much higher wages than they did before.

It does seem grossly unjust when the bare statement is made that the competent pig-iron handler, for instance, who has been so trained that he piles $3\frac{6}{10}$ times as much iron as the incompetent man formerly did, should receive an increase of only 60 per cent. in wages.

It is not fair, however, to form any final judgment until all of the elements in the case have been considered. At the first glance we see only two parties to the transaction, the workmen and their employers. We overlook the third great party, the whole people, — the consumers, who buy the product of the first two and who ultimately pay both the wages of the workmen and the profits of the employers.

The rights of the people are therefore greater than those of either employer or employé. And this third great party should be given its proper share of any gain. In fact, a glance at industrial history shows that in the end the whole people receive the greater part of the benefit coming from industrial improvements. In the past hundred years, for example, the greatest factor tending toward increasing the output, and thereby the prosperity of the civilized world, has been the introduction of machinery to replace hand labor. And without doubt the greatest gain through this change has come to the whole people — the consumer.

Through short periods, especially in the case of patented apparatus, the dividends of those who have introduced new machinery have been greatly increased, and in many cases, though unfortunately not universally, the employés have obtained materially higher wages, shorter hours, and better working conditions. But in the end the major part of the gain has gone to the whole people.

And this result will follow the introduction of scientific management just as surely as it has the introduction of machinery.

To return to the case of the pig-iron handler. We must assume, then, that the larger part of the gain which has come from his great increase in output will in the end go to the people in the form of cheaper pig-iron. And before deciding upon how the balance is to be divided between the workmen and the employer, as to what is just and fair compensation for the man who does the piling and what should be left for the company as profit, we must look at the matter from all sides.

First. As we have before stated, the pig-iron handler is not an extraordinary man difficult to find, he is merely a man more or less of the type of the ox, heavy both mentally and physically.

Second. The work which this man does tires him no more than any healthy normal laborer is tired by a proper day's work. (If this man is overtired by his work, then the task has been wrongly set and this is as far as possible from the object of scientific management.)

Third. It was not due to this man's initiative or originality that he did his big day's work, but to the knowledge of the science of pig-iron handling developed and taught him by some one else.

Fourth. It is just and fair that men of the same general grade (when their all-round capacities are considered) should be paid about the same wages when they are all working to the best of their abilities. (It would be grossly unjust to other laborers, for instance, to pay this man $3\frac{6}{10}$ as high wages as other men of his general grade receive for an honest full day's work.)

Fifth. As is explained (page 74), the 60 per cent. increase in pay which he received was not the result of an arbitrary judgment of a foreman or superintendent, it was the result of a long series of careful experiments impartially made to determine what compensation is really for the man's true and best interest when all things are considered.

Thus we see that the pig-iron handler with his 60 per cent. increase in wages is not an object for pity but rather a subject for congratulation.

After all, however, facts are in many cases more convincing than opinions or theories, and it is a significant fact that those workmen who have come under this system during the past thirty years have invariably been satisfied with the increase in pay which they have received, while their employers have been equally pleased with their increase in dividends.

The writer is one of those who believes that more and more will the third party (the whole people), as it becomes acquainted with the true facts, insist that justice shall be done to all three parties. It will demand the largest efficiency from both employers and employés. It will no longer tolerate the type of employer who has his eye on dividends alone, who refuses to do his full share of the work and who merely cracks his whip over the heads of his workmen and attempts to drive them into harder work for low pay. No more will it tolerate tyranny on the part of labor which demands one increase after another in pay and shorter hours while at the same time it becomes less instead of more efficient.

And the means which the writer firmly believes will be adopted to bring about, first, efficiency both in employer and employé and then an equitable division of the profits of their joint efforts will be scientific management, which has for its sole aim the attainment of justice for all three parties through impartial scientific investigation of all the elements of the problem. For a time both sides will rebel against this advance. The workers will resent any interference with their old rule-of-thumb methods, and the management will resent being asked to take on new duties and burdens; but in the end the people through enlightened public opinion will force the new order of things upon both employer and employé.

It will doubtless be claimed that in all that has been said no new fact has been brought to light that was not known to some one in the past. Very likely this is true. Scientific management does not necessarily involve any great invention, nor the discovery of new or startling facts. It does, however, involve a certain *combination* of elements which have not existed in the past, namely, old knowledge so collected, analyzed, grouped, and classified into laws and rules that it constitutes a science; accompanied by a complete change in the mental attitude of the working men as well as of those on the side of the management, toward each other, and toward their respective duties and responsibilities. Also, a new division of the duties between the two sides and intimate, friendly cooperation to an extent that is impossible under the philosophy of the old management. And even all of this in many cases could not exist without the help of mechanisms which have been gradually developed.

It is no single element, but rather this whole combination, that constitutes scientific management, which may be summarized as:

Science, not rule of thumb.

Harmony, not discord.

Cooperation, not individualism.

Maximum output, in place of restricted output.

The development of each man to his greatest efficiency and prosperity.

The writer wishes to again state that: "The time is fast going by for the great personal or individual achievement of any one man standing alone and without the help of those around him. And the time is coming when all great things will be done by that type of cooperation in which each man performs the function for which he is best suited, each man preserves his own individuality and is supreme in his particular function, and each man at the same time loses none of his originality and proper personal initiative, and yet is controlled by and must work harmoniously with many other men."

The examples given above of the increase in output realized under the new management fairly represent the gain which is possible. They do not represent extraordinary or exceptional cases, and have been selected from among thousands of similar illustrations which might have been given.

Let us now examine the good which would follow the general adoption of these principles.

The larger profit would come to the whole world in general.

The greatest material gain which those of the present generation have over past generations has come from the fact that the average man in this generation, with a given expenditure of effort, is producing two times, three times, even four times as much of those things that are of use to man as it was possible for the average man in the past to produce. This increase in the productivity of human effort is, of course, due to many causes, besides the increase in the personal dexterity of the man. It is due to the discovery of steam and electricity, to the introduction of machinery, to inventions, great and small, and to the progress in science and education. But from whatever cause this increase in productivity has come, it is to the greater productivity of each individual that the *whole country* owes its greater prosperity.

Those who are afraid that a large increase in the productivity of each workman will throw other men out of work, should realize that the one element more than any other which differentiates civilized from uncivilized countries — prosperous from poverty-stricken peoples — is that the average man in the one is five or six times as productive as the other. It is also a fact that the chief cause for the large percentage of the unemployed in England (perhaps the most virile nation in the world), is that the workmen of England, more than in any other civilized country, are deliberately restricting their output because they are possessed by the fallacy that it is against their best interest for each man to work as hard as he can.

The general adoption of scientific management would readily in the future double the productivity of the average man engaged in industrial work. Think of what this means to the whole country. Think of the increase, both in the necessities and luxuries of life, which becomes available for the whole country, of the possibility of shortening the hours of labor when this is desirable, and of the increased opportunities for education, culture, and recreation which this implies. But while the whole world would profit by this increase in production, the manufacturer and the workman will be far more interested in the especial local gain that comes to them and to the people immediately around them. Scientific management will mean, for the employers and the workmen who adopt it — and particularly for those who adopt it first — the elimination of almost all causes for dispute and disagreement between them. What constitutes a fair day's work will be a question for scientific investigation, instead of a subject to be bargained and haggled over. Soldiering will cease because the object for soldiering will no longer exist. The great increase in wages which accompanies this type of management will largely eliminate the wage question as a source of dispute. But more than all other causes, the close, intimate cooperation, the constant personal contact between the two sides, will tend to diminish friction and discontent. It is difficult for two people whose interests are the same, and who work side by side in accomplishing the same object, all day long, to keep up a quarrel.

The low cost of production which accompanies a doubling of the output will enable the companies who adopt this management, particularly those who adopt it first, to compete far better than they were able to before, and this will so enlarge their markets that their men will have almost constant work even in dull times, and that they will earn larger profits at all times.

This means increase in prosperity and diminution in poverty, not only for their men but for the whole community immediately around them.

As one of the elements incident to this great gain in output, each workman has been systematically trained to his highest state of efficiency, and has been taught to do a higher class of work than he was able to do under the old types of management; and at the same time he has acquired a friendly mental attitude toward his employers and his whole working conditions, whereas before a considerable part of his time was spent in criticism, suspicious watchfulness, and sometimes in open warfare. This direct gain to all of those working under the system is without doubt the most important single element in the whole problem.

Is not the realization of results such as these of far more importance than the solution of most of the problems which are now agitating both the English and American peoples? And is it not the duty of those who are acquainted with these facts, to exert themselves to make the whole community realize this importance?

*Taylor's Testimony
Before the
Special House Committee*

Taylor's Testimony
Before the
Special House Committee

A REPRINT OF THE PUBLIC DOCUMENT

Hearings Before Special Committee of the House of Representatives to Investigate the Taylor and Other Systems of Shop Management Under Authority of House Resolution 90

Taylor's Testimony Before the Special House Committee[1]

Thursday, January 25, 1912.

The committee met at 10.40 o'clock a. m., Hon. William B. Wilson (chairman) presiding.

TESTIMONY OF MR. FREDERICK WINSLOW TAYLOR

The witness was duly sworn by the chairman.

The Chairman. Will you please give your name and address to the stenographer, Mr. Taylor?

Mr. Taylor. Frederick Winslow Taylor, Highland Avenue, Chestnut Hill, Philadelphia, Pa.

The Chairman. Mr. Taylor, are you the author or compiler of the system of shop management generally known as the "Taylor system"?

Mr. Taylor. I have had a very great deal to do with the development of the system of management which has come to be called by certain people the "Taylor system," but I am only one of many men who have been instrumental in the development of

---

[1] Reprint of public document, *Hearings Before Special Committee of the House of Representatives to Investigate the Taylor and Other Systems of Shop Management Under Authority of H. Res. 90;* Vol. III, pp. 1377—1508.

this system. I wish to state, however, that at no time have I personally called the system the "Taylor system," nor have I ever advocated the desirability of calling it by that name. I have constantly protested against it being branded either with my name or the name of any other man, and I believe it has been a very great injury to the cause that it has been branded with any man's name. I think it should be properly called by some generic term which could be and ought to be acceptable to the whole country. Many self-respecting and able managers object to working under the brand of any man's name, whereas there is no management that could properly object to working under the name, we will say, of "scientific management."

The Chairman. In developing and collating the different parts of this system and in introducing it in different establishments, by what name have you designated it?

Mr. Taylor. The first general designation was a "piece-rate system," because the prominent feature —the feature which at that time interested me most—was a new and radically different type of piecework than anything introduced before. I afterwards pointed out, however, that piecework was really one of the comparatively unimportant elements of our system of management. The next paper written by me on the subject was called "Shop management," and in that paper the task idea—the idea of setting a measured standard of work for each man to do each day—was the most prominent feature, and for some time after this the system was called the "task system." The word "task," however, had a severe sound and did not at all adequately represent the sentiment of the system; it sounded as though you were treating men severely, whereas the whole idea underlying our system is justice and not severity. So it was recognized that this designation was not the proper one, but at the time no better name appeared. Finally the name was agreed upon which I think is correct and which does represent the system better than any other name yet suggested, namely, "scientific management."

The Chairman. Would you state, for the information of the committee, how you developed this system, when you developed it, where you developed it, and what the essential features of it are when developed, and state it in your own way?

Mr. Taylor. Mr. Chairman, before beginning with the early steps which were taken and which led toward the development of scientific management, I should like to attempt to make it clear what the essence of scientific management is; what may be called the atmosphere surrounding it; the sentiments which accompany scientific management when real scientific management comes to exist, and which are appropriate to it; I wish to make clear those sentiments, on the one hand, which come to be most important for those on the management's side, and those sentiments, on the other hand, which come to be the essence and most important to the men working under scientific management, because a mere statement of details and of various steps taken one after another in developing the system, unless one understands the goal toward which they are converging, is apt to be misleading rather than enlightening.

The most important fact which is connected with the working people of this country and which has been forced upon my attention possibly more during the past year than it has in former years, is the fact that the average workingman believes it to be for his interest and for the interest of his fellow workmen to go slow instead of going fast, to restrict output instead of turning out as large a day's work as is practicable.

Now, I find that this fallacy is practically universal with workingmen, and in using the term "workingmen" I have in mind only that class of workmen who are engaged in what may be called cooperative industries, in which several men work together. To illustrate, I have not in mind the coachman, the gardener, or the isolated workman of any kind. I do not mean to say that men outside the cooperative trades believe it to be for their best interest and for the best interest of their fellow workmen to go slow, but I do say that those engaged in cooperative trades generally so believe. Therefore, in using the word "workmen" I hope it will be understood that I am referring simply to that group of men cooperatively engaged, and that is rather a small group of men in any community. We who are engaged in cooperative industry have somehow gotten the impression that the whole world is engaged in the same sort of work, but the class of which I speak forms a rather small minority, but, nevertheless, a very important element of the community.

When you get almost any workingman to talking with you intimately and saying exactly what he believes and feels without reserve; I mean when he speaks without feeling that he is going to meet with an antagonistic opinion not in sympathy with him; to put this in still a third way, when you get that man to telling his real views, he will almost always state that he cannot see how it could be for the interest of his particular trade—that is, for the interest of those men associated with him, and with whose work he is familiar—to very greatly increase their output per day.

The question the workman will ask you, if you have his confidence, is: "What would become of those of us in my particular trade who would be thrown out of work in case we were all to greatly increase our output each day?" Each such man in a particular working group feels that in his town or section or particular industry there is, in the coming year, only about so much work to be done. As far as he can see, if he were to double his output, and if the rest of the men were to double their output tomorrow or next week or next month or next year, he can see no other outcome except that one-half of the workmen engaged with him would be thrown out of work.

That is the honest viewpoint of the average workman in practically all trades. And let me say here that this is a strictly honest view; it is no fake view; there is no hypocrisy about it. This is a firm conviction on the part of almost all workingmen. Holding those views and acting upon them, the workmen cannot be blamed for impressing upon other workmen their conviction that it is not for their mutual interest to greatly increase the output in their particular trade. And as a result they almost all come to the conclusion that it would be humane, it would be a kindly thing, it would be acting merely in the best interests of their brothers, to restrict output rather than to materially increase their output.

Now, I think that is the view of the great majority of the workingmen of this country, and I do not blame them for it. I think I may say that for the almost universality with which this view is found among workingmen, and still more for the fact that this view is growing instead of diminishing, that the men who are not themselves working in cooperative industry and who belong, we will say, taking a single example, to the literary classes, men who have the leisure time for study and investigation and the opportunity for knowing better, are mainly to blame. Some one is surely to blame for the fact that workingmen hold this view, because it is a fallacy which some one should have taken the trouble to point out long ago. This view is directly the opposite of the truth. This view is false from beginning to end, and I say again that for this fallacy on the part of the working people the men who have the leisure and the opportunity to educate themselves, the men whose duty it is—or ought to be—to see that the community is properly educated and told the truth, are mainly to blame. I know of very few men in this country who have taken the trouble to bring out the truth of this fact and make it clear to the working people.

On the contrary, the men who are immediately in contact with the workmen—most of all the labor leaders—are teaching the workmen just the opposite of the facts in this respect, and yet I want to say right here, gentlemen, that while I shall have to say quite a little in the way of blame as to the views and acts of certain labor leaders during my talk, in the main I look upon them as strictly honest, upright, straightforward men. I think you will find as many good men among them as you will in any class, but you will also find many misguided men among them, men whose prejudices are carrying them away in the wrong direction, just as you will find with men of other classes. And please note here that I am using the words "class" or "classes" throughout in the sense of groups of men and women with somewhat similar aims in life, and not at all with the "upper and lower class" distinctions which are sometimes given to these words. So that when I say the labor leaders are misdirecting their followers, are giving them wrong views, are teaching wrong doctrines to their men, I say this with no idea of imputing wrong motives to labor leaders. They themselves are as ignorant of the underlying truths of political economy as the workmen whom they are teaching. I say this quite advisedly because I have talked with a great many of them and I find that they are as firmly convinced of the truth of this fallacy as to the restriction of output as the workmen themselves. Therefore, I repeat again, the teaching of this doctrine by almost all labor leaders is the result of honest conviction and not of any less praiseworthy motive.

And yet, in spite of the fact that nearly all labor leaders are teaching this doctrine, and that almost no one in this country is giving much, if any, time to counteracting the evil effects—and they are tremendous—of this fallacy, that it is for the interest of the workman to go slow. In spite of this fact, I may say that all that is necessary to do to prove the direct contrary of this fallacy is to investigate the facts of any trade, whatever that trade may be. I do not care what trade you go into, get back to the basic facts, the fundamental truths connected with that trade, and you will find that every time there has been an increased output per individual workman in that trade produced by any cause that it has made more work in the trade and has never diminished the number of workmen in the trade. All you have to do is to go back into the history of any trade and look up the facts and you will find it to be true; that in no case has the permanent effect of increasing the output per individual in the trade been that of throwing men out of work, but the effect has always been to make work for more men.

Now, that is the history of every trade, but in spite of that fact the world at large, both on the workman's side and on the manufacturer's side believes this fallacy (and I find a great many men who ought to know better completely misinformed on the side of the management). And yet this is a fallacy, and a blighting fallacy, as far as the interests of the workingmen and the interests of the whole country are concerned. Now, I feel it important or desirable to give just one illustration to show that an increase in output does not throw men out of work, and I could give thousands, simply thousands of such illustrations.

Take any trade, go back through the history of it, and see whether increase of output on the part of the workman has resulted in throwing men out of work. That is what people generally believe; that is what these working people who have testified here believe. They believe if they were to increase their output it would result in throwing a lot of them out of their jobs. And I have had much sympathy with the workingmen who have testified before your committee, because I feel that they firmly believe that it would not be for their best interests to turn out a larger output. I believe these men are honestly mistaken, just as the rest of the world has been honestly mistaken in many other instances.

Let us examine the actual facts in one trade—the cotton trade, for instance. It is as well known, perhaps, and as well understood as any trade in the whole list. The power loom was invented some time between 1780 and 1790, I think it was; I am not quite sure about that date, but it was somewhere about that time. It was very slow in coming into use. Somewhere about the year 1840—the exact date is immaterial, and I give that as about the time of the occurrence—there were in round numbers 5,000 cotton weavers in Manchester, England. About that time these weavers became convinced that the power loom was going to win out, that the hand looms which they were operating were doomed. And they knew that the power loom would turn out per man about three times the output. That is a general figure. I do not wish to say that this ratio is exact, but in any case it is nearly so. Those men knew the possibilities of the power loom and realized that when it was introduced it would turn out a very much larger output per man than was being then turned out by the hand loom.

Now, what could they see? They were certain, those men were honestly certain, and it was a natural conviction on their part, that nothing could happen through the introduction of this power loom except that after it was in, after it was fully installed and doing three times the work that the hand loom did, that instead of there being 5,000 weavers in Manchester they would be reduced to 1,500 or 2,000, and that 3,000 weavers would be thrown out of a job. Now, those men felt fully convinced of that; with them there was no doubt about it; it was a matter of certainty, and they did in kind just what all of us would be apt to do in kind if we were convinced that three-fifths of our working body were to have our means of livelihood taken away from us. What I mean to say is that, broadly speaking, we would adopt the same general policy of opposition that they adopted. I am not advocating violence, arson, or any of the wrong things that were done by these men when I say that we would in a general way have done, broadly speaking, what they did. We would have opposed the introduction of any such policy by every means in our power. What the Manchester weavers did was to break into the establishments where these power looms were being installed. They smashed up the looms. They burned down the buildings in which they were being used. They beat up the scabs using them, and they did almost everything that was in their power to prevent the introduction of the power loom.

And even after that exhibition of fearful violence, gentlemen, I do not hesitate to say that I do not feel very bitterly toward those men. I believe that they were misguided. I feel a certain sympathy for them, not in their violence—I do not endorse that for one moment—but I cannot help but feel a certain sympathy for the men who believe, with absolute certainty, that their means of livelihood is being taken away from them. You cannot help but feel sympathy for men who believe that, even if you thoroughly disapprove of their acts. I do not want to be misquoted in this. These men did murder, violence, and arson. I do not believe in anything of that sort under any circumstances.

Now, gentlemen, the power loom came into use just as every labor-saving device that is a real labor-saving device is sure to come at all times. In spite of any opposition that may come from any source whatever, I do not care what the source is, I do not care how great the opposition, or what it may be, any truly labor-saving device will win out. All that you have to do to find proof of this is to look at the history of the industrial world. And, gentlemen, scientific management is merely the equivalent of a labor-saving device; that is all it is; it is a means, and a very proper and right means, of making men more efficient than they now are, and without imposing materially greater burdens on them than they now have, and if scientific management is a device for doing that it will win out in spite of all the labor opposition in the world; in spite of any opposition that may be brought to bear against it from any quarter whatever, from any class of people, or from the whole people, it will win out. If scientific management is right, and I believe it is right; if it is a labor-saving device for enabling men to do more work with no greater effort on their part, then it is going to win out.

Now, let us see what happened from the introduction of the power loom in 1840, or thereabouts. Did it throw men out of work; did it make work for a less number of men? In Manchester, England, now— and, again, the figures I am giving are merely the broadest kind of general figures, as I am not personally familiar with the cotton industry. The data I have has been given to me by a man who is familiar with it, but I do not want to quibble over the exact figures, as they are not material. It is the broad general facts that count. In Manchester, England, today, the average weaver turns out, I am told, from 8 to 10 times the yardage of cotton cloth formerly turned out by the old hand weaver; the man who does his work with this modern machinery turns out 8 to 10 times the yardage formerly turned out by the hand weaver. The man who told me of the conditions said these figures were well within the limit. In Manchester, England, in 1840, there were

5,000 operatives, and in Manchester, today there are 265,000 operatives. Now, in the light of those figures has the introduction of the power loom, has the introduction of labor-saving machinery thrown men out of work?

What has happened in the cotton industry is typical of what happens in every industry, it makes no difference what that industry is. Broadly speaking, all that you have to do is to bring wealth into the world, and the world uses it. Now, real wealth, as you all know, has but very little to do with money; money is the least important element in wealth. The wealth of the world comes from two sources—from what comes out of the ground or from beneath the surface of the earth, on the one hand, and what is produced by man on the other hand. And the broad fact is that all you have to do is to bring wealth into the world and the world uses it. This is just what happened in the cotton industry.

If you will multiply the figures given in the Manchester illustration you will see that in each day now in Manchester there are 400 or 500 yards of cotton cloth coming out for every single yard that came out each day in 1840, whereas the population of England certainly has not more than doubled; I do not know exactly, but my impression is that it has not more than doubled since 1840. Suppose we even granted that it has trebled and the fact would still be astounding that there now comes out of Manchester, England, 400 to 500 yards of cotton cloth for every single yard that came out in 1840. The true meaning of this great production is that just that much more wealth is being unloaded on the world. This is the fundamental meaning of increase in output in all trades, namely, that additional wealth is coming into the world. Such wealth is real wealth, for it consists of those things which are most useful to man; those things that man needs for his everyday happiness, for his prosperity, and his comfort. The meaning of increased output, whether it be in one trade or another, is always the same, the world is just receiving that much more wealth.

Let us see, now, in a definite way what the increased output of cotton goods means to the American workman. None of us probably appreciate now that in 1840 the ordinary cotton shirt or dress made, for example, from Manchester cottons was a luxury to be worn only by the middle classes, as the English describe it, and that cotton goods were worn by the poor people only as a rare luxury. Now the cotton shirt and the cotton dress, cotton goods generally, have become an absolute daily necessity of all classes of mankind all over the civilized world. And this magnificent result (more magnificent for the working people than for any other portion of the community) has been brought about solely by this great increase in output so stubbornly fought against by the cotton weavers in 1840. It is in those changes which directly affect the poor—which give them a higher standard of living and make from the luxuries of one generation the necessities of the next that we can best see the meaning of an increase in the wealth of the world. And the most important fact of this whole subject is that any association of men, whether it be a group of workmen or a group of capitalists or manufacturers, a manufacturer's association, or whatever it may be, any men who deliberately restrict the output in any industry are robbing the people. And they rob the people of the wealth that justly belongs to them, whether they restrict output honestly, believing it to be for the interest of their trade, or dishonestly for any other reason. There is one point along this line which I want to make clear, gentlemen—that is, that many people believe the ridiculous nonsense that the wealth of the world is enjoyed by the rich. The fact is, that of the real wealth of the world, of the real necessities of life, of practically all the good things of this world, nineteen-twentieths are consumed and used by the working people, and only about one-twentieth by the rich people. Therefore that group of men who prevent wealth from coming into the world are robbing the working people of this nineteen-twentieths and the rich people of but one-twentieth. In fact I doubt if they are robbing the rich people at all. That, after all, is the essence of the whole matter—the robbing of the poor through restriction of output—and I want to try and make it clear that I believe it is quite as much a crime for a manufacturer to restrict output for the sake of holding up prices as it is for the workman to restrict output for this or any other reason.

I don't mean to say for one instant that times may not come in every industry when it is wise to restrict output temporarily, but when that is true it is due merely to a lack of balance in the output of the world and lack of proper poise in industrial conditions. It is perfectly clear that there is such a thing as overproduction; that is no myth, but overproduction, in 99 cases out of 100, properly translated, means a lack of balance, a lack of evenness in production, a failure to maintain a fair balance between the necessities of life and production. It is a special condition, not a normal one. The world doesn't want, for example, 20 times the cotton goods that it has used in the past manufactured all at once. If there then were to be a fair balance maintained at all times between the various necessities of life and the amount of their production, then it would not be necessary to restrict output at any time. It is true, however, that the world seems to get out of kilter at certain fairly regular times; these periods appear to come at intervals of about 20 years. At such times we wake up to find that the world has attempted to start more new enterprises than there is available capital to handle these enterprises with. This condition is not confined to this country, but all over the world and in every class of trade and industry; men make their estimates in a reckless way about new things they will attempt. They start so many new enterprises and on such a large scale that the world's capital and credit is insufficient to carry them through, and then there is a panic. The whole world becomes overanxious, and there follows a period of depression.

No, I do not mean to say that overproduction does not at times exist and should be checked, but I do mean to say that, as a guiding policy—that is, a permanent policy on the part of workingman or manufacturer to restrict the world's output to just so much and no more is mere robbery; it is deliberate robbery of the poor people of those things to which they are entitled and which they can get only from the real wealth of the world.

Now, gentlemen, the firm conviction on the part of workmen that an increase in output on their part would inevitably result in throwing many of their brother workmen out of work is only one of the two great reasons why the working people are, generally speaking, restricting their output by deliberately going slow instead of working at proper speed. I am now going to discuss the second great reason why workmen deliberately turn out a small instead of a large output. For this second cause I doubt whether either the manufacturer or the workman is directly to blame. I feel that any blame for this second cause should attach to the faulty system of management in general use; certainly the workmen cannot be blamed. Now, we will say you are manufacturing this article which I hold in my hand, a fountain pen, and we will assume that it is possible for one man to make that pen—to do all the work himself; I will assume this in order to have a simple case, for we know that it is not possible for one man alone to make it.

We will say that the workman is employed on daywork—that is, he is paid by the day, not by piecework; and is turning out 10 of these pens a day and is paid $2.50 a day for his work. If he has a foreman who is wide awake and interested both in the workman and the company he is working for, as he ought to be, that foreman will probably suggest to the workman that instead of making this pen on daywork that he should make it on piecework, manufacture it by the piece; in other words, that he should be paid 25 cents each for the 10 pens that he makes each day, and so be allowed to earn $2.50 a day, just as he has earned in the past, the only change being from day's wages to piecework. Now, the foreman's object and the workman's object in changing from daywork to piecework is, on the one hand, to enable the workman to get higher wages, and, on the other hand, to get an increased output for the factory. At the end of, perhaps, a year, through the energy of the workman, through his ingenuity and the help of his foreman, through the advice he gets by talking with other workmen, instead of turning out 10 pens a day he finds himself turning out 20 a day. Now, if the foreman amounts to anything, if he is at all a decent kind of a fellow, he feels very glad of the fact that the workman is earning $5 a day where before he only earned $2.50, and he is also pleased that the company is getting such an increase in output from its plant that it is also making more money. It must be understood that this increase in the output will enable the company to earn more money, in spite of the fact that it is paying the same wages per pen that were originally paid. That foreman, if he is any kind of a man, must feel very happy over this state of things. Now, gentlemen, something of this sort happens; I have seen it happen a great many times: There are some members on the board of directors of the company who think that at certain intervals it is necessary or desirable for them to look over the pay roll and see how things are going. And I think that I may say that to the horror of some of those directors, they find that this workman making pens is earning $5 a day, where before he only got $2.50 a day. That is all those directors can see to it. Now, there are just as good men and as conscientious men on the boards of directors of our companies as anywhere else in the world, no better and no worse; yet from a lack of understanding of all sides of the problem they feel genuinely a certain horror at finding that one of their workmen is getting $5 a day where before he only got $2.50. And I have heard them say, and I do not think it is at all an uncommon view for them to hold, "We are spoiling the labor market in this part of the country by paying such wages." What they fear is that if workmen in their part of the country come to receive $5 a day, while those of their competitors are paid only $2.50, that they will be unable to compete. And as a result they order their foreman to see that he doesn't "continue to spoil the labor market in that part of the country." Now, the foreman, acting on the orders of the board of directors, cuts the price per pen down until the workman finds himself turning out 20 pens a day where before he only turned out 10, and is receiving perhaps $2.50, or at most $2.75 or $3, when before he was receiving $2.50 a day.

Now, gentlemen, I have no sympathy whatever with the blackguarding that workmen are receiving from a good deal of the community; there are a great many people who look upon them as greedy, selfish, grasping, and even worse, but I don't sympathize with this view in the least. They are not different in the least from any other class in the community; they are no more grasping and selfish, nor are they less so than other classes of people. It may be a debatable question as to whether they are or are not more grasping than other people. There is one thing, however, we can be perfectly sure of and that is, whatever else they are or they are not, they are not fools. And let me tell you that a workman, after having received one cut of that sort in his wages as a reward for turning out a larger day's work, is a very extraordinary man if he doesn't adopt soldiering and deliberately going slow instead of fast as a permanent policy so as to keep his employer from speeding him up and then cutting his piecework price. I soldiered when I was a workman, and I believe that even many of the most sensible workmen, understanding the conditions as I have outlined them, will inevitably adopt the policy of going slow. Under those conditions it would take an exceedingly broadminded man to do anything else than adopt soldiering as his permanent policy. I will not say that this soldiering is the best policy for the workman to adopt, even for his own best interest in the long run, but I do say that I do not blame him for doing it. In spite of the miserable policy of cutting piecework prices when men increase their output, I believe that those workmen who do not adopt the policy of restricting output and going slow, i.e., soldiering, will in the end be far better off than those who soldier. Certainly, this whole situation is no fault of theirs; they didn't introduce the system which makes soldiering seem to be necessary, and if blame rests anywhere it certainly does not rest with the working people, but somewhere else.

Now, the first thing that I want to make clear, then, before starting in to describe what scientific management, or, as you, Mr. Chairman, have called it, the "Taylor system," is (if you will allow me, however, I will substitute the term scientific management for the "Taylor system"), with the understanding that the two are equivalent in the future—the fact that I wish to make clear is, first, that this restriction of output, that this going slow on the part of the workman is an almost universal fact in this country, and that from the workmen's point of view there is ample justification for the policy which, in the main, they have adopted. That is what I wish to make clear as a foundation for what I shall say later. Now, let me first, in the broadest kind of way outline or describe what I look upon as the essence of scientific management.

There are many elements of scientific management, many details connected with scientific management, that it is utterly impossible to go into details in a hearing of this kind; but I want to try and make clear before going much further into the history of the development of scientific management—I want to make clear what may be called the essence of it so that when I use the words "scientific management," you men who are listening may have a clear, definite idea of what is in my own mind, because I know that what is in your mind when the words "scientific management" are used has a totally different meaning from what is in my mind, and I want you to know what is in my mind when I use these words. I want to clear the deck, sweep away a good deal of rubbish first by pointing out what scientific management is not. I think that will clear the deck a good deal.

Scientific management is not any efficiency device, not a device of any kind for securing efficiency; nor is it any bunch or group of efficiency devices. It is not a new system of figuring costs; it is not a new scheme of paying men; it is not a piecework system; it is not a bonus system; it is not a premium system; it is no scheme for paying men; it is not holding a stop watch on a man and writing things down about him; it is not time study; it is not motion study nor an analysis of the movements of men; it is not the printing and ruling and unloading of a ton or two of blanks on a set of men and saying, "Here's your system; go use it." It is not divided foremanship or functional foremanship; it is not any of the devices which the average man calls to mind when scientific management is spoken of. The average man thinks of one or more of these things when he hears the words "scientific management" mentioned, but scientific management is not any of these devices. I am not sneering at cost-keeping systems, at time study, at functional foremanship, nor at any new and improved scheme of paying men, nor at any efficiency devices, if they are really devices that make for efficiency. I believe in them; but what I am emphasizing is that these devices in whole or in part are not scientific management; they are useful adjuncts to scientific management, so are they also useful adjuncts of other systems of management.

Now, in its essence, scientific management involves a complete mental revolution on the part of the workingman engaged in any particular establishment or industry—a complete mental revolution on the part of these men as to their duties toward their work, toward their fellow men, and toward their employers. And it involves the equally complete mental revolution on the part of those on the management's side—the foreman, the superintendent, the owner of the business, the board of directors—a complete mental revolution on their part as to their duties toward their fellow workers in the management, toward their workmen, and toward all of their daily problems. And without this complete mental revolution on both sides scientific management does not exist.

That is the essence of scientific management, this great mental revolution. Now, later on, I want to show you more clearly what I mean by this great mental revolution. I know that perhaps it sounds to you like nothing but bluff—like buncombe—but I am going to try and make clear to you just what this great mental revolution involves, for it does involve an immense change in the minds and attitude of both sides, and the greater part of what I shall say today has relation to the bringing about of this great mental revolution. So that whether the details may be interesting or uninteresting, what I hope you will see is that this great change in attitude and viewpoint must produce results which are magnificent for both sides, just as fine for one as for the other. Now, perhaps I can make clear to you at once one of the very great changes in outlook which come to the workmen, on the one hand, and to those in the management on the other hand.

I think it is safe to say that in the past a great part of the thought and interest both of the men, on the side of the management, and of those on the side of the workmen in manufacturing establishments has been centered upon what may be called the proper division of the surplus resulting from their joint efforts, between the management on the one hand, and the workmen on the other hand. The management have been looking for as large a profit as possible for themselves, and the workmen have been looking for as large wages as possible for themselves, and that is what I mean by the division of the surplus. Now, this question of the division of the surplus is a very plain and simple one (for I am announcing no great fact in political economy or anything of that sort). Each article produced in the establishment has its definite selling price. Into the manufacture of this article have gone certain expenses, namely, the cost of materials, the expenses connected with selling it, and certain indirect expenses, such as the rent of the building, taxes, insurance, light and power, maintenance of machinery, interest on the plant, etc. Now, if we deduct these several expenses from the selling price, what is left over may be called the surplus. And out of this surplus comes the profit to the manufacturer on the one hand, and the wages of the workmen on the other hand. And it is largely upon the division of this surplus that the attention of the workman and of the management has been centered in the past. Each side has had its eye upon this surplus, the working man wanting as large a share in the form of wages as he could get, and the management wanting as large a share in the form of profits as it could get; I think I am safe in saying that in the past it has been in the division of this surplus that the great labor troubles have come between employers and employees.

Frequently, when the management have found the selling price going down they have turned toward a cut in the wages—toward reducing the workman's share of the surplus—as their way of getting out whole, of preserving their profits intact. While the workman (and you can hardly blame him) rarely feels willing to relinquish a dollar of his wages, even in dull times, he wants to keep all that he has had in the past, and when busy times come again very naturally he wants to get more. Thus it is over this division of the surplus that most of the troubles have arisen; in the extreme cases this has been the cause of serious disagreements and strikes. Gradually the two sides have come to look upon one another as antagonists, and at times even as enemies—pulling apart and matching the strength of the one against the strength of the other.

The great revolution that takes place in the mental attitude of the two parties under scientific management is that both sides take their eyes off of the division of the surplus as the all-important matter, and together turn their attention toward increasing the size of the surplus until this surplus becomes so large that it is unnecessary to quarrel over how it shall be divided. They come to see that when they stop pulling against one another, and instead both turn and push shoulder to shoulder in the same direction, the size of the surplus created by their joint efforts is truly astounding. They both realize that when they substitute friendly cooperation and mutual helpfulness for antagonism and strife they are together able to make this surplus so enormously greater than it was in the past that there is ample room for a large increase in wages for the workmen and an equally great increase in profits for the manufacturer.. This, gentlemen, is the beginning of the great mental revolution which constitutes the first step toward scientific management. It is along this line of complete change in the mental attitude of both sides; of the substitution of peace for war; the substitution of hearty brotherly cooperation for contention and strife; of both pulling hard in the same direction instead of pulling apart; of replacing suspicious watchfulness with mutual confidence; of becoming friends instead of enemies; it is along this line, I say, that scientific management must be developed.

The substitution of this new outlook—this new viewpoint—is of the very essence of scientific management, and scientific management exists nowhere until after this has become the central idea of both sides; until this new idea of cooperation and peace has been substituted for the old idea of discord and war.

This change in the mental attitude of both sides toward the "surplus" is only a part of the great mental revolution which occurs under scientific management. I will later point out other elements of this mental revolution. There is, however, one more change in viewpoint which is absolutely essential to the existence of scientific management. Both sides must recognize as essential the substitution of exact scientific investigation and knowledge for the old individual judgment or opinion, either of the workman or the boss, in all matters relating to the work done in the establishment. And this applies both as to the methods to be employed in doing the work and the time in which each job should be done.

Scientific management cannot be said to exist, then, in any establishment until after this change has taken place in the mental attitude of both the management and the men, both as to their duty to cooperate in producing the largest possible surplus and as to the necessity for substituting exact scientific knowledge for opinions or the old rule-of-thumb or individual knowledge.

These are the two absolutely essential elements of scientific management.

What has scientific management accomplished? It has been introduced in a great number and variety of industries in this country, to a greater or less degree, and in those companies which have come under scientific management it is, I think, safe and conservative to say that the output of the individual workman has been, on the average, doubled. This doubling of the output has enabled the manufacturer to earn a larger profit, because it has cheapened the cost of manufacture; and, in addition to enabling the manufacturer to earn a larger profit, it has in many cases—in fact, in most cases—resulted in a very material lowering of the selling price of the article. Through this lowering of the selling price the whole public, the buyer and user, of the joint product of the labor and machinery have profited by getting what they buy cheaper. This is the greatest interest that the general public has in scientific management—that in the end they will get more for their money than they are now getting—in other words, that scientific management will in the end enable us all to live better than we are now living. Through scientific management, then, the manufacturer has already profited, and the general public has also profited.

The greatest gain has come, however, in my judgment, to the workmen who have been working under scientific management. They have received from 30 to 100 per cent higher wages than they received in the past; and, in addition, I do not recall a single case in which they have ever worked longer hours than they did before, but I do recall many instances in which the hours of work were shortened. Perhaps the greatest gain, however,—and I say it without hesitation—is not the increase in wages received by the workmen, but the fact that those who are working under scientific management have come to look upon their employers as their best friends instead of their enemies. They have come to realize that friendship and cooperation are better than war.

Now, this, of course, is a mere assertion. By way of proving this fact, however, I wish to state that until this last year, during the 30 years that scientific management has been gradually developed—has been in process of evolution—there has never been a single strike of employees working under scientific management—never one in all the 30 years in which it has been used.

Scientific management has been introduced in competitive industries. Among their competitors, situated in many cases right alongside of them, who have not adopted scientific management, there have been repeated strikes. Yet even during the very difficult period of changing from the old type of management to the new, until last year, there has never been a strike among the men working under the principles of scientific management, while in corresponding establishments not working under scientific management there have been repeated strikes.

Thereupon, at 12 o'clock noon, the committee took a recess until 2 o'clock.

After Recess

The committee reconvened 2.05 o'clock p. m., pursuant to taking a recess, Hon. William B. Wilson (chairman) presiding.

The Chairman. You may go ahead, Mr. Taylor.

Mr. Taylor. It must be realized that during the many years that scientific management has been in process of evolution that much of the mechanism—

which has improperly come to be looked upon by many people as the essence of scientific management —has been adopted and used by those who were in no way engaged in working under the principles of scientific management. And that the false use, if I may speak of it in this way, of elements which have been associated with scientific management have led to strikes. I shall try to point out that many elements of what may be called the mechanism of scientific management are powerful when used by those on the management's side. These elements are powerful both for good and for bad, and it is impossible to be assured that even useful elements shall always be used in the right way. So that, in a number of cases, men who were out of sympathy with scientific management and yet who were using the elements which have been in the eyes of the public associated with scientific management have brought on strikes by using these elements entirely without any relation to the real, fundamental, and essential principles of scientific management. In order that the essential difference between the principles of scientific management and those of the older type of management may be made more clear, it seems to me desirable to first point out, or indicate, what I think you gentlemen will all recognize as representing the best of the older type of management.

If you have a company, say, employing from 500 to 1,000 men you will have among the employees of this company perhaps 15 or 20 different trades. Now, the men working at these different trades have probably learned all that they know, one may almost say, through tradition; that is, trades are now learned, not from books but just as they were 100 years ago; apprentices learn by watching and observing the way other men work, by imitating the best workmen, and by asking questions of those immediately around them. The apprentice learns by reading a little, by some teaching on the part of the foreman and superintendent, but mainly by imitating the best methods of those workmen with whom he comes closely into contact. Trades, then, are learned now practically as they were in the Middle Ages. They are transmitted from hand to eye and comparatively little is learned from books. I think I may truthfully say that during the two apprenticeships I served, one as a pattern maker and one as a machinist, I did not spend more than two and a half hours in reading books about my work. Of course there are many more books and more useful books published now about the different trades than there were 37 years ago; but, still, my impression is that the same fact remains true. I have had the object lesson of watching my own son, who left college at the end of his freshman year and is working a year in a machine shop under the sad, baleful conditions of scientific management as they have been pictured by some of the witnesses before this committee, in which he is obliged to do a severe task every day. I have given this boy as many books as I could on the machininst's trade, but I do not think he has yet spent an hour reading the books we have put before him; so that my opinion remains the same about the present-day apprentice as it was about the old one;

that is, that he is learning almost all that he gets through the old traditional channels.

Notwithstanding this fact the knowledge which every journeyman has of his trade is his most valuable possession. It is his great life's capital, and none the less valuable—perhaps even more valuable—from the fact that it is attained in the old-fashioned traditional way rather than through such study as is to be had at school or college. In my judgment, then, the manager who really understands the problem which is before him must appreciate that the most important thing for him to do under the old type of management which is in common use is to get what may be called the initiative of his workmen, and by this I mean the workman's hard work, his good will, his ingenuity, his determination to do everything that he can to further his employer's interest. Now, owing to the fact, as I have tried to explain at the opening of my testimony, that practically all of the workingmen of this country are fully convinced that it is for their interest to go slow and to restrict output instead of turning out a maximum output, no manager who really understands conditions as they exist in our shops would dream that he could get the true initiative of his workmen unless he did something more and better for them than is done by employers in the average shop—unless he gave his workmen some special incentive, some reason, for wishing to do more work than is done in the ordinary shop. Because, as I have already stated, the average workman is engaged during a very considerable part of his time in watching the clock to be sure that he doesn't work so fast as to to spoil a piecework rate; to be sure that he is not doing what he would look upon as an injustice to himself and his fellow workmen.

There are a few manufacturers, perhaps not more than one manufacturer in a hundred, however, who are large enough minded and whose hearts are kindly enough disposed to lead them to honestly desire that their employees should be better off than the employees of their competitors; to lead them to try and arrange matters so that their employees can earn higher wages than the employees of their competitors. And if these employers will only persist long enough in deliberately paying their men higher wages than are paid to the workmen of their competitors, it has been my observation that invariably the workmen respond by giving them their real initiative, by working hard and faithfully, by using their ingenuity to see how they can turn out as much work as possible, instead of using their ingenuity, as they ordinarily do, to convince their employers that they are working hard and yet not work hard enough to spoil any piecework job.

Now, this special case, this rare case, in which the management deliberately treat their employees far better than the employees of their competitors are treated, to my mind represents the best of the older types of management. And I again assert that any manufacturer who will only persist long enough in treating his employees in this way will succeed in getting their true initiative. I have known a good many employers to set out to adopt this scheme of paying higher wages than their competitors and become discouraged because their employees did not immediately respond by doing their share under this new arrangement. It must be remembered, however, that workmen are naturally and very properly suspicious of their employers. If they have lived long in this world, they have seen or heard of a great many tricks being played by employers. Now, again, gentlemen, I do not wish to be quoted as saying that all employers are tricky, but I do wish to say that, in my judgment, employers are just as tricky as workmen are tricky, neither more nor less so.

All of you men here who are workmen know that there are a whole lot of tricky workmen, and all you men here who are employers know that there are a whole lot of tricky employers; not that any very large portion of workmen are tricky, and not that a large portion of employers are tricky men, but tricky men are there just the same, on both sides. You cannot blame, therefore, any set of workmen for being slow in responding to even this kindly treatment; what they suspect is—and they can almost all point to some personal experience or to some friend's experience to warrant their suspicion—what they suspect is that this is merely a trick on the part of their employer to get them to work at a higher rate of speed and then, through some infernal excuse or reason or flimflam game, that ultimately the piecework price will be cut down and they will find themselves working at a high rate of speed for the same old pay.

Thereupon, at 2.28 o'clock p. m. the committee took a recess for 30 minutes.

After Recess

The committee reconvened at 2.58 o'clock p. m., pursuant to taking a recess, Hon. William B. Wilson (chairman) presiding.

Mr. Taylor. What I want to try to prove to you and make clear to you is that the principles of scientific management when properly applied, and when a sufficient amount of time has been given to make them really effective, must in all cases produce far larger and better results, both for the employer and the employees, than can possibly be obtained under even this very rare type of management which I have been outlining, namely, the management of "initiative and incentive," in which those on the management's side deliberately give a very large incentive to their workmen, and in return the workmen respond by working to the very best of their ability at all times in the interest of their employers.

I want to show you that scientific management is even far better than this rare type of management.

The first great advantage which scientific management has over the management of initiative and incentive is that under scientific management the initiative of the workmen—that is, their hard work, their good will, their ingenuity—is obtained practically with absolute regularity, while under even the best of the older type of management this initiative is only obtained spasmodically and somewhat irregularly. This obtaining, however, of the initiative of the workmen is the lesser of the two great causes which make scientific management better for both sides than the older type of management. By far the greater gain under scientific management comes from the new, the very great, and the extraordinary burdens and duties which are voluntarily assumed by those on the management's side.

These new burdens and new duties are so unusual and so great that they are to the men used to managing under the old school almost inconceivable. These duties and burdens voluntarily assumed under scientific management, by those on the management's side, have been divided and classified into four different groups and these four types of new duties assumed by the management have (rightly or wrongly) been called the "principles of scientific management."

The first of these four groups of duties taken over by the management is the deliberate gathering in on the part of those on the management's side of all of the great mass of traditional knowledge, which in the past has been in the heads of the workmen, and in the physical skill and knack of the workmen, which he has acquired through years of experience. The duty of gathering in of all this great mass of traditional knowledge and then recording it, tabulating it, and, in many cases, finally reducing it to laws, rules, and even to mathematical formulae, is voluntarily assumed by the scientific managers. And later, when these laws, rules, and formulae are applied to the everyday work of all the workmen of the establishment, through the intimate and hearty cooperation of those on the management's side, they invariably result, first, in producing a very much larger output per man, as well as an output of a better and higher quality; and, second, in enabling the company to pay much higher wages to their workmen; and, third, in giving to the company a larger profit. The first of these principles, then, may be called the development of a science to replace the old rule-of-thumb knowledge of the workmen; that is, the knowledge which the workmen had, and which was, in many cases, quite as exact as that which is finally obtained by the management, but which the workmen nevertheless in nine hundred and ninety-nine cases out of a thousand kept in their heads, and of which there was no permanent or complete record.

A very serious objection has been made to the use of the word "science" in this connection. I am much amused to find that this objection comes chiefly from the professors of this country. They resent the use of the word science for anything quite so trivial as the ordinary, every-day affairs of life. I think the proper answer to this criticism is to quote the definition recently given by a professor who is, perhaps, as generally recognized as a thorough scientist as any man in the country—President McLaurin, of the Institute of Technology, of Boston. He recently defined the word science as "classified or organized knowledge of any kind." And surely the gathering in of knowledge which, as previously stated, has existed, but which was in an unclassified condition in the minds of workmen, and then the reducing of this knowledge to laws and rules and formulae, certainly represents the organization and classification of knowledge, even though it may not meet with the approval of some people to have it called science.

The second group of duties which are voluntarily assumed by those on the management's side, under scientific management, is the scientific selection and then the progressive development of the workmen. It becomes the duty of those on the management's side to deliberately study the character, the nature, and the performance of each workman with a view to finding out his limitations on the one hand, but even more important, his possibilities for development on the other hand; and then, as deliberately and as systematically to train and help and teach this workman, giving him, wherever it is possible, those opportunities for advancement which will finally enable him to do the highest and most interesting and most profitable class of work for which his natural abilities fit him, and which are open to him in the particular company in which he is employed. This scientific selection of the workman and his development is not a single act; it goes on from year to year and is the subject of continual study on the part of the management.

The third of the principles of scientific management is the bringing of the science and the scientifically selected and trained workmen together. I say "bringing together" advisedly, because you may develop all the science that you please, and you may scientifically select and train workmen just as much as you please, but unless some man or some men bring the science and the workman together all your labor will be lost. We are all of us so constituted that about three-fourths of the time we will work according to whatever method suits us best; that is, we will practice the science or we will not practice it; we will do our work in accordance with the laws of the science or in our own old way, just as we see fit unless some one is there to see that we do it in accordance with the principles of the science. Therefore I use advisedly the words "bringing the science and the workman together." It is unfortunate, however, that this word "bringing" has rather a disagreeable sound, a rather forceful sound; and, in a way, when it is first heard it puts one out of touch with what we have come to look upon as the modern tendency. The time for using the word "bringing," with a sense of forcing, in relation to most matters, has gone by; but I think that I may soften this word down in its use in this particular case by saying that nine-tenths of the trouble with those of us who have been engaged in helping people to change from the older type of management to the new management—that is, to scientific management—that nine-tenths of our trouble has been to "bring" those on the management's side to do their fair share of the work and only one-tenth of our trouble has come on the workman's side. Invariably we find very great opposition on the part of those on the management's side to do their new duties and comparatively little opposition on the part of the workmen to cooperate in doing their new duties. So that the word "bringing" applies much more forcefully to those on the management's side than to those on the workman's side.

The fourth of the principles of scientific management is perhaps the most difficult of all of the four principles of scientific management for the average man to understand. It consists of an almost equal division of the actual work of the establishment between the workmen, on the one hand, and the management, on the other hand. That is, the work which under the old type of management practically all was done by the workman, under the new is divided into two great divisions, and one of these divisions is deliberately handed over to those on the management's side. This new division of work, this new share of the work assumed by those on the management's side, is so great that you will, I think, be able to understand it better in a numerical way when I tell you that in a machine shop, which, for instance, is doing an intricate business—I do not refer to a manufacturing company, but, rather, to an engineering company; that is, a machine shop which builds a variety of machines and is not engaged in manufacturing them, but, rather, in constructing them—will have one man on the management's side to every three workmen; that is, this immense share of the work—one-third—has been deliberately taken out of the workman's hands and handed over to those on the management's side. And it is due to this actual sharing of the work between the two sides more than to any other one element that there has never (until this last summer) been a single strike under scientific management. In a machine shop, again, under this new type of management there is hardly a single act or piece of work done by any workman in the shop which is not preceded and followed by some act on the part of one of the men in the management. All day long every workman's acts are dovetailed in between corresponding acts of the management. First, the workman does something, and then a man on the management's side does something; then the man on the management's side does something, and then the workman does something; and under this intimate, close, personal cooperation between the two sides it becomes practically impossible to have a serious quarrel.

Of course I do not wish to be understood that there are never any quarrels under scientific management. There are some, but they are the very great exception, not the rule. And it is perfectly evident that while the workmen are learning to work under this new system, and while the management is learning to work under this new system, while they are both learning, each side to cooperate in this intimate way with the other, there is plenty of chance for disagreement and for quarrels and misunderstandings, but after both sides realize that it is utterly impossible to turn out the work of the establishment at the proper rate of speed and have it correct without this intimate, personal cooperation, when both sides realize that it is utterly impossible for either one to be successful without the intimate, brotherly cooperation of the other, the friction, the disagreements, and quarrels are reduced to a minimum. So I think that scientific management can be justly and truthfully characterized as management in which harmony is the rule rather than discord.

There is one illustration of the application of the principles of scientific management with which all of us are familiar and with which most of us have been familiar since we were small boys, and I think this instance represents one of the best illustrations of the application of the principles of scientific management. I refer to the management of a first-class American baseball team. In such a team you will find almost all of the elements of scientific management.

You will see that the science of doing every little act that is done by every player on the baseball field has been developed. Every single element of the game of baseball has been the subject of the most intimate, the closest study of many men, and, finally, the best way of doing each act that takes place on the baseball field has been fairly well agreed upon and established as a standard throughout the country. The players have not only been told the best way of making each important motion or play, but they have been taught, coached, and trained to it through months of drilling. And I think that every man who has watched first-class play, or who knows anything of the management of the modern baseball team, realizes fully the utter impossibility of winning with the best team of individual players that was ever gotten together unless every man on the team obeys the signals or orders of the coach and obeys them at once when the coach gives those orders; that is, without the intimate cooperation between all members of the team and the management, which is characteristic of scientific management.

Now, I have so far merely made assertions; I have merely stated facts in a dogmatic way. The most important assertion I have made is that when a company, when the men of a company and the management of a company have undergone the mental revolution that I have referred to earlier in my testimony, and that when the principles of scientific management have been applied in a correct way in any particular occupation or industry that the results must, inevitably, in all cases, be far greater and better than they could possibly be under the best of the older types of management, even under the especially fine management of "initiative and incentive," which I have tried to outline.

I want to try and prove the above-stated fact to you gentlemen. I want to try now and make good in this assertion. My only hope of doing so lies in showing you that whenever these four principles are correctly applied to work, either large or small, to work which is either of the most elementary or the most intricate character, that inevitably results follow which are not only greater, but enormously greater, than it is possible to accomplish under the old type of management. Now, in order to make this clear I want to show the application of the four principles first to the most elementary, the simplest kind of work that I know of, and then to give a series of further illustrations of one class of work after another, each a little more difficult and a little more intricate than the work which preceded it, until I shall finally come to an illustration of the application of these same principles to about the most intricate type of mechanical work that I know of. And in all of these illustrations I hope that you will look for and see the application of the four principles I have described. Other elements of the stories may interest you, but the thing that I hope you will see and have before you in all cases is the effect of the four following elements in each particular case: First, the development of the science, i.e., the gathering in on the part of those on the management's side of all the knowledge which in the past has been kept in the heads of the workmen; second, the scientific selection and the progressive development of the workmen; third, the bringing of the science and the scientifically selected and trained men together; and, fourth, the constant and intimate cooperation which always occurs between the men on the management's side and the workmen.

I ordinarily begin with a description of the pig-iron handler. For some reason, I don't know exactly why, this illustration has been talked about a great deal, so much, in fact, that some people seem to think that the whole of scientific management consists in handling pig iron. The only reason that I ever gave this illustration, however, was that pig-iron handling is the simplest kind of human effort; I know of nothing that is quite so simple as handling pig-iron. A man simply stoops down and with his hands picks up a piece of iron, and then walks a short distance and drops it on the ground. Now, it doesn't look as if there was very much room for the development of a science; it doesn't seem as if there was much room here for the scientific selection of the man nor for his progressive training, nor for cooperation between the two sides; but, I can say, without the slightest hesitation, that the science of handling pig-iron is so great that the man who is fit to handle pig-iron as his daily work cannot possibly understand that science; the man who is physically able to handle pig-iron and is sufficiently phlegmatic and stupid to choose this for his occupation is rarely able to comprehend the science of handling pig-iron; and this inability of the man who is fit to do the work to understand the science of doing his work becomes more and more evident as the work becomes more complicated, all the way up the scale. I assert, without the slightest hesitation, that the high class mechanic has a far smaller chance of ever thoroughly understanding the science of his work than the pig-iron handler has of understanding the science of his work, and I am going to try and prove to your satisfaction, gentlemen, that the law is almost universal—not entirely so, but nearly so—that the man who is fit to work at any particular trade is unable to understand the science of that trade without the kindly help and cooperation of men of a totally different type of education, men whose education is not necessarily higher but a different type from his own.

I dare say most of you gentlemen are familiar with pig-iron handling and with the illustration I have used in connection with it, so I won't take up any of your time with that. But I want to show you how these principles may be applied to some one of the lower classes of work. You may think I am a little highfalutin when I speak about what may be called the atmosphere of scientific management, the relations that ought to exist between both sides, the intimate and friendly relations that should exist between employee and employer. I want, however, to emphasize this as one of the most important features of scientific management, and I can hardly do so without going into detail, without explaining minutely the duties of both sides, and for this reason I want to take some of your time in explaining the application of these four principles of scientific management to one of the cheaper kinds of work, for instance, to shoveling. This is one of the simplest kinds of work, and I want to give you an illustration of the application of these principles to it.

Now, gentlemen, shoveling is a great science compared with pig-iron handling. I dare say that most of you gentlemen know that a good many pig-iron handlers can never learn to shovel right; the ordinary pig-iron handler is not the type of man well suited to shoveling. He is too stupid; there is too much mental strain, too much knack required of a shoveler for the pig-iron handler to take kindly to shoveling.

You gentlemen may laugh, but that is true, all right; it sounds ridiculous, I know, but it is a fact. Now, if the problem were put up to any of you men to develop the science of shoveling as it was put up to us, that is, to a group of men who had deliberately set out to develop the science of doing all kinds of laboring work, where do you think you would begin? When you started to study the science of shoveling I make the assertion that you would be within two days—just as we were within two days—well on the

way toward development of the science of shoveling.
At least you would have outlined in your minds those
elements which required careful, scientific study in
order to understand the science of shoveling. I do
not want to go into all of the details of shoveling, but
I will give you some of the elements, one or two of
the most important elements of the science of shoveling; that is, the elements that reach further and have
more serious consequences than any other. Probably
the most important element in the science of shoveling is this: There must be some shovel load at which
a first-class shoveler will do his biggest day's work.
What is that load? To illustrate: When we went to
the Bethlehem Steel Works and observed the shovelers in the yard of that company, we found that each
of the good shovelers in that yard owned his own
shovel; they preferred to buy their own shovels
rather than to have the company furnish them. There
was a larger tonnage of ore shoveled in that works
than of any other material and rice coal came next
in tonnage. We would see a first-class shoveler go
from shoveling rice coal with a load of $3\frac{1}{2}$ pounds
to the shovel to handling ore from the Massaba
Range, with 38 pounds to the shovel. Now, is $3\frac{1}{2}$
pounds the proper shovel load or is 38 pounds the
proper shovel load? They cannot both be right.
Under scientific management the answer to this question is not a matter of anyone's opinion; it is a question for accurate, careful, scientific investigation.

Under the old system you would call in a first-rate
shoveler and say, "See here, Pat, how much ought
you to take on at one shovel load?" And if a couple

of fellows agreed, you would say that's about the right load and let it go at that. But under scientific management absolutely every element in the work of every man in your establishment, sooner or later, becomes the subject of exact, precise, scientific investigation and knowledge to replace the old, "I believe so," and "I guess so." Every motion, every small fact becomes the subject of careful, scientific investigation.

What we did was to call in a number of men to pick from, and from these we selected two first-class shovelers. Gentlemen, the words I used were "first-class shovelers." I want to emphasize that. Not poor shovelers. Not men unsuited to their work, but first-class shovelers. These men were then talked to in about this way, "See here, Pat and Mike, you fellows understand your job all right; both of you fellows are first-class men; you know what we think of you; you are all right now; but we want to pay you fellows double wages. We are going to ask you to do a lot of damn fool things, and when you are doing them there is going to be some one out alongside of you all the time, a young chap with a piece of paper and a stop watch and pencil, and all day long he will tell you to do these fool things, and he will be writing down what you are doing and snapping the watch on you and all that sort of business. Now, we just want to know whether you fellows want to go into that bargain or not? If you want double wages while that is going on all right, we will pay you double; if you don't all right, you needn't take the job unless you want to; we just called you in to see whether you want to work this way or not.

"Let me tell you fellows just one thing: If you go into this bargain, if you go at it, just remember that on your side we want no monkey business of any kind; you fellows will have to play square; you fellows will have to do just what you are supposed to be doing; not a damn bit of soldiering on your part; you must do a fair day's work; we don't want any rushing, only a fair day's work and you know what that is as well as we do. Now, don't take this job unless you agree to these conditions, because if you start to try to fool this same young chap with the pencil and paper he will be onto you in 15 minutes from the time you try to fool him, and just as surely as he reports you fellows as soldiering you will go out of this works and you will never get in again. Now, don't take this job unless you want to accept these conditions; you need not do it unless you want to; but if you do, play fair."

Well, these fellows agreed to it, and, as I have found almost universally to be the case, they kept their word absolutely and faithfully. My experience with workmen has been that their word is just as good as the word of any other set of men that I know of, and all you have to do is to have a clear, straight, square understanding with them and you will get just as straight and fair a deal from them as from any other set of men. In this way the shoveling experiment was started. My remembrance is that we first started them on work that was very heavy, work requiring a very heavy shovel load. What we did was to give them a certain kind of heavy material ore, I think, to handle with a certain size of shovel. We sent these two men into different parts of the yard, with two different men to time and study them, both sets of men being engaged on the same class of work. We made all the conditions the same for both pairs of men, so as to be sure that there was no error in judgment on the part of either of the observers and that they were normal, first-class men.

The number of shovel loads which each man handled in the course of the day was counted and written down. At the end of the day the total tonnage of the material handled by each man was weighed and this weight was divided by the number of shovel loads handled, and in that way, my remembrance is, our first experiment showed that the average shovel load handled was 38 pounds, and that with this load on the shovel the man handled, say, about 25 tons per day. We then cut the shovel off, making it somewhat shorter, so that instead of shoveling a load of 38 pounds it held a load of approximately 34 pounds. The average, then, with the 34 pound load, of each man went up, and instead of handling 25 he had handled 30 tons per day. These figures are merely relative, used to illustrate the general principles, and I do not mean that they were the exact figures. The shovel was again cut off, and the load made approximately 30 pounds, and again the tonnage ran up, and again the shovel load was reduced, and the tonnage handled per day increased, until at about 21 or 22 pounds per shovel we found that these men were doing their largest day's work. If you cut the shovel load off still more, say until it averages 18 pounds instead of 21½, the tonnage handled per day will begin to fall off, and at 16 pounds it will be still lower, and so on right down. Very well; we now have developed the scientific fact that a workman well suited to his job, what we call a first-class shoveler, will do his largest day's work when he has a shovel load of 21½ pounds.

Now, what does that fact amount to? At first it may not look to be a fact of much importance, but let us see what it amounted to right there in the yard of the Bethlehem Steel Co. Under the old system, as I said before, the workmen owned their shovels, and the shovel was the same size whatever the kind of work. Now, as a matter of common sense, we saw at once that it was necessary to furnish each workman each day with a shovel which would hold just 21½ pounds of the particular material which he was called upon to shovel. A small shovel for the heavy material, such as ore, and a large scoop for light material, such as ashes. That meant, also, the building of a large shovel room, where all kinds of laborers' implements were stored. It meant having an ample supply of each type of shovel, so that all the men who might be called upon to use a certain type in any one day could be supplied with a shovel of the size desired that would hold just 21½ pounds. It meant, further, that each day each laborer should be given a particular kind of work to which he was suited, and that he must be provided with a particular shovel suited to that kind of work, whereas in the past all the laborers in the yard of the Bethlehem

Steel Co. had been handled in masses, or in great groups of men, by the old-fashioned foreman, who had from 25 to 100 men under him and walked them from one part of the yard to another. You must realize that the yard of the Bethlehem Steel Co. at that time was a very large yard. I should say that it was at least 1½ or 2 miles long and, we will say, a quarter to a half mile wide, so it was a good large yard; and in that yard at all times an immense variety of shoveling was going on.

There was comparatively little standard shoveling which went on uniformly from day to day. Each man was likely to be moved from place to place about the yard several times in the course of the day. All of this involved keeping in the shovel room 10 or 15 kinds of shovels, ranging from a very small flat shovel for handling ore up to immense scoops for handling rice coal, and forks with which to handle coke, which, as you know, is very light. It meant the study and development of the implement best suited to each type of material to be shoveled, and assigning, with the minimum of trouble, the proper shovel to each one of the four to six hundred laborers at work in that yard. Now, that meant mechanism, human mechanism. It meant organizing and planning work at least a day in advance. And, gentlemen, here is an important fact, that the greatest difficulty which we met with in this planning did not come from the workmen. It came from the management's side. Our greatest difficulty was to get the heads of the various departments each day to inform the men in the labor office what kind of work and how much of it was to be done on the following day.

This planning the work one day ahead involved the building of a labor office where before there was no such thing. It also involved the equipping of that office with large maps showing the layout of the yards so that the movements of the men from one part of the yard to another could be laid out in advance, so that we could assign to this little spot in the yard a certain number of men and to another part of the yard another set of men, each group to do a certain kind of work. It was practically like playing a game of chess in which four to six hundred men were moved about so as to be in the right place at the right time. And all this, gentlemen, follows from the one idea of developing the science of shoveling; the idea that you must give each workman each day a job to which he is well suited and provide him with just that implement which will enable him to do his biggest day's work. All this, as I have tried to make clear to you, is the result that followed from the one act of developing the science of shoveling.

In order that our workmen should get their share of the good that came from the development of the science of shoveling and that we should do what we set out to do with our laborers,—namely, pay them 60 per cent higher wages than were paid to any similar workmen around that whole district. Before we could pay them these extra high wages it was necessary for us to be sure that we had first-class men and that each laborer was well suited to his job, because the only way in which you can pay wages

60 per cent higher than other people pay and not overwork your men is by having each man properly suited and well trained to his job. Therefore, it became necessary to carefully select these yard laborers; and in order that the men should join with us heartily and help us in their selection it became necessary for us to make it possible for each man to know each morning as he came in to work that on the previous day he had earned his 60 per cent premium, or that he had failed to do so. So here again comes in a lot of work to be done by the management that had not been done before. The first thing each workman did when he came into the yard in the morning—and I may say that a good many of them could not read and write—was to take two pieces of paper out of his pigeonhole; if they were both white slips of paper, the workman knew he was all right. One of those slips of paper informed the man in charge of the tool room what implement the workman was to use on his first job and also in what part of the yard he was to work. It was in this way that each one of the 600 men in that yard received his orders for the kind of work he was to do and the implement with which he was to do it, and he was also sent right to the part of the yard where he was to work, without any delay whatever. The old-fashioned way was for the workmen to wait until the foreman got good and ready and had found out by asking some of the heads of departments what work he was to do, and then he would lead the gang off to some part of the yard and go to work. Under the new method each man gets his orders almost automati-

cally; he goes right to the tool room, gets the proper implement for the work he is to do, and goes right to the spot where he is to work without any delay. The second piece of paper, if it was a white piece of paper, showed this man that he had earned his 60 per cent higher wages; if it was a yellow piece of paper the workman knew that he had not earned enough to be a first-class man, and that within two or three days something would happen, and he was absolutely certain what this something would be. Every one of them knew that after he had received three or four yellow slips a teacher would be sent down to him from the labor office. Now, gentlemen, this teacher was no college professor. He was a teacher of shoveling; he understood the science of shoveling; he was a good shoveler himself, and he knew how to teach other men to be good shovelers, This is the sort of man who was sent out of the labor office. I want to emphasize the following point, gentlemen: The workman, instead of hating the teacher who came to him—instead of looking askance at him and saying to himself, "Here comes one of those damn nigger drivers to drive me to work"— looked upon him as one of the best friends he had around there. He knew that he came out there to help him, not to nigger drive him. Now, let me show you what happens. The teacher comes, in every case, not to bulldoze the man, not to drive him to harder work than he can do, but to try in a friendly, brotherly way to help him, so he says, "Now, Pat, something has gone wrong with you. You know no workman who is not a high-priced workman can stay on this gang, and you will have to get off of it if we can't find out what is the matter with you. I believe you have forgotten how to shovel right. I think that's all there is the matter with you. Go ahead and let me watch you awhile. I want to see if you know how to do the damn thing, anyway."

Now, gentlemen, I know you will laugh when I talk again about the science of shoveling. I dare say some of you have done some shoveling. Whether you have or not, I am going to try to show you something about the science of shoveling, and if any of you have done much shoveling, you will understand that there is a good deal of science about it.

There is a good deal of refractory stuff to shovel around a steel works; take ore, or ordinary bituminous coal, for instance. It takes a good deal of effort to force the shovel down into either of these materials from the top of the pile, as you have to when you are unloading a car. There is one right way of forcing the shovel into materials of this sort, and many wrong ways. Now, the way to shovel refractory stuff is to press the forearm hard against the upper part of the right leg just below the thigh, like this (indicating), take the end of the shovel in your right hand and when you push the shovel into the pile, instead of using the muscular effort of your arms, which is tiresome, throw the weight of your body on the shovel like this (indicating); that pushes your shovel in the pile with hardly any exertion and without tiring the arms in the least. Nine out of ten workmen who try to push a shovel in a pile of that sort will use the strength of their arms, which involves more than twice the necessary exertion. Any of you men who don't know this fact just try it. This is one illustration of what I mean when I speak of the science of shoveling, and there are many similar elements of this science. Now, this teacher would find, time and time again, that the shoveler had simply forgotten how to shovel; that he had drifted back to his old wrong and inefficient way of shoveling, which prevented him from earning his 60 per cent higher wages. So he would say to him, "I see all that is the matter with you is that you have forgotten how to shovel; you have forgotten what I showed you about shoveling some time ago. Now, watch me," he says, "this is the way to do the thing." And the teacher would stay by him two, three, four, or five days, if necessary, until he got the man back again into the habit of shoveling right.

Now, gentlemen, I want you to see clearly that, because that is one of the characteristic features of scientific management; this is not nigger driving; this is kindness; this is teaching; this is doing what I would like mighty well to have done to me if I were a boy trying to learn how to do something. This is not a case of cracking a whip over a man and saying, "Damn you, get there." The old way of treating with workmen, on the other hand, even with a good foreman, would have been something like this: "See here, Pat, I have sent for you to come up here to the office to see me; four or five times now you have not earned your 60 per cent increase in wages; you know that every workman in this place has got to earn 60 per cent more wages than they pay in any other place around here, but you're no good and that's all there is to it; now, get out of this." That's the old way. "You are no good; we have given you a fair chance; get out of this," and the workman is pretty lucky if it isn't "get to hell out of this," instead of "get out of this."

The new way is to teach and help your men as you would a brother; to try to teach him the best way and show him the easiest way to do his work. This is the new mental attitude of the management toward the men, and that is the reason I have taken so much of your time in describing this cheap work of shoveling. It may seem to you a matter of very little consequence, but I want you to see, if I can, that this new mental attitude is the very essence of scientific management; that the mechanism is nothing if you have not got the right sentiment, the right attitude in the minds of the men, both on the management's side and on the workman's side. Because this helps to explain the fact that until this summer there has never been a strike under scientific management.

The men who developed the science of shoveling spent, I should say, four or five months studying the subject and during that time they investigated not only the best and most efficient movements that the men should make when they are shoveling right, but they also studied the proper time for doing each of the elements of the science of shoveling. There are many other elements which go to make up this science, but I will not take up your time describing them.

Now, all of this costs money. To pay the salaries

of men who are studying the science of shoveling is an expensive thing. As I remember it there were two college men who studied this science of shoveling and also the science of doing many other kinds of laboring work during a period of about three years; then there were a lot of men in the labor office whose wages had to be paid, men who were planning the work which each laborer was to do at least a day in advance; clerks who worked all night so that each workman might know the next morning when he went to work just what he had accomplished and what he had earned the day before; men who wrote out the proper instructions for the day's work for each workman. All of this costs money; it costs money to measure or weigh up the materials handled by each man each day. Under the old method the work of 50 or 60 men was weighed up together; the work done by a whole gang was measured together. But under scientific management we are dealing with individual man and not with gangs of men. And in order to study and develop each man you must measure accurately each man's work. At first we were told that this would be impossible. The former managers of this work told me "You cannot possibly measure up the work of each individual laborer in this yard; you might be able to do it in a small yard, but our work is of such an intricate nature that it is impossible to do it here."

I want to say that we had almost no trouble in finding some cheap way of measuring up each man's work, not only in that yard but throughout the entire plant.

But all of that costs money, and it is a very proper question to ask whether it pays or whether it doesn't pay, because, let me tell you, gentlemen, at once, and I want to be emphatic about it, scientific management has nothing in it that is philanthropic; I am not objecting to philanthropy, but any scheme of management which has philanthropy as one of its elements ought to fail; philanthropy has no part in any scheme of management. No self-respecting workman wants to be given things, every man wants to earn things, and scientific management is no scheme for giving people something they do not earn. So, if the principles of scientific management do not pay, then this is a miserable system. The final test of any system is, does it pay?

At the end of some three and a half years we had the opportunity of proving whether or not scientific management did pay in its application to yard labor. When we went to the Bethlehem Steel Co. we found from 400 to 600 men at work in that yard, and when we got through 140 men were doing the work of the 400 to 600, and these men handled several million tons of material a year.

We were very fortunate to be able to get accurate statistics as to the cost of handling a ton of materials in that yard under the old system and under the new. Under the old system the cost of handling a ton of materials had been running between 7 and 8 cents, and all you gentlemen familiar with railroad work know that this is a low figure for handling materials. Now, after paying for all the clerical work which was necessary under the new system for the time study and the teachers, for building and running the labor office and the implement room, for constructing a telephone system for moving men about the yard, for a great variety of duties not performed under the old system, after paying for all these things incident to the development of the science of shoveling and managing the men the new way, and including the wages of the workmen, the cost of handling a ton of material was brought down from between 7 and 8 cents to between 3 and 4 cents, and the actual saving, during the last six months of the three and one-half years I was there, was at the rate of $78,000 a year. That is what the company got out of it; while the men who were on the labor gang received an average of sixty per cent more wages than their brothers got or could get anywhere around that part of the country. And none of them were overworked, for it is no part of scientific management ever to overwork any man; certainly overworking these men could not have been done with the knowledge of anyone connected with scientific management, because one of the first requirements of scientific management is that no man shall ever be given a job which he cannot do and thrive under through a long term of years. It is no part of scientific management to drive anyone. At the end of three years we had men talk to and investigate all of these yard laborers and we found that they were almost universally satisfied with their jobs.

Of course certain men are permanent grouches and when we run across that kind we all know what to expect. But, in the main, they were the most satisfied and contented set of laborers I have ever seen anywhere; they lived better than they did before, and most of them were saving a little money; their families lived better, and as to having any grouch against their employers, those fellows, every one, looked upon them as the best friends they ever had, because they taught them how to earn 60 per cent more wages than they had ever earned before. This is the round-up of both sides of this question. If the use of the system does not make both sides happier, then it is no good.

To give you one illustration of the application of scientific management to a rather high class of work, gentlemen, bricklaying, so far as I know, is one of the oldest of the trades, and it is a truly extraordinary fact that bricks are now laid just about as they were 2,000 years before Christ. In England they are laid almost exactly as they were then; in England the scaffold is still built with timbers lashed together—in many cases with the bark still on it—just as we see that the scaffolds were made in old stone-cut pictures of bricklaying before the Christian era. In this country we have gone beyond the lashed scaffold, and yet in most respects it is almost literally true that bricks are still laid as they were 4,000 years ago. Virtually the same trowel, virtually the same brick, virtually the same mortar, and, from the way in which they were laid, according to one of my friends, who is a brick work contractor and a student of the subject, who took the trouble to take down some bricks laid 4,000 years ago to study the way in which the mortar was spread, etc., it appears that they even spread the mortar in the same way then as we do now. If, then, there is any trade in which one would say that the principles of scientific management would produce but small results, that the development of the science would do little good, it would be in a trade which thousands and thousands of men through successive generations had worked and had apparently reached, as far as methods and principles were concerned, the highest limit of efficiency 4,000 years ago. In bricklaying this would seem to be true since practically no progress has been made in this art since that time. Therefore, viewed broadly, one would say that there was a smaller probability that the principles of scientific management could accomplish notable results in this trade than in almost any other.

Mr. Frank Gilbreth is a man who in his youth worked as a bricklayer; he was an educated man and is now a very successful contractor. He said to me, some years ago, "Now, Taylor, I am a contractor, putting up all sorts of buildings, and if there is one thing I know it is bricklaying; I can go out right now, and I am not afraid to back myself, to beat any man I know of laying bricks for ten minutes, both as to speed and accuracy; you may think I am blowing, but that is one way I got up in the world. I cannot stand it now for more than ten minutes; I'm soft; my hands are tender, I haven't been handling bricks for years, but for ten minutes I will back myself against anyone. I want to ask you about this scientific management; do you think it can be applied to bricklaying? Do you believe that these things you have been shouting about (at that time it was called the 'task system'), do you believe these principles can be applied to bricklaying?" "Certainly," I said, "some day some fellow will make the same kind of study about bricklaying that we have made of other things, and he will get the same results." "Well," he said, "if you really think so, I will just tell you who is going to do it, his name is Frank Gilbreth."

I think it was about three years later that he came to me and said: "Now, I'm going to show you something about bricklaying. I have spent three years making a motion and time study of bricklaying, and not I alone did it; my wife has also spent almost the same amount of her time studying the problems of bricklaying, and I think she has made her full share of the progress which has been made in the science of bricklaying." Then he said, "I will show you just how we went to work at it. Let us assume that I am now standing on the scaffold in the position that the bricklayer occupies when he is ready to begin work. The wall is here on my left, the bricks are there in a pile on the scaffold to my right, and the mortar is here on the mortar-board alongside of the bricks. Now, I take my stand as a bricklayer and am ready to start to lay bricks, and I said to myself, 'What is the first movement that I make when I start to lay bricks?' I take a step to the right with the right foot. Well, is that movement necessary? It took me a year and a half to cut out that motion—that step to the right—and I will tell you later how I cut it out. Now, what motion do I make next? I stoop down to the floor to the pile of bricks and disentangle a brick from the pile and pick it up off the pile. 'My God,' I said, 'that is nothing short of barbarous.' Think of it! Here I am a man weighing over 250 pounds, and every time I stoop down to pick up a brick I lower 250 pounds of weight down two feet so as to pick up a brick weighing 4 pounds, and then raise my 250 pounds of weight up again, and all of this to lift up a brick weighing 4 pounds. Think of this waste of effort. It is monstrous. It took me—it may seem to you a pretty long while—but it took a year and a half of thought and work to cut out that motion; when I finally cut it out, however, it was done in such a simple way that anyone in looking at the method which I adopted would say, 'There is no invention in that, any fool could do that; why did you take a year and a half to do a little thing like that?' Well, all I did was to put a table on the scaffold right alongside of me here on my right side and put the bricks and mortar on it, so as to keep them at all times at the right height, thus making it unnecessary to stoop down in picking them up. This table was placed in the middle of the scaffold with the bricklayer on one side of it, and with a walkway on the other side along which the bricks were brought by wheelbarrow or by hod to be placed on the table without interfering with the bricklayer or even getting in his way." Then Mr. Gilbreth made his whole scaffold adjustable, and a laborer was detailed to keep all of the scaffolds at all times at such a height that as the wall goes up the bricks, the mortar, and the men will occupy that position in which the work can be done with the least effort.

Mr. Gilbreth has studied out the best position for

each of the bricklayer's feet and for every type of bricklaying the exact position for the feet is fixed so that the man can do his work without unnecessary movements. As a result of further study both on the part of Mr. and Mrs. Gilbreth, after the bricks are unloaded from the cars and before bringing them to the bricklayer they are carefully sorted by a laborer and placed with their best edges up on a simple wooden frame, constructed so as to enable him to take hold of each brick in the quickest time and in the most advantageous position. In this way the bricklayer avoids either having to turn the brick over or end for end to examine it before laying it, and he saves also the time taken in deciding which is the best edge and end to place on the outside of the wall. In most cases, also, he saves the time taken in disentangling the brick from a disorderly pile on the scaffold. This "pack of bricks," as Mr. Gilbreth calls his loaded wooden frames, is placed by the helper in its proper position on the adjustable scaffold close to the mortar box.

We have all been used to seeing bricklayers tap each brick after it is placed on its bed of mortar several times with the end of the handle of the trowel so as to secure the right thickness for the joint. Mr. Gilbreth found that by tempering the mortar just right the bricks could be readily bedded to the proper depth by a downward pressure of the hand which lays them. He insisted that the mortar mixers should give special attention to tempering the mortar and so save the time consumed in tapping the brick.

In addition to this he taught his bricklayers to

make simple motions with both hands at the same time, where before they completed a motion with the right hand before they followed it later with one made by the left hand. For example, Mr. Gilbreth taught his bricklayers to pick up a brick in the left hand at the same time that he takes a trowel of mortar with the right hand. This work with two hands at the same time is, of course, made possible by substituting a deep mortar box for the old mortar-board, on which the mortar used to spread out so thin that a step or two had to be taken to reach it, and then placing the mortar box and the brick pile close together and at the proper height on his new scaffold.

Now, what was the practical outcome of all this study? To sum it up he finally succeeded in teaching his bricklayers, when working under the new method, to lay bricks with five motions per brick, while with the old method they used 18 motions per brick. And, in fact, in one exceedingly simple type of bricklaying he reduced the motions of his bricklayers from 18 to 2 motions per brick. But in the ordinary bricklaying he reduced the motions from 18 to 5. When he first came to me, after he had made this long and elaborate study of the motions of bricklayers, he had accomplished nothing in a practical way through this study, and he said, "You know, Fred, I have been showing all my friends these new methods of laying bricks and they say to me, 'Well, Frank, this is a beautiful thing to talk about, but what in the devil do you think it amounts to? You know perfectly well the unions have forbidden their members to lay more than so many bricks per day; you know they won't allow this thing to be carried out." But Gilbreth said, "Now, my dear boy, that doesn't make an iota of difference to me. I'm just going to see that the bricklayers do the right thing. I belong to the bricklayers' union in Boston, and the next job that I get in Boston this thing goes through. I'm not going to do it in any underhanded way. Everyone knows that I have always paid higher wages than the union scale in Boston. I've got a lot of friends at the head of the unions in Boston, and I'm not afraid of having any trouble."

He got his job near Boston, and he went to the leaders of the union and told them just what you can tell any set of sensible men. He said to them, "I want to tell you fellows some things that you ought to know. Most of my contracts around here used to be brick jobs; now, most of my work is in reinforced concrete or some other type of construction, but I am first and last a bricklayer; that is what I am interested in, and if you have any sense you will just keep your hands off and let me show you bricklayers how to compete with the reinforced concrete men. I will handle the bricklayers myself. All I want of you leaders is to keep your hands off and I will show you how bricklayers can compete with reinforced concrete or any other type of construction that comes along."

Well, the leaders of the union thought that sounded all right, and then he went to the workmen and said to them, "No fellow can work for me for less than $6.50 a day—the union rate was $5 a day—but every man who gets on this job has got to lay bricks my way; I will put a teacher on the job to show you all my way of laying bricks and I will give every man plenty of time to learn, but after a bricklayer has had a sufficient trial at this thing, if he won't do my way or cannot do my way, he must get off the job." Any number of bricklayers were found to be only too glad to try the job, and I think he said that before the first story of the building was up he had the whole gang trained to work in the new way, and all getting their $6.50 a day when before they only received $5 per day; I believe those are the correct figures; I am not absolutely sure about that, but at least he paid them a very liberal premium above the average bricklayer's pay.

It is one of the principles of scientific management to ask men to do things in the right way, to learn something new, to change their ways in accordance with the science, and in return to receive an increase of from 30 to 100 per cent in pay, which varies according to the nature of the business in which they are engaged.

Thereupon, at 4.55 o'clock p. m., the committee adjourned until 11 o'clock a. m. Friday, January 26, 1912.

FRIDAY, JANUARY 26, 1912

The committee met at 11 o'clock a. m., Hon. W. B. Wilson (chairman) presiding.

There were also present Representatives Redfield and Tilson.

Mr. Taylor. After Mr. Gilbreth had trained his complete force of bricklayers so that they were all working the new instead of the old way, a very great and immediate increase in the output per man occurred. So that during the latter part of the construction of this building the bricklayers—and I wish it distinctly understood that all of these men were union bricklayers; Mr. Gilbreth himself has for years insisted on having what is known as the closed shop on his work—who were engaged in building a 12-inch wall with drawn joints on both sides—which you gentlemen who understand bricklaying will recognize as a difficult wall to build; a 12-inch wall with drawn joints on both sides—these bricklayers averaged 350 bricks per man per hour, whereas the most rapid union rate up to that time had been 120 bricks per man per hour. And you will recognize, gentlemen, that this is due principally to the very great simplification of the work brought about thru Mr. Gilbreth's three years' of analysis and study of the art of bricklaying, which enabled him to reduce the number of motions made by the workman in laying a brick from 18 per brick to 5 per brick.

The immense gain which has been made through this study will be realized when it is understood that in one city in England the union bricklayers on this type of work have limited their output to 275 bricks per day per man, when on municipal work, and 375 bricks per day per man when on private work.

I want to make it clear to you gentlemen that this great increase in output on the part of Mr. Gilbreth's bricklayers could only be brought about, and was brought about, through the application of the four principles of scientific management to which I referred yesterday in my testimony.

In the first place, it is perfectly clear that unless Mr. Gilbreth had developed the science of bricklaying himself this could not have been done.

In the second place, unless the management coöperated in the most hearty way in the scientific selection of the workmen, and then in his progressive development—that is, first choosing the workmen (picking out those men who were able and willing to adopt the new methods in bricklaying), and then teaching them the new movements—this result could not have been realized.

You will appreciate this fact when you know (as those of you who are familiar with bricklaying know) that practically the whole of a wall must go up at the same rate of speed; that it is impossible for the man working on the middle of the wall, for instance, to put his work up faster than the men working on either side of him. If he did, you would have the most horrible looking wall imaginable, unsightly, and with broken joints. Therefore, the whole wall must go up uniformly, and yet under the old system of management no one bricklayer has the authority to compel other men to adopt new methods and cooperate with him doing work faster.

Now, I have not the slightest doubt that during the last 4,000 years all the methods that Mr. Gilbreth developed have many, many times suggested themselves to the minds of bricklayers. I do not believe Mr. Gilbreth was the first man to invent those methods, and yet if any man or men had invented Gilbreth's improvements and methods prior to the time that the principles of scientific management were understood and accepted, no useful results could have come from them, because the adoption of Gilbreth's methods demands a degree of cooperation, coupled with a kind of leadership on the management's side, which is entirely impossible with the independent individualism which characterizes the old type of management. Under the old system a resourceful man might persuade some, or even most of your bricklayers to adopt the new and scientific methods, but one stubborn man, by refusing to join with the rest, could prevent a realization of any great increase in output. It therefore requires in the development of these methods that the management shall assume the responsibility for seeing that each workman either learns an entirely new method of doing his work or else gets off the job. This is something which no management ever thought of doing in the past.

In short, it requires the hearty cooperation of the management at all points with the workmen, and the voluntary assumption on the part of the management of new duties which they never did before. To make this point clear, it requires the management to appoint men to go around and keep the scaffolding at a proper height, all day long, and to keep the bricklayers supplied with the right kind of brick, systematically placed near them with their right edge up, etc. Every care must be taken by the management to see that the mortar is tempered exactly for the particular kind of work which is to be done.

Mr. Gilbreth puts on special men to see that all conditions under which his men work shall be the best that are known and that these perfect conditions shall be maintained at all times.

I want to emphasize the fact that it is due to the application of what I have pointed out as the four principles of scientific management that Mr. Gilbreth has accomplished his large results, namely:

First. The development—by the management, not the workmen—of the science of bricklaying, with rigid rules for each motion of every man, and the perfection and standardization of all implements and working conditions.

Second. The careful selection and subsequent training of the bricklayers into first-class men, and the elimination of all men who refuse to, or are unable to adopt, the best methods.

Third. Bringing the first-class bricklayer and the science of bricklaying together, through the constant help and watchfulness of the management, and through paying each man a large daily bonus for working fast and doing what he is told to do.

Fourth. An almost equal division of the work and responsibility between the workman and the management. All day long the management work almost side by side with the men, helping, encouraging, and smoothing the way for them, while in the past they stood at one side, gave the men but little help, and threw on to them almost the entire responsibility as to methods, implements, speed, and harmonious co-operation.

Now, before I start on the last illustration—that

is, the illustration of the application of these principles to the work of a machine shop—it may perhaps be better for me to explain the first steps that were taken toward scientific management, because that will help you to understand how the science of cutting metals came to be developed. I defer entirely to your judgment, gentlemen, on that matter. If, on the contrary, it be your desire that I shall go ahead at once with machine-shop illustration, I will do so, and afterwards proceed with a description of how scientific management first started.

The Chairman. Proceed in your own way.

Mr. Taylor. Thank you. In 1878 I came to the Midvale Steel Works as a day laborer, after having served two apprenticeships as a pattern maker and a machinist. I came then as a laborer because I could not get work at my trade. Work at that time was very dull—it was toward the end of the long period of depression following the panic of 1873. I was assigned to work on the floor of the machine shop. Soon after I went there the clerk of the shop got mixed up in his accounts and they thought he was stealing—I never could quite believe that he was; I thought it was merely a mix-up—and they put me in to take his place, simply because I was able to do clerical work.

I did this clerical work all right, although it was distasteful to me, and after having trained another clerk to do the work of the shop I asked permission of the foreman to work as a machinist. They gave me a job on the lathe, because I had made good as a clerk when they needed one, and I worked for some time with the lathe gang.

Shortly after this they wanted a gang boss to take charge of the lathes and they appointed me to this position.

Now, the machine shop of the Midvale Steel Works was a piecework shop. All the work practically was done on piecework, and it ran night and day—five nights in the week and six days. Two sets of men came on, one to run the machines at night and the other to run them in the daytime.

We who were the workmen of that shop had the quantity output carefully agreed upon for everything that was turned out in the shop. We limited the output to about, I should think, one-third of what we could very well have done. We felt justified in doing this, owing to the piecework system— that is, owing to the necessity for soldiering under the piecework system—which I pointed out yesterday.

As soon as I became gang boss the men who were working under me and who, of course, knew that I was onto the whole game of soldiering or deliberately restricting output, came to me at once and said, "Now, Fred, you are not going to be a damn piecework hog, are you?" I said, "If you fellows mean you are afraid I am going to try to get a larger output from these lathes" I said, "Yes; I do propose to get more work out." I said, "You must remember I have been square with you fellows up to now and worked with you. I have not broken a single rate. I have been on your side of the fence. But now I have accepted a job under the management of this company and I am on the other side of the fence, and I will tell you perfectly frankly that I am going to try to get a bigger output from those lathes." They answered, "Then, you are going to be a damn hog."

I said, "Well, if you fellows put it that way, all right." They said, "We warn you, Fred, if you try to bust any of these rates, we will have you over the fence in six weeks." I said, "That is all right; I will tell you fellows again frankly that I propose to try to get a bigger output off these machines."

Now, that was the beginning of a piecework fight that lasted for nearly three years, as I remember it —two or three years—in which I was doing everything in my power to increase the output of the shop, while the men were absolutely determined that the output should not be increased. Anyone who has been through such a fight knows and dreads the meanness of it and the bitterness of it. I believe that if I had been an older man—a man of more experience—I should have hardly gone into such a fight as this—deliberately attempting to force the men to do something they did not propose to do.

We fought on the management's side with all the usual methods, and the workmen fought on their side with all their usual methods. I began by going to the management and telling them perfectly plainly, even before I accepted the gang boss-ship, what would happen. I said, "Now these men will show you, and show you conclusively, that, in the first place, I know nothing about my business; and that in the second place, I am a liar, and you are being fooled, and they will bring any amount of evidence to prove these facts beyond a shadow of a doubt." I said to the management, "The only thing I ask you, and I must have your firm promise, is that when I say a thing is so you will take my word against the word of any 20 men or any 50 men in the shop." I said, "If you won't do that, I won't lift my finger toward increasing the output of this shop." They agreed to it and stuck to it, although many times they were on the verge of believing that I was both incompetent and untruthful.

Now, I think it perhaps desirable to show the way in which that fight was conducted.

I began, of course, by directing some one man to do more work than he had done before, and then I got on the lathe myself and showed him that it could be done. In spite of this, he went ahead and turned out exactly the same old output and refused to adopt better methods or to work quicker until finally I laid him off and got another man in his place. This new man—I could not blame him in the least under the circumstances—turned right around and joined the other fellows and refused to do any more work than the rest. After trying this policy for a while and failing to get any results I said distinctly to the fellows, "Now, I am a mechanic; I am a machinist. I do not want to take the next step, because it will be contrary to what you and I look upon as our interest as machinists, but I will take it if you fellows won't compromise with me and get more work off of these lathes, but I warn you if I have to take this step it will be a durned mean one." I took it.

I hunted up some especially intelligent laborers who were competent men, but who had not had the opportunity of learning a trade, and I deliberately taught these men how to run a lathe and how to work fast and right. Every one of these laborers promised me, "Now if you will teach me the machinist trade, when I learn to run a lathe I will do a fair day's work," and every solitary man, when I had taught them their trade, one after another turned right around and joined the rest of the fellows and refused to work one bit faster.

That looked as if I were up against a stone wall, and for a time I was up against a stone wall. I did not blame even these laborers in my heart; my sympathy was with them all of the time, but I am telling you the facts as they then existed in the machine shops of this country and, in truth, as they still exist.

When I had trained enough of these laborers so that they could run the lathes, I went to them and said, "Now, you men to whom I have taught a trade are in a totally different position from the machinists who were running these lathes before you came here. Every one of you agreed to do a certain thing for me if I taught you a trade, and now not one of you will keep his word. I did not break my word with you, but every one of you has broken his word with me. Now, I have not any mercy on you; I have not the slightest hesitation in treating you entirely differently from the machinists." I said, "I know that very heavy social pressure has been put upon you outside the works to keep you from carrying out your agreement with me, and it is very difficult for you to stand out against this pressure, but you ought not to have made your bargain with me if you did not intend to keep your end of it. Now, I am going to cut your rate in two tomorrow and you are going to work for half price from now on. But all you will have to do is to turn out a fair day's work and you can earn better wages than you have been earning."

These men, of course, went to the management, and protested I was a tyrant, and a nigger driver, and for a long time they stood right by the rest of the men in the shop and refused to increase their output a particle. Finally, they all of a sudden gave right in and did a fair day's work.

I want to call your attention, gentlemen, to the bitterness that was stirred up in this fight before the men finally gave in, to the meanness of it, and the contemptible conditions that exist under the old piecework system, and to show you what it leads to. In this contest, after my first fighting blood which was stirred up through strenuous opposition had subsided, I did not have any bitterness against any particular man or men. My anger and hard feelings were stirred up against the system; not against the men. Practically all of those men were my friends, and many of them are still my friends. As soon as I began to be successful in forcing the men to do a fair day's work, they played what is usually the winning card. I knew that it was coming. I had predicted to the owners of the company what would happen when we began to win, and had warned them that they must stand by me; so that I had the backing of the company in taking effective steps to checkmate the final move of the men. Every time I broke a rate or forced one of the new men whom I had trained to work at a reasonable and proper speed, some one of the machinists would deliberately break some part of his machine as an object lesson to demonstrate to the management that a fool foreman was driving the men to overload their machines until they broke. Almost every day ingenious accidents were planned, and these happened to machines in different parts of the shop, and were, of course, always laid to the fool foreman who was driving the men and the machines beyond their proper limit.

Fortunately, I had told the management in advance that this would happen, so they backed me up fully. When they began breaking their machines, I said to the men, "All right; from this time on, any accident that happens in this shop, every time you break any part of a machine you will have to pay part of the cost of repairing it or else quit. I don't care if the roof falls in and breaks your machine, you will pay all the same." Every time a man broke anything I fined him and then turned the money over to the mutual benefit association, so that in the end it came back to the men. But I fined them, right or wrong. They could always show every time an accident happened that it was not their fault and that it was an impossible thing for them not to break their machine under the circumstances. Finally, when they found that these tactics did not produce the desired effect on the management, they got sick and tired of being fined, their opposition broke down, and they promised to do a fair day's work.

After that we were good friends, but it took three years of hard fighting to bring this about. I was a young man in years, but I give you my word I was a great deal older than I am now with worry, meanness, and contemptibleness of the whole damn thing. It is a horrid life for any man to live, not to be able to look any workman in the face all day long without seeing hostility there and feeling that every man around is his virtual enemy. These men were a nice lot of fellows and many of them were my friends outside of the works. This life was a miserable one, and I made up my mind either to get out of the business entirely, and go into some other line of work, or to find some remedy for this unbearable condition.

When I came to think over the whole matter, I realized that the thing which we on the management's side lacked more than anything else was exact knowledge as to how long it ought to take the workman to do his work. I knew how to do the work about as well as the rest of the workmen (many of them were better mechanics than I was, but on the whole I knew well enough how the work ought to be done in the shop). I could take any workman and show him how to run his lathe, but when it came to telling a man how long it ought to take him to do his work there was no foreman who at that time could do this with any degree of accuracy even if he knew ten times as much about the time problem as I did. You will remember, of course, that the chief object of the men in soldiering was to keep their foreman ignorant of how fast the work could be done. Realizing this deficiency on my part, I asked permission from Mr. William Sellers, the president of the Midvale Steel Company to make a series of careful scientific experiments to find out how quickly the various kinds of work that went into the shop ought to be done.

Now, these experiments were started along a variety of lines. One of the types of investigation which was started at that time was that which has come to be generally known as "motion study" or "time study." A young man was given a stop watch and ruled and printed blanks like those shown after page 160 of the red bound book written by me, entitled "Shop Management," which is in the hands of your committee. This man for two years and one half, I think, spent his entire time in analyzing the motions of the workmen in the machine shop in relation to all the machine work going on in the shop —all the operations, for example, which were performed while putting work into and taking work out from the machines were analyzed and timed. I refer to the details of all such motions as are repeated over and over again in machine shops. I dare say you gentlemen realize that while the actual work done in the machine shops of this country is infinite in its variety, and that while there are millions and millions of different operations that take place, yet these millions of complicated or composite operations can be analyzed intelligently and readily resolved into a comparatively small number of simple elementary operations, each of which is repeated over and over again in every machine shop. As a sample of these elementary operations which occur in all machine shops, I would cite picking up a bolt and clamp and putting the bolt head into the slot of a machine, then placing a distance piece under the back end of the clamp and tightening down the bolt. Now, this is one of the series of simple operations that take place in every machine shop hundreds of times a day. It is clear that a series of motions such as this can be analyzed, and the best method of making each of these movements can be found out, and then a time study can be made to determine the exact time which a man should take for each job when he does his work right, without any hurry and yet who does not waste time. This was the general line of one of the investigations which we started at that time.

At the same time, another series of investigations was started which I shall describe later, and which resulted in developing the art or science of cutting metals.

Before starting to describe these experiments, however, I want to make it clear to you that these scientific experiments, namely, accurate motion and time study of men and a study of the art of cutting metals, which were undertaken to give the foreman of the machine shop of the Midvale Steel Works knowledge which was greatly needed by him, in order to prevent soldiering and the strife that goes with it, marked the first steps which were taken in the evolution of what is called scientific management. These steps were taken in an earnest endeavor to correct what I look upon as one of the crying evils of the older systems of management. And I think that I may say that every subsequent step which was taken and which has resulted in the development of scientific management was in the same way taken, not as the result of some preconceived theory by any one man or any number of men, but in an equally earnest endeavor to correct some of the perfectly evident and serious errors of the older type of management. Thus scientific management has been an evolution in which many men have had their part, and I feel that this fact should be emphasized. Personally I am profoundly suspicious of any new theory, my own as well as any other man's theory, and until a theory has been proved to be correct from practical experience, it is safe to say that in nine cases out of ten it is wrong.

Scientific management, then, is no new or untried theory. Far from being a mere theory, on the contrary, the theory of scientific management has only come to be a matter of interest and of investigation during the past few years, whereas this type of management itself has been in process of evolution during a period of about 30 years, through actual use in shops, through being tried out, experimented with, and improved in the most practical way by hundreds, almost thousands of men. Scientific management, then, is not a theory, but is the practical result of a long evolution.

The illustrations of shoveling and bricklaying which I have given you have thus far been purposely confined to the more elementary types of work, so that a very strong doubt must still remain as to whether this kind of cooperation is desirable in the case of more intelligent mechanics, that is, in the case of men who are more capable of generalization, and who would therefore be more likely, of their own volition, to choose the more scientific and better methods. The following illustration will be given for the purpose of demonstrating the fact that in the higher classes of work the scientific laws which are developed are so intricate that the high-priced mechanic needs—even more than the cheap laborer—the cooperation of men better educated than himself in finding the laws, and then in selecting, developing, and training him to work in accordance with these laws. This illustration should make perfectly clear my original proposition that in practically all of the mechanic arts the science which underlies each workman's act is so great and amounts to so much that the workman who is best suited to actually doing the work is incapable, either through lack of education or through insufficient mental capacity, of understanding this science.

A doubt, for instance, will remain in your minds—in the case of an establishment which manufactures the same machine year in and year out in large quantities and in which, therefore, each mechanic repeats the same limited series of operations over and over again—whether the ingenuity of each workman and the help which he from time to time receives from his foreman will not develop such superior methods and such a personal dexterity that no scientific study which could be made would result in a material increase in efficiency.

A number of years ago a company employing in one of their departments about 300 men, which had been manufacturing the same machine for 10 to 15 years, sent for my friend Mr. Barth to report as to whether any gain could be made in their work through the introduction of scientific management. Their shops had been run for many years under a good superintendent and with excellent foremen and workmen on piece work. The whole establishment was, without doubt, in better physical condition than the average machine shop in this country. The superintendent was distinctly displeased when Mr. Barth told him that through the adoption of scientific management the output, with the same number of men and machines, could be more than doubled. He said that he believed that any such statement was mere boasting, absolutely false, and instead of inspiring him with confidence he was disgusted that anyone would make such an impudent claim. He, however, readily assented to Mr. Barth's proposition that he should select any one of the machines whose output he considered as representing the average of the shop, and that Mr. Barth should then demonstrate on this machine that through scientific methods its output could be more than doubled.

The machine selected by the superintendent fairly represented the work of the shop. It had been run for 10 or 12 years past by a first-class mechanic, who was more than equal in his ability to the average workmen in the establishment. In a shop of this sort, in which similar machines are made over and over again, the work is necessarily greatly subdivided, so that no one man works upon more than a comparatively small number of parts during the year. A careful record was therefore made, in the presence of both parties, of the time actually taken in finishing each of the parts which this man worked upon. The total time required by the old-fashioned skilled lathe hand to finish each piece, as well as the exact speeds and feeds which he took, were noted, and a record was kept of the time which he took in setting the work in the machine and in removing it. After obtaining in this way a statement of what represented a fair average of the work done in the shop, Mr. Barth applied to this one machine the principles of scientific management.

The first thing that Mr. Barth did was to study the proper speed at which this machine ought to be run. I am well within the limit, gentlemen, in saying that not one machine in twenty in the average shop in this country is properly speeded. This may seem incredible, and yet I make this statement with a great deal of confidence, because the Tool Builders' Association of the United States—the men who manufacture the machine tools of this country—last spring asked me to address their annual convention. I told them, just as I have told you, that not one in twenty of the machines in their shops was properly speeded; and I added, "You gentlemen know whether I am telling the truth or not, and I challenge anyone who thinks I am wrong in this statement to go into his own shop and let me show him how far wrong the speeds of his machines are." Not a man took up this challenge. And these tool builders are the men who make and sell the machines used in our machine shops.

I have here four quite elaborate slide rules, which have been developed especially to make a rapid study of machine tools. The one which I have marked "A" takes care of all the belting problems connected with machine tools. The one marked "B" solves all of the problems connected with gearing. The slide rule marked "C" determines accurately the pressure which the chip or shaving which is being cut from the metal exerts on the top of the tool. The one marked "D" shows just how fast the lathe or other metal-cutting machine ought to run while the tool is taking any given kind of cut.

By means of these four quite elaborate slide rules, which have been especially made for the purpose of determining the all-round capacity of metal-cutting machines, Mr. Barth made a careful analysis of every element of this machine in its relation to the work in hand. Its pulling power at its various speeds, its feeding capacity, and its proper speeds were determined by means of the slide rules, and changes were then made in the countershaft and driving pulleys so as to run the lathe at its proper speed. Tools, made of high-speed steel and of the proper shapes were properly dressed, treated, and ground. It should be understood, however, that in this case the high-speed steel which had heretofore been in general use in the shop was also used in Mr. Barth's demonstration. Mr. Barth then made a large special slide rule, by means of which the exact speeds and feeds were indicated at which each kind of work could be done in the shortest possible time in this particular lathe. After preparing in this way so that the workman should work according to the new method, one after another, pieces of work were finished in the lathe, corresponding to the work which had been done in the preliminary trials, and the gain in time made through running the machine according to scientific principles ranged from two and one-half times the speed in the slowest instance to nine times the speed in the highest.

Thereupon, at 12 o'clock noon, a recess was taken until 2 o'clock p. m.

After Recess

The Committee met at 2 o'clock p. m., Hon. William B. Wilson (chairman) presiding.

Mr. Taylor. The change from rule-of-thumb management to scientific management involves, however, not only a study of what is the proper speed for doing the work and a remodeling of the tools and the implements in the shop, but also a complete change in the mental attitude of all the men in the shop toward their work and toward their employers. The physical improvements in the machines necessary to insure large gains and the motion study followed by minute study with a stop watch of the time in which each workman should do his work can be made comparatively quickly. But the change in the mental attitude and in the habits of the 300 or more workmen can be brought about only slowly and through a long series of object lessons, which finally demonstrates to each man the great advantage which he will gain by heartily cooperating in his everyday work with the men in the management. Within three years, however, in this shop the output had been more than doubled per man and per machine. The men had been carefully selected and in almost all cases promoted from a lower to a higher order of work and so instructed by their teachers—the functional foremen—that they were able to earn higher wages than ever before. The average increase in the daily earnings of each man was about 35 per cent, while at the same time the sum total of the wages paid for doing a given amount of work was lower than before. This increase in the speed of doing the work, of course, involved a substitution of the quickest hand methods for the old independent rule-of-thumb methods and an elaborate analysis of the hand work done by each man. By hand work is meant such work as depends upon the manual dexterity and speed of a workman and which is independent of the work done by the machine. The time saved by scientific hand work was in many cases greater even than that saved in machine work.

It seems important to fully explain the reason why, with the aid of a slide rule, and after having studied the art of cutting metals, it was possible for the scientifically equipped man, Mr. Barth, who had never before seen these particular jobs, and who had never worked on this machine, to do work from two and one-half to nine times as fast as it had been done before by a good mechanic who had spent his whole time for some 10 to 12 years in doing this very work upon this particular machine. In a word, this was possible because the art of cutting metals involves a true science of no small magnitude, a science, in fact, so intricate that it is impossible for any machinist who is suited to running a lathe year in and year out either to understand it or to work according to its laws without the help of men who have made this their specialty. Men who are unfamiliar with machine-shop work are prone to look upon the manufacture of each piece as a special problem, independent of any other kind of machine work. They are apt to think, for instance, that the problems connected with making the parts of an engine require the especial study, one may say almost the life study, of a set of engine-making mechanics, and that these problems are entirely different from those which would be met with in machining lathe or planer parts. In fact, however, a study of those elements which are peculiar either to engine parts or to lathe parts is trifling compared with the great study of the art, or science, of cutting metals, upon a knowledge of which rests the ability to do really fast machine work of all kinds.

The real problem is how to remove chips fast from a casting or a forging, and how to make the piece smooth and true in the shortest time, and it matters but little whether the piece being worked upon is part, say, of a marine engine, a printing press, or an automobile. For this reason, the man with the slide rule, familiar with the science of cutting metals, who had never before seen this particular work, was able completely to distance the skilled mechanic who had made the parts of this machine his specialty for years.

It is true that whenever intelligent and educated men find that the responsibility for making progress in any of the mechanic arts rests with them, instead of upon the workmen who are actually laboring at the ·trade, that they almost invariably start on the road which leads to the development of a science where in the past has existed mere traditional or rule-of-thumb knowledge. When men whose education has given them the habit of generalizing and everywhere looking for laws, find themselves confronted with a multitude of problems, such as exist in every trade and which have a general similarity one to another, it is inevitable that they should try to gather those problems into certain logical groups, and then search for some general laws or rules to guide them in their solution. As I have tried to point out, however, the underlying principles of the management of "initiative and incentive"—that is, the underlying philosophy of this management—necessarily leaves the solution of all of these problems in the hands of each individual workman, while the philosophy of scientific management places their solution in the hands of the management. The workman's whole time is each day taken in actually doing the work with his hands, so that, even if he had the necessary education and habits of generalizing in his thought, he lacks the time and the opportunity for developing these laws, because the study of even a simple law involving, say, time study requires the cooperation of two men, the one doing the work while the other times him with a stop watch. And even if the workman were to develop laws where before existed only rule-of-thumb knowledge, his personal interest would lead him almost inevitably to keep his discoveries secret so that he could, by means of this special knowledge, personally do more work than other men and so obtain higher wages.

Under scientific management, on the other hand, it becomes the duty and also the pleasure of those who are engaged in the management not only to develop laws to replace rule-of-thumb, but also to teach impartially all of the workmen who are under them the quickest ways of working. The useful results obtained from these laws are always so great that any company can well afford to pay for the time and the experiments needed to develop them. Thus, under scientific management, exact scientific knowledge and methods are everywhere, sooner or later, sure to replace rule-of-thumb, whereas under the old type of management working in accordance with scientific laws is an impossibility.

The development of the art or science of cutting metals is an apt illustration of this fact. In the early eighties, about the time that I started to make the investigations above referred to to determine the proper movements to be made by machinists in putting their work into and removing it from machines and time required to do this work, I also obtained the permission of Mr. William Sellers, the president of the Midvale Steel Co., to make a series of experiments to determine what angles and shapes of tools were the best for cutting steel, and also to try to determine the proper cutting speed for steel. At the time that these experiments were started it was my belief that they would not last longer than six months, and, in fact, if it had been known that a longer period than this would be required, the permission to spend a considerable sum of money in making them would not have been forthcoming.

A 66-inch diameter vertical boring mill was the first machine used in making these experiments, and large locomotive tires, made out of hard steel of uniform quality, were day after day cut up into chips in gradually learning how to make, shape, and use the cutting tools so that they would do faster work. At the end of six months sufficient practical information had been obtained to far more than repay the cost of materials and wages which had been expended in experimenting. And yet the comparatively small number of experiments which had been made served principally to make it clear that the actual knowledge attained was but a small fraction of that which still remained to be developed and which was badly needed by us in our daily attempt to direct and help the machinists in their work.

Experiments in this field were carried on, with occasional interruptions, through a period of about 26 years, in the course of which 10 different experimental machines were especially fitted up to do this work. Between 30,000 and 50,000 experiments were carefully recorded, and many other experiments were made of which no record was kept. In studying these laws more than 800,000 pounds of steel and iron was cut up into chips with the experimental tools, and it is estimated that from $150,000 to $200,000 was spent in the investigation.

Work of this character is intensely interesting to anyone who has any love for scientific research. It should be fully appreciated that the motive power which kept these experiments going through many years and which supplied the money and the opportunity for their accomplishment was not an abstract search after scientific knowledge, but was the very practical fact that we lacked the exact information which was needed every day in order to help our machinists to do their work in the best way and in the quickest time.

All of these experiments were made to enable us to answer correctly the two questions which face every machinist each time that he does a piece of work in a metal-cutting machine, such as a lathe, planer, drill press, or milling machine. These two questions are:

In order to do the work in the quickest time, at what cutting speed shall I run my machine? and what feed shall I use?

These questions sound so simple that they would appear to call for merely the trained judgment of any good mechanic. In fact, however, after working 26 years, it has been found that the answer in every case involves the solution of an intricate mathematical problem, in which the effect of 12 independent variables must be determined.

Each of the 12 following variables has an important effect upon the answer. The figures which are given with each of the variables represent the effect of this element upon the cutting speed. For example, after the first variable (A) I quote:

> The proportion is as 1 in the case of semi-hardened steel or chilled iron to 100 in the case of a very soft low-carbon steel.

The meaning of this quotation is that soft steel can be cut one hundred times as fast as the hard steel or chilled iron. The ratios which are given, then, after each of these elements indicate the wide range of judgment which practically every machinist has been called upon to exercise in the past in determining the best speed at which to run his machine and the best feed to use.

(A) The quality of the metal which is to be cut, i. e. its hardness or other qualities which affect the cutting speed. The proportion is as 1 in the case of semi-hardened steel or chilled iron to 100 in the case of very soft, low-carbon steel.

(B) The chemical composition of the steel from which the tool is made, and the heat treatment of the tool. The proportion is as 1 in tools made from tempered carbon steel to 7 in the best highspeed tools.

(C) The thickness of the shaving, or the thickness of the spiral strip or band of metal which is to be removed by the tool. The proportion is as 1 with thickness of shaving three-sixteenths of an inch to $3\frac{1}{2}$ with thickness of shaving one sixty-fourth of an inch.

(D) The shape or contour of the cutting edge of the tool. The proportion is as 1 in a thread tool to 6 in a broad-nosed cutting tool.

(E) Whether a copious stream of water or other cooling medium is used on the tool. The proportion is as 1 for tool running dry to 1.41 for tool cooled by a copious stream of water.

(F) The depth of the cut. The proportion is as 1 with one-half inch depth of cut to 1.36 with one-eighth inch depth of cut.

(G) The duration of the cut, i. e., the time which a tool must last under pressure of the shaving without being re-ground. The proportion is as 1 when tool is to be ground every one and one-half hours to 1.20 when tool is to be ground every 20 minutes.

(H) The lip and clearance angles of the tool. The proportion is as 1 with lip angle of 68° to 1.023 with lip angle of 61°.

(J) The elasticity of the work and of the tool on account of producing chatter. The proportion is as 1 with tool chattering to 1.15 with tool running smoothly.

(K) The diameter of the casting or forging which is being cut.

(L) The pressure of the chip or shaving upon the cutting surface of the tool.

(M) The pulling power and the speed and feed changes of the machine.

It may seem preposterous to many people that it should have required a period of 26 years to investigate the effect of these 12 variables upon the cutting speed of metals. To those, however, who have had personal experience as experimenters it will be appreciated that the great difficulty of the problem lies in the fact that it contains so many variable elements.

And, in fact, the great length of time consumed in making each single experiment was caused by the difficulty of holding 11 variables constant and uniform throughout the experiment, while the effect of the twelfth variable was being investigated. Holding the 11 variables constant was far more difficult than the investigation of the twelfth element.

As, one after another, the effect upon the cutting speed of each of these variables was investigated, in order that practical use could be made of this knowledge, it was necessary to find a mathematical formula which expressed in concise form the laws which had been obtained. As examples of the 12 formulae which were developed, the 3 following are given.

$$P = 45{,}000 \ D^{14/15} F^{3/4}$$

$$V = \frac{90}{T^{1/8}}$$

$$V = \frac{11.9}{F^{0.665} \left(\dfrac{48}{3}D\right)^{0.2373 \ + \ \frac{2.4}{18+24D}}}$$

After these laws had been investigated and the various formulae which mathematically expressed them had been determined there still remained the difficult task of how to solve one of these complicated mathematical problems quickly enough to make this knowledge available for everyday use. If a good mathematician who had these formulae before him were to attempt to get the proper answer (i. e. to get the correct cutting speed and feed by working in the ordinary way), it would take him from two to six hours, say, to solve a single problem; far longer to solve the mathematical problem than would be taken in most cases by the workman in doing the whole job in his machine.

Thus a task of considerable magnitude which faced us was that of finding a quick solution of this problem, and as we made progress in its solution the whole problem was from time to time presented by me to one after another of the noted mathematicians in this country. They were offered any reasonable fee for a rapid, practical method to be used in its solution. Some of these men merely glanced at it; others, for the sake of being courteous, kept it before them for some two or three weeks. They all gave us practically the same answer, that in many cases it was possible to solve mathematical problems which contained 4 variables and in some cases problems with 5 or 6 variables, but that it was manifestly impossible to solve a problem containing 12 variables in any other way than by the slow process of "trial and error."

A quick solution was, however, so much of a necessity in our everyday work of running machine shops that in spite of the small encouragement received from the mathematicians we continued at irregular periods, through a term of 15 years, to give a large amount of time searching for a simple solution. Four or five men at various periods gave practically their whole time to this work (among these men were Mr. Sinclair, Mr. Gault, and Mr. Barth) and finally, while we were at the Bethlehem Steel Co. the slide rule was developed, which is illustrated on folder No. 11 of the paper "On the art of cutting metals," which is in the hands of your committee and is described in detail in the paper presented by Mr. Carl G. Barth to the American Society of Mechanical Engineers, entitled "Slide rules for the machine shop, as a part of the Taylor system of management" (Vol. XXV of The Transactions of the American Society of Mechanical Engineers). By means of this slide rule one of these intricate problems can be solved in less than half a minute by any good mechanic, whether he understands anything about mathematics or not, thus making available for everyday practical use the years of experimenting on the art of cutting metals.

This is a good illustration of the fact that some way can be always be found of making practical, everyday use of complicated scientific data which appears to be beyond the experience and the range of the technical training of ordinary practical men. These slide rules have been for years in constant daily use by machinists having no knowledge of mathematics.

A glance at the intricate mathematical formulae which represent the laws of cutting metals should clearly show the reason why it is impossible for any machinist, without the aid of these laws and who depends upon his personal experience, correctly to guess at the answer to the two questions:

What speed shall I use?
What feed shall I use?

even though he may repeat the same piece of work many times.

To return to the case of the machinist who had

been working for 10 to 12 years in machining the same pieces over and over again, there was but a remote chance in any of the various kinds of work which this man did that he should hit upon the one best method of doing each piece of work out of the hundreds of possible methods which lay before him. In considering this typical case it must also be remembered that the metal-cutting machines throughout our machine shops have practically all been speeded by their makers by guesswork and without the knowledge obtained through a study of the art of cutting metals. As I have said before, in the machine shops systemized by us we have found that there is not one machine in twenty which is speeded by its makers at anywhere near the correct cutting speed. So that, in order to compete with the science of cutting metals the machinist, before he could use proper speeds, would first have to put new pulleys on the countershaft of his machine and also make in most cases changes in the shapes and treatment of his tools, etc. Many of these changes are matters entirely beyond his control, even if he knows what ought to be done.

If the reason is clear to you why the rule-of-thumb knowledge obtained by the machinist who is engaged on repeat work cannot possibly compete with the true science of cutting metals, it should be even more apparent why the high-class mechanic, who is called upon to do a great variety of work from day to day, is even less able to compete with this science. The high-class mechanic who does a different kind of work each day, in order to do each job in the quickest time, would need, in addition to a thorough knowledge of the art of cutting metals, a vast knowledge and experience in the quickest way of doing each kind of handwork. And by calling to mind the gain which was made by Mr. Gilbreth through his motion and time study in laying bricks, you will appreciate the great possibilities for quicker methods of doing all kinds of handwork which lie before every tradesman after he has the help which comes from a scientific motion and time study of his work.

For nearly 30 years past time-study men connected with the management of machine shops have been devoting their whole time to a scientific motion study, followed by accurate time study with a stop watch of all elements connected with the machinist's work. When, therefore, the teachers, who form one section of the management, and who are cooperating with the workingmen, are in possession both of the science of cutting metals and of equally elaborate motion-study and time-study science connected with this work, it is not difficult to appreciate why even the highest-class mechanic is unable to do his best work without constant daily assistance from his teachers.

Now, gentlemen, what I have been trying to illustrate is the effect which the development of a great science has upon the workman's daily life. The sciences of shoveling and of bricklaying are comparatively small, and yet their effect upon the workman is great. The science of cutting metals required 26 years of constant effort to develop, and what I have been trying to show you is that when a large science, such as this, is applied to the work of a first-class mechanic, even though he be a man having a good high-school education, that the effect of science upon the work of this man is quite as great as the effect of the smaller science, such as that of bricklaying, upon a less intellectual and less well-educated man.

You will remember that Mr. Barth, with the knowledge obtained from the science of cutting metals, was able to show the high-class mechanic how to do work from two and one-half to nine times as fast as he had formerly done it, and this with no greater effort to himself than he had exerted before.

Now, gentlemen, the development of the science of cutting metals is merely typical of what is going to take place in all of the great industries of this country during the next twenty to thirty years. Already bleaching has been taken out of the old rule-of-thumb methods and developed into a science, and the dyeing business is now being studied scientifically, and right at this minute probably 10 to 15 other large and important sciences are receiving the same minute, painstaking study which will ultimately result in developing a science where now exists mere traditional rule-of-thumb knowledge. And in each of these cases results will be accomplished which are fairly comparable with those achieved under the science of cutting metals·

The development of a science sounds like a formidable undertaking, and in fact, anything like a thorough study of a science such as that of cutting metals necessarily involves many years of work. The science of cutting metals, however, represents in its complication, and in the time required to develop it, almost an extreme case in the mechanic arts. Yet even in this very intricate science within a few months after starting enough knowledge had been obtained to much more than pay for the work of experimenting. This holds true in the case of practically all scientific development in the mechanic arts. The first laws developed for cutting metals were crude and contained only a partial knowledge of the truth, yet this imperfect knowledge was vastly better than the utter lack of exact information or the very imperfect rule-of-thumb which existed before, and it enables the workmen, with the help of the management, to do far quicker and better work.

For example, a very short time was needed to discover one or two types of tools which, though imperfect as compared with the shapes developed years afterwards, were superior to all other shapes and kinds in common use. These tools were adopted as standard and made possible an immediate increase in the speed of every machinist who used them. These types were superseded in a comparatively short time by still other tools which remained standard until they in turn made way for later improvements.

The science which exists in most of the mechanic arts is, however, far simpler than the science of cutting metals. In almost all cases, in fact, the laws or rules which are developed are so simple that the average man would hardly dignify them with the name of a science. In most trades the science is developed through a comparatively simple analysis and time study of the movements required by the workmen to do some small part of his work, and this study is usually made by a man equipped merely with a stop watch and a properly ruled notebook. Hundreds of these "time study men" are now engaged in developing elementary scientific knowledge where before existed only rule-of-thumb. Even the motion study of Mr. Gilbreth in bricklaying involves a much more elaborate investigation than that which occurs in most cases. The general steps to be taken in developing a simple law of this class are as follows:

First. Find, say, 10 to 15 different men (preferably in as many separate establishments and different parts of the country) who are especially skillful in doing the particular work to be analyzed.

Second. Study the exact series of elementary operations or motions which each of these men uses in doing the work which is being investigated, as well as the implements each man uses.

Third. Study with a stop watch the time required to make each of these elementary movements and then select the quickest way of doing each element of the work.

Fourth. Eliminate all false movements, slow movements, and useless movements.

Fifth. After doing away with all unnecessary movements, collect into one series the quickest and best movements, as well as the best implements.

This new method, involving that series of motions which can be made quickest and best, is then substituted in place of the 10 or 15 inferior series which were formerly in use. This best method becomes standard and remains standard, to be taught first to the teachers (or functional foremen) and by them to every workman in the establishment until it is superseded by a quicker and better series of movements. In this simple way one element after another of the science is developed.

In the same way each type of implement used in a trade is studied. Under the philosophy of the management of "initiative and incentive" each workman is called upon to use his own best judgment so as to do the work in the quickest time, and from this results, in all cases, a large variety in the shapes and types of implements which are used for any specific purpose. Scientific management requires, first, a careful investigation of each of the many modifications of the same implement, developed under rule-of-thumb; and second, after a time study has been made for speed attainable with each of these implements that the good points of several of them shall be united in a single standard implement, which will enable the workman to work faster and with greater ease than he could before. This one implement, then, is adopted as standard in place of the many different kinds before in use, and it remains standard for all workmen to use until superseded by an implement which has been shown, through motion and time study, to be still better.

With this explanation it will be seen that the development of a science to replace rule-of-thumb is in most cases by no means a formidable undertaking and that it can be accomplished by ordinary, every-day men without any elaborate scientific training;

but that, on the other hand, the successful use of even the simplest improvement of this kind calls for records, system, and cooperation where in the past existed only individual effort.

Now, what I want to bring out and make clear to you is that under scientific management there is nothing too small to become the subject of scientific investigation. Every single motion of every man in the shop sooner or later becomes the subject of accurate, careful study to see whether that motion is the best and quickest that can be used, and as you see, this is a new mental attitude assumed by the employer which differs radically from the old. The old idea, both of employer and employee, was to leave all of these details to someone's judgment. The new idea is that everything requires scientific investigation, and that is what I am trying to make clear to you.

There are a number of facts connected with scientific management which I think can be better brought out under cross-examination than by direct statement.

The Chairman. Well, if you have concluded your direct statement, Mr. Taylor, we will adjourn the committee until 11 o'clock tomorrow morning, when we will proceed with the cross-examination.

SATURDAY, JANUARY 27, 1912

The committee met at 11 o'clock a. m.

Present: Messrs. William B. Wilson (chairman), and John Q. Tilson.

The Chairman. The committee will be in order.

Mr. Taylor, did you serve your apprenticeship as a machinist in the Midvale plant?

Mr. Taylor. No, sir; I served my apprenticeship in a small shop. It was under the management of the firm of Ferrell & Jones, a shop in which steam pumps were made and a variety of miscellaneous machinery, but yet a very small shop.

The Chairman. How long did you serve as an apprentice?

Mr. Taylor. I started in 1874 and finished in 1878, the end of 1878.

The Chairman. Making four years?

Mr. Taylor. Four years of work; yes, sir.

The Chairman. How old were you when you began your apprenticeship?

Mr. Taylor. About 18 years old.

The Chairman. You were a journeyman machinist when you went to the Midvale plant, were you?

Mr. Taylor. Yes; I may say, Mr. Chairman, that my father had some means, and owing to the fact that I worked during my first year of apprenticeship for nothing, the second year for $1.50 a week, the third year for $1.50 a week, and the fourth year for $3 a week, I was given, perhaps, special opportunities to progress from one kind of work to another; that is, I told the owners of the establishment that I wanted an opportunity to learn fast rather than wages, and for that reason, I think, I had specially good opportunities to progress. I am merely saying that to explain why in four years I was able to get through with my apprenticeship as a pattern maker and as a machinist. That is a very short time, as you will realize. I may add that I do not think I was a very high order of journeyman when I started in.

The Chairman. How long did you work as a journeyman machinist at the Midvale plant before you were promoted to the position of gang foreman?

Mr. Taylor. My remembrance is not very clear in the matter, but I should not think it was more than two months.

The Chairman. How long had you worked as a journeyman machinist before that at this other plant?

Mr. Taylor. That is the first work I had after I got through with my apprenticeship.

The Chairman. You went right from there to the Midvale plant as a journeyman machinist?

Mr. Taylor. Yes, sir.

The Chairman. And worked at the Midvale plant two months as a journeyman machinist before you were promoted to the position of gang foreman?

Mr. Taylor. Gang boss; yes.

The Chairman. During the time that you were working as a journeyman machinist you worked exactly as the other man in the plant worked?

Mr. Taylor. Oh, yes; absolutely.

The Chairman. You found there a disposition on the part of the workmen to soldier?

Mr. Taylor. We all soldiered; it is safe to say that there was not a man in the shop that did not soldier.

The Chairman. Yourself included?

Mr. Taylor. Certainly, sir.

The Chairman. You did not while there do any

greater amount of work than the other machinists?

Mr. Taylor. Well, there may have been a shade of difference between my work and that of the rest of the men. I will not say that I did work harder. Possibly I did a little more work, but it was not enough to cause my brother workmen to feel that I was breaking rates and making a hog of myself, as they would put it then.

The Chairman. But you were there long enough and worked with them long enough to feel that the workmen were soldiering?

Mr. Taylor. I absolutely knew it; there was no question about it. I saw the same thing, Mr. Chairman, all through my apprenticeship, from the time I started as an apprentice until I got through; the thing was practically universal in the shop.

The Chairman. And when you became a gang foreman, having this information, you determined to take strong measures to break up that soldiering?

Mr. Taylor. I determined to try to get a larger output from the machines, but I do not think I had in mind what measures I was going to take; at first I do not think I had any policy clearly in mind. I thought at first that I would be able to persuade a lot of my friends to do more work, but I soon found that was out of the question.

The Chairman. Did you find during that time that the workmen themselves admitted that they were soldiering?

Mr. Taylor. Of course they did.

The Chairman. They admitted that to the foreman?

Mr. Taylor. I do not know what they admitted to the other foreman (the old gentleman, as we called him; the old man was an old English gentleman of more than 70 years of age). I really do not know what they admitted to him; but all through the time that I was their foreman or their gang boss and was trying to get them to do a larger day's work there was no denying the matter at all with me; they knew that I knew it, and they justified it, and so did I justify it, Mr. Chairman, in view of prevailing conditions, and my sympathies were with them through-out the whole performance. Now, that may sound like an anomaly, but I am telling you the fact. My sympathies were with the workman, and my duty lay to the people by whom I was employed. My sympathies were so great that when, as I have told you before, they came to me for personal advice as a friend and asked me in a serious, sober way, "Fred, if you were in my place, would you do what you are asking me to do, turn out a bigger output?" my answer was, as I have said in the record before, "If I were in your place, I would do just what you are doing; I would fight against this as hard as any of you are; only," I said, "I would not make a fool of myself; when the time comes that you see that I have succeeded, or the men on our side have succeeded, in forcing or compelling you to do a larger day's work, I would not then make a fool of myself. When that time comes I would work up to proper speed." I told them that over and over again. Our official relations were of the most strained and most disagreeable and contemptible nature, but my personal relations with most of the men throughout that fight were agreeable.

The Chairman. Let me find out whether your conception of what is meant by the term "soldiering" and my conception are the same. Do you mean by the term "soldiering" a failure on the part of the workman to do as much work as he could do without physical or mental injury to himself?

Mr. Taylor. Would it not be better for me to quote from what I have written on the matter? What I have written has been very carefully prepared to express my exact views.

The Chairman. I just wanted to get your conception as to what constitutes soldiering. If that fits your conception, of course we will be glad to hear it.

Mr. Tilson. What we want is your present idea of that term; and if it is expressed in your book, we will be glad to have it.

Mr. Taylor. It is expressed in my book better than I could state it extemporaneously; I could state it in a shorter way, but I do not want to have people coming back at me and misrepresenting my real views because of any brief extemporaneous statement that I may make. There are several kinds of soldiering, and they are described in my book; if you want a full definition of soldiering, I beg to refer to my book.

The Chairman. We would like to have your whole view about soldiering.

Mr. Taylor. Well, I will read from my book as follows:

On the part of the men the greatest obstacle to the attainment of this standard is the slow pace which they adopt, or the loafing or "soldiering," marking time, as it is called. This loafing or soldiering proceeds from two causes. First, from the natural instinct and tendency of man to take it easy, which may be called natural soldiering. Second, from more intricate second thought and reasoning caused by their relations with other men, which may be called systematic soldiering.

I might add that in England it is called "hanging it out" and in Scotland "ca' cannie," and every man in England, let me tell you, hangs it out, and every man in Scotland will ca' cannie.

(Reading:)
There is no question that the tendency of the average man (in all walks of life) is toward working at a slow, easy gait, and that it is only after a good deal of thought and observation on his part or as a result of example, conscience, or external pressure that he takes a more rapid pace.

There are, of course, men of unusual energy, vitality, and ambition who naturally choose the fastest gait, set up their own standards, and who will work hard, even though it may be against their best interests. But these few uncommon men only serve by affording a contrast to emphasize the tendency of the average.

This common tendency to "take it easy" is greatly increased by bringing a number of men together on similiar work and at a uniform standard rate of pay by the day. Under this plan the better men gradually but surely slow down their gait to that of the poorest and least efficient. When a naturally energetic man works for a few days beside a lazy one, the logic of the situation is unanswerable: "Why should I work hard when that lazy fellow gets the same pay that I do and does only half as much work?"

A careful time study of men working under these conditions will disclose facts which are ludicrous as well as pitiable.

To illustrate: The writer has timed a naturally energetic workman who, while going and coming from work would walk at a speed of from 3 to 4 miles per hour, and not infrequently trot home after a day's work. On arriving at his work he would immediately slow down to a speed of about one mile an hour. When, for example, wheeling a loaded wheelbarrow he would go at a good fast pace even up hill in order to be as short a time as possible under load, and immediately on the return walk slow down to a mile an hour, improving every opportunity for delay short of actually sitting down. In order to be sure not to do more than his lazy neighbor he would actually tire himself in his effort to go slow

These men were working under a foreman of good reputation and one highly thought of by his employer who, when his attention was called to this state of things, answered: "Well, I can keep them from sitting down, but the devil can't make them get a move on while they are at work."

The natural laziness of men is serious, but by far the greatest evil from which both workmen and employers are suffering, is the systematic soldiering which is almost universal under all of the ordinary schemes of management and which results from a careful study on the part of the workmen of what they think will promote their best interests.

The writer was very much interested recently to hear one small but experienced golf caddy boy of 12 explaining to a green caddy who had shown special energy and interest the necessity of going slow and lagging behind his man when he came up to the ball, showing him that since they were paid by the hour, the faster they went the less money they got, and finally telling him that if he went too fast the other boys would give him a licking.

This represents a type of systematic soldiering

which is not, however, very serious, since it is done with the knowledge of the employer, who can quite easily break it up if he wishes.

The greater part of the systematic soldiering, however, is done by the men with the deliberate object of keeping their employers ignorant of how fast work can be done.

So universal is soldiering for this purpose that hardly a competent workman can be found in a large establishment, whether he works by the day or on piecework, contract work or under any of the ordinary systems of compensating labor, who does not devote a considerable part of his time to studying just how slowly he can work and still convince his employer that he is going at a good pace.

The causes for this are, briefly, that practically all employers determine upon a maximum sum which they feel it is right for each of their classes of employees to earn per day, whether their men work by the day or by the piece.

Each workman soon finds out about what this figure is for his particular case, and he also realizes that when his employer is convinced that a man is capable of doing more than he has done, he will find sooner or later some way of compelling him to do it with little or no increase of pay.

Employers derive their knowledge of how much of a given class of work can be done in a day from either their own experience, which has frequently grown hazy with age, from casual and unsystematic observation of their men, or at best from records which are kept, showing the quickest time in which each job has been done. In many cases the employer will feel almost certain that a given job can be done faster than it has been, but he rarely cares to take the drastic measures necessary to force men to do it in the quickest time, unless he has an actual record, proving conclusively how fast the work can be done.

It evidently becomes for each man's interest, then, to see that no job is done faster than it has been in the past. The younger and less experienced men are taught this by their elders, and all possible persuasion and social pressure is brought to bear upon the greedy and selfish men to keep them from making new records which result in temporarily increasing their wages, while all those who come after them are made to work harder for the same old pay.

Under the best daywork of the ordinary type, when accurate records are kept of the amount of work done by each man and of his efficiency, and when each man's wages are raised as he improves, and those who fail to rise to a certain standard are discharged and a fresh supply of carefully selected men are given work in their places, both the natural loafing and systematic soldiering can be largely broken up. This can be done, however, only when the men are thoroughly convinced that there is no intention of establishing piecework even in the remote future, and it is next to impossible to make men believe this when the work is of such a nature that they believe piecework to be practicable. In most cases their fear of making a record which will be used as a basis for piecework will cause them to soldier as much as they dare.

It is, however, under piecework that the art of systematic soldiering is thoroughly developed. After a workman has had the price per piece of the work he is doing lowered two or three times as a result of his having worked harder and increased his output he is likely to entirely lose sight of his employer's side of the case and to become imbued with a grim determination to have no more cuts if soldiering can prevent it. Unfortunately for the character of the workman, soldiering involves a deliberate attempt to mislead and deceive his employer, and thus upright and straightforward workmen are compelled to become more or less hypocritical. The employer is soon looked upon as an antagonist, if not as an enemy, and the mutual confidence which should exist between a leader and his men—the enthusiasm, the feeling that they are all working for the same end and will share in the results—is entirely lacking.

The feeling of antagonism under the ordinary piecework system becomes in many cases so marked on the part of the men that any proposition made by their employers, however reasonable, is looked upon with suspicion. Soldiering becomes such a fixed habit that the men will frequently take pains to restrict the product of the machines which they are running when even a large increase in output would involve no more work on their part.

The Chairman. Now, with that definition of soldiering before us I want to ask whether I understood your direct testimony correctly to be that after you became foreman you ultimately succeeded in breaking up that soldiering, destroying the loafing, and removing the slow pace which you had found existing both in this automatic and systematic form, and thereby increased productivity?

Mr. Taylor. Yes, sir; to a large extent, but not entirely. I did not succeed in entirely breaking up the soldiering; I did not expect to succeed in that. As I told you before, we had the work in that shop laid out so that I think we were doing about one-third of a full day's work, and I succeeded in doubling the output of those men on the whole, I should say. It is many years ago and I make this statement in round numbers.

The Chairman. But you had succeeded in increasing the pace to such an extent that you did increase the productivity?

Mr. Taylor. Doubled it.

The Chairman. Never having worked yourself at that increased pace, would you think it possible for you to determine the soreness of muscle or the tiredness of brain which the increased pace brought to the workmen?

Mr. Taylor. I had many times done work at full speed, just as practically all of the workmen in the shop had worked at full speed. They all did work at full speed. We would not have known what full speed was unless we had worked at full speed, but we invariably did that when there was no one around to watch us and when there would be no record kept of it which could be used to break a rate to our own disadvantage. In this way we all knew what the right pace was, and then we settled upon what we thought the company ought to have in the way of work.

The Chairman. Is it not a fact when you speeded up for a comparatively short time and did the work rapidly that you thereby determined the length of time in which the work could be done rather than the length of time in which it should be done?

Mr. Taylor. Mr. Chairman, in my statement of what I believed was a proper day's work for that shop I stated what ought to be done and what could be done—what ought to be done as a fair day's work—that is, what could be done and kept up through a long term of years without any injury to the man, but what, on the contrary, would develop him—make him stronger, happier, and more contented in doing it. It was a perfectly proper pace and a pace such as you and I would be willing to take.

The Chairman. But that conclusion was arrived at by observation on your part, was it not, rather than by actual experience?

Mr. Taylor. By working myself and noting that I was not hurried; that I was perfectly contented; that I did not feel driven. It was personal experience and the experience of my friends who were working on their jobs in the same way. It was not watching anyone else so much as it was our own personal experience, and then we interchanged our views.

The Chairman. Would not the fact that your people were in better financial circumstances than the average workingman remove from your mind the same fear of ultimate exhaustion that would be continually in the mind of the workman who was dependent entirely upon his day's wages for his living?

Mr. Taylor. Well, I never had in mind ultimate exhaustion. I never had such a thing in my mind, and I do not think any of us in that shop had any fear of ultimate exhaustion. I never heard anyone talk about it. There was no fear that I ever heard expressed of anyone being overworked in that shop. That was not the fear.

The Chairman. Is it not true that a workman must provide for himself through his earning capacity for his entire lifetime; or, if from any cause he fails to provide for himself through his earnings he becomes a public charge and what is known as a pauper?

Mr. Taylor. Certainly, sir.

The Chairman. Would it not naturally, then, be in the mind of the workingman who has no other resources except his earnings from day to day that he must conserve his earning power so as to last him through the longest possible period of his life?

Mr. Taylor. It certainly should be, Mr. Chairman. Perhaps I could make the matter clearer to you by telling you that in machine work—running machine tools—it is next to impossible to overwork a man. In working on the average machine tool, of necessity the greater part of the day is spent by the man standing at his machine doing nothing except watch his machine work. I think I would be safe in saying that not more than three hours of actual physical work would be the average that any machinist would have to do in running his machine—not more than three hours' actual physical work in the day. The rest of the time the machine is working, and he simply stands there watching it. So there is no fear of overwork in the machine shop. Perhaps I can make it clearer to you by telling you that I worked the whole winter of 1895, I think it was, in running a machine myself. I went back and ran a machine for the whole winter in making a series of experiments in developing the "art of cutting metals," which I described to you in my direct testimony, and during this time I worked more steadily on that lathe than I had ever worked in my whole lifetime as a workman. I worked the same hours as the other workmen, and I tell you it was the easiest and happiest year I have had since I got out of my apprenticeship—that year of going back and working on a lathe. I worked hard from the machinist's standpoint and harder than I had ever worked before in my life as a mechanic. I was known to be a manager, and the men knew I was in there conducting some of the series of experiments that I have told you about on the art of cutting metals, and yet some of the men came to me and begged me not to set too fast a pace or the other fellows might have their rate cut as a result.

I give you my word, Mr. Chairman, that during that winter there was never a day that I was overworked, and I was physically soft; I was a comparatively middle-aged man and had not done any work by hand for 12 or 14 years, and yet I was not in the slightest degree overworked.

The Chairman. Is it not the purpose of the advocates of scientific management to apply it to all classes of work whether it is machine work or any other kind of work?

Mr. Taylor. It certainly is, sir.

The Chairman. So that the explanation which you have made would only apply to those cases where machines are used and where physical and mental energy is not required in handling the machines?

Mr. Taylor. It might apply to some other cases; it certainly would apply to the cases you speak of. But I know of a good many kinds of handwork, that is, work done without any machine, in which it is next to impossible to overwork, such, as for instance, very light, delicate work in which the muscular effort is so slight that it is next to impossible for a man to overwork himself physically. In work of this type he might overwork himself mentally or become tired mentally, but not physically.

The Chairman. Now, having removed, to some extent, the soldiering which occurred and thereby cheapened the cost of production, by what method does the public at large get the benefit of that cheapened cost of production?

Mr. Taylor. Usually the manufacturer who is manufacturing his goods, we will say at half the price he did formerly, wishes to enlarge his sales and so lowers the price in order to get a greater proportion of the business, and in that way the public profits by the lowering of the cost; that is the usual course.

Mr. Tilson. If everybody used the same system and thereby reduced the cost of production his competitors in business would force him to sell cheaper to the public, would they not?

Mr. Taylor. Yes, sir.

The Chairman. I am trying to bring out the inception of this thing. If an establishment reduces the cost by this process would the owner of the establishment sell the goods produced in the shop at any lower rate than the rate that was necessary to enable him to undersell his competitor and secure the trade?

Mr. Taylor. Naturally, he would not; in nine cases out of ten he would lower his price just enough to get the order. And you gentlemen who have had to do with the selling side of business know that the sales department is exceedingly slow in lowering prices, that is, making cuts in prices; they will usually wait until they get a big order before they cut at all, and so the process of lowering the price to the public is usually a slow one.

The Chairman. So that until other establishments introduced the system and thereby cut the cost of production competition between the manufacturers would not be sufficiently keen to enable the public to receive the entire benefit?

Mr. Taylor. Mr. Chairman, I think in the course of your question you used the term "introduced the system." I wish it clearly understood that everything I have said up to now during this cross examination bears no relation whatever to scientific management; it refers to just the opposite; it refers to the most unscientific management; it is the beastly management of the past that I have been referring to, and this has nothing to do with scientific management. All that I have had to say has relation to the brutal thing that I had to deal with in the early days, while in charge of the shop of the Midvale Steel Works, and that system was just the opposite of scientific management. I was trying to place before you the horror of the older system of management; it was the horror of this system which started me to take the first steps which, as time went on, finally produced the evolution of scientific management. I want that clearly understood. No one dislikes the older system of management more than I do.

The Chairman. However, if I understood your testimony correctly, you found this soldiering going on in this establishment and you took the methods which you have described to abolish that soldiering?

Mr. Taylor. Yes, sir.

The Chairman. And growing out of the experience thus arrived at you undertook to develop a scientific system by which the method of production could be improved, including, among other things, the automatic removal of soldiering by the system itself?

Mr. Taylor. My whole object was to remove the cause for antagonism between the boss and the men who were under him; to try to make both sides friends in the place of tactical enemies. Now, under this old system those men were my personal friends, but when we came to business, the moment that we went thru the gate of that place we were enemies— we were bitter enemies. I was trying to drive them and they were not going to be driven. I told you my early experience in the machine shop perfectly frankly, so as to try and make clear to you the sad and unfortunate mental attitude that accompanies the older type of management.

The Chairman. Now, having developed this system of management by which the advocates of it declare the cost of production is reduced we have already gotten to the point when it is introduced in one shop the owner of which in selling the product will simply sell low enough to secure the trade, and I want to get to the point at which the public at large receives all the benefits that can possibly come thru it.

Mr. Taylor. The time when the public at large gets the benefit?

The Chairman. Yes.

Mr. Taylor. That occurs with absolute certainty when dull times come along, if not before. In the iron and steel business—in the early years of the iron and steel business—whenever dull times came along, so far as my knowledge of it went, with few exceptions, prices fell to such a point that it was not a question of how much money you could make, but how little you must lose. The owners of the steel works and iron works practically all recognized that they must lose a certain amount of money in dull years, and the only question was how small they could make that loss. The competition was so keen during the dull years in the iron and steel business that it brought about this result; on the other hand, when busy times came along, when a good year came again, I have known them to earn right off 50 per cent in profits, and in that way largely make up the losses which came in dull times.

The Chairman. Now, assuming a case like the Midvale steel plant, where, I understand, this system was developed; assuming that the Midvale steel plant had scientific management and thereby reduced the cost of production, when a dull period came would not the fact that the Midvale Steel Co. had this reduced cost of production as compared with other competitors enable them to secure a very much larger share of the contracts, a proportionately larger share of the contracts and the work than they had formerly secured?

Mr. Taylor. That would be the theory, Mr. Chairman, but, as a rule, I think it has been true that your competitor meets your cuts in prices and he is willing to go to the verge of ruination in meeting your cuts, even though he loses more money than you are losing. Even though you may be making a little bit of money while he is losing a great deal of money, he, generally speaking, meets your cut; and that is a very unfortunate part of the competitive feature of industry. That has been an unfortunate feature and has led in the past to the survival of the fittest and to driving of many of the weaker companies to the wall.

The Chairman. Would it not be true, however, under the circumstances described, that if the competitors still continued to hold their share of the business, assuming that the Midvale Steel Co. were selling at cost and not under cost, it would only be a question of time until the entire capital of the competitors would be used up?

Mr. Taylor. If the dull times went on through a long enough term of years that would be true, but, fortunately, in most cases they did not continue for a great length of time. Fortunately, the dull times, during which you had to sell at low cost, did not last long enough so that many people were entirely ruined, although many of them came out battered and scarred, in bad financial condition, and overloaded with debt, and so on.

The Chairman. Now, assuming that they have not been driven to the wall by the dull times, those who are competitors of the Midvale Steel Co., which we are using as an illustration, and industrial activity and prosperity recurs, would not the same condition, so far as the benefits to the people who are concerned, exist after the restoration of industrial activity as existed prior to the industrial depression, unless the other establishments also introduced a system by which the cost would be reduced?

Mr. Taylor. If I understand you right, I think it would, sir, but I do not know that it is altogether clear in my mind just what you mean. I think I should agree with you that the conditions would return approximately to where they were before the dull times came on. I think that has been the history of it.

The Chairman. Now, it has taken, as I understand, 30 years of development to reach the stage in which scientific management now exists. I believe you made that statement, Mr. Taylor, or words to that effect?

Mr. Taylor. To be exact, I should say 29 years, I can mark the starting of it; it started in 1882; in the fall of 1882, if I remember rightly, the first steps were taken and that would be, perhaps, 29 years and 2 or 3 months.

The Chairman. Now, Mr. Taylor, is it not a fact that when any great improvement in machinery takes place or any system is introduced that requires less men to produce the same material, and while the public ultimately will receive the benefit of the improvement, that until it reaches the time when the public does secure the entire benefit there is a disturbed condition in the trades affected by the improvements and that a readjustment must take place and that the workmen who have been working in that trade or industry have to bear the entire burden until readjustment does take place?

Mr. Taylor. I think a careful study of the history of the introduction of labor-saving machinery would indicate that the larger part of the benefits from the introduction of new machinery first come to the employers or capitalists and that the workmen who were running the new machines, on the whole, have not, upon the immediate introduction of new machinery, profited to the extent to which they ought to have profited in an increase in wages and a betterment of conditions; that is, not immediately; but without question, ultimately not only those workmen who are working at the particular trade affected, but all of the collateral workmen affected by it do profit and profit immensely through increased production, which brings more wealth into the world for them to use; but the immediate effect has been that the workmen running the machine have not profited as they should have profited, in my judgment, through the introduction of labor-saving machinery.

And right here I want to point out the essential difference between scientific management and the management of the past. I have never heard that through the introduction of labor-saving machinery any manufacturer, under the old system of management, has insisted, as a part of the introduction of the labor-saving machinery, that his men should be paid from 30 to 100 per cent higher wages than are being paid to the same type of workmen working in similar industries in the immediate neighborhood. Manufacturers have in the past, on the contrary, been very careful to pay their men no higher wages than were paid in competitive industries right around them. In contrast to this, all of those men who are interested in the introduction of scientific management insist that the workmen shall get from 30 to 100 per cent higher wages as their share of this new scheme. The workmen get this great increase in wages right off; they do not have to ask for it—it is voluntarily and gladly given to them. And you will realize that under the old system of management an increase, say of 50 per cent, in wages could only come as a result of six or eight successful strikes, and that the average workman under the old system would not reach the goal in a lifetime. Now, if you will genuinely investigate—I am not speaking of you personally, Mr. Chairman, because anything you investigate is genuinely investigated, but some of the witnesses who have testified before this committee have not genuinely investigated it—the history of the introduction of scientific management, you will find that it is the truth that the 30 per cent to 100 per cent increase in wages which the workman receives as his share has been carefully awarded him right off; and that marks the difference in the history of the introduction of labor-saving contrivances of all kinds, such as new machinery and improved processes, on the one hand, and the introduction of this new labor-saving device on the other hand, namely, scientific management—a study of the motions of men and the simplification of their movements and acts. The introduction of labor-saving machinery has rarely been accompanied by a direct increase in wages, while the introduction of scientific management has always netted the workman an increase of 30 per cent to 100 per cent in wages.

The Chairman. Stating a hypothetical case, Mr. Taylor, there are something over 700,000 coal miners in the United States, producing approximately 500,-000,000 tons of coal; suppose that by the introduction of scientific management or the improvement of machinery, or by any other process, you were able to create conditions whereby 400,000 men produced the 500,000,000 tons of coal, would not the 300,000 men thereby temporarily displaced have to be provided for in some other way until a complete readjustment had taken place?

Mr. Taylor. Most certainly, providing those men were thrown out of a job all at once; but the history of the introduction of labor-saving machinery, as well as the history of the introduction of scientific management, indicates that in no industry is it possible to make any sudden change. In the case of scientific management, if you will read what I have written about it, I have carefully emphasized the fact that even in the most elementary work to make this great change is a question of not a month, not of a year, but two or three years, even in the most elementary work, and that in an intricate establishment it is a matter of not less than five years before a great increase in the output per man can be made. While the change in the type of management is going on, and while the increase in output per man grows and the cost gradually goes down, the history of the world shows that the world uses more and more of the new materials created. The introduction of labor-saving machinery does not tend to throw men out of work; that is not the history of the industrial world, nor even the history of any individual industry, and I challenge you gentlemen to state a case in which it is not true that the introduction of labor-saving machinery in the end has made work for more men, instead of throwing men out of work. The history of all industries indicates that labor-saving machinery, which enables a man to turn out a larger output, makes work for more men in those industries, and it would do the same thing in the coal trade as in any other trade.

The Chairman. I believe it is generally admitted

on all sides that the ultimate cheapening of the cost of production results in a greater consumption of the article and consequently a greater amount of production of the article, but is it not true that that increased consumption is itself a matter of growth; that it does not come suddenly?

Mr. Taylor. Yes, Mr. Chairman, that is true; but a study of industrial history indicates that consumption grows about as fast as production; that is the history of the world, I think. And, Mr. Chairman, as a matter of interest, I would call your attention to a very remarkable book on the law of wages which deals with statistics in the coal trade. This book was recently sent to me, and I have been reading it during the past few days; it shows statistically the effect of the introduction of labor-saving machinery on the wages of workmen in the coal trade, showing that the larger the amount of labor-saving machinery used in the industry the higher the wages. It is a most interesting book called "The Law of Wages," and it was published quite recently. Its author is Mr. A. L. Moore. I think you will be greatly interested in it, particularly in the conclusions or summaries of the last chapter; it is the most illuminating book statistically on the effect of various elements on wages that I have been able to get hold of.

The Chairman. Notwithstanding the fact that production keeps pace with consumption and consumption, to a certain extent, keeps pace with production, is it not true that when labor-saving machinery is introduced in any industry or any improvement in method introduced which reduces the number of men necessary to produce a given amount of material until the readjustment takes place, that a great many workmen are thrown out of employment and must be absorbed in some other lines until the growth in that line takes them back again?

Mr. Taylor. Yes; I think that is almost universally true. I think, however, it mainly comes about in this way; that the workmen who for years were accustomed to working in a certain way find that the new method of doing the work is irksome to them or sometimes that they are unable to do the work in the new way. These men find themselves not only seriously inconvenienced but they are sometimes brought to actual suffering from this cause; I think the introduction of labor-saving machinery is always accompanied by some unfortunate occurrences of that sort.

The Chairman. Now, then, what method has been developed or evolved by scientific management for taking care of the workmen thus displaced until the readjustment has taken place?

Mr. Taylor. I think I may say that in those establishments in which scientific management has been introduced there is not a single case that I can recall in which, after scientific management was introduced, there were less men employed than before. Not a single case, that is, in which the total number of men employed in the establishment were less than before. Sometimes many of the men who under the old system of management were workmen have been transferred from the working side to the management side, you understand, and in that case there may have been fewer workmen employed. By workmen, I mean those who are actually doing the work with their hands. But in this case the men who formerly did the work with their hands have been transferred to the management side, they have become teachers, guiders, and helpers. However, I do not think I can mention a single case in which there have been fewer men employed. I believe that in our arsenals, when scientific management will have been introduced, there will be more men at work than formerly; and I believe that in our navy yards the same result will follow. I believe that workmen from the arsenals and the navy yards who have appeared before your committee are laboring under an entire misapprehension as to the results which will follow the introduction of scientific management into the arsenals and into the navy yards, though scientific management has not been, and is not being introduced in the navy yards, according to Secretary Meyer. The results will be just the same there as everywhere else. I say there will be more men employed in the navy yards.

The Chairman. Then it is your belief that if this system of scientific management was universally adopted that no readjustment would be necessary so far as the employment of men is concerned?

Mr. Taylor. Mr. Chairman, there is a very great readjustment which necessarily follows from the very principles of scientific management. As I tried to outline at the beginning of my testimony, these principles involve a very careful study on the part of the management of the capacity and possibilities of each workman, and an entire change in that man's work if it becomes necessary, and it is necessary in most cases, in order to give each man the type of work to which he is best suited. So that scientific management does involve a series of very great changes in the workmen. I know of no system in which the changes are so great, but they almost all involve better conditions and more prosperity for the workmen; they are nine-tenths in the direction of good; they mean better work, higher wages, and more interesting work; those changes tend to make the workmen more efficient and make them into higher types of men. There are changes in plenty, but they are all to the good.

The Chairman. Is it not true that a number of men who have been eliminated from certain classes because they were considered not to be best suited for that class of work have been principally taken care of by virtue of the fact that the system in itself is only applied in a comparatively small percentage of the work to be done?

Mr. Taylor. Do you mean a comparatively small percentage of the work to be done in the world?

The Chairman. In the community at large.

Mr. Taylor. No, sir. If you will ask me about specific cases that you have in mind, I will tell you what happened to the men who were laid off. For instance, it may be in your mind to know what became of the 400 or 600 workmen in the yard of the Bethlehem Steel Co. that I spoke to you about and who were reduced finally to 140 men. There is a specific case.

The Chairman. In order that you may know what is running in my mind, I will say that I am not so particularly interested in any specific case as I am interested in what would be the general condition if this system was generally applied, and knowing from observation and experience the readjustment that has to take place when labor-saving machinery is introduced and knowing about the hardships that have to be borne by the workmen pending the readjustment, I wanted to find out—and that is what all this line of questioning has been leading up to—whether this scientific management has evolved any method by which the workmen could be taken care of during the period of readjustment.

Mr. Taylor. I have tried to explain that, Mr. Chairman, by saying that under scientific management we make a definite and careful study of each workman in the place; men are appointed in all of these establishments whose chief duty is to make this study of the workmen, of their possibilities and their character, and then to deliberately train each of those workmen to do that work for which he is best fitted. Under this system, then, instead of treating them brutally, they are treated as kindly as we know how. The only case that is at all usual, in which men suffer under this system, is this: there are certain men in all establishments who are lazy—one may say incorrigibly lazy. Now when such a man as that is found every effort is made to induce him to cease to be lazy and to work as he ought to work, and generally you are successful in this if you will only keep at the man long enough. I have in mind now several cases in which the worst shirkers under the old system have been finally trained men and developed into foremen, under scientific management, because under persistent, firm but kindly treatment, and with hope of advancement before them, they became such energetic men and developed such an interest in their work. But there are a few men who remain, you might say, incorrigibly lazy, and when those men are proved to be unchangeable shirkers they have to get out of the establishment in which scientific management is being introduced. Scientific management has no place for them.

Thereupon the committee adjourned to meet Tuesday, January 30, 1912, at 2 o'clock p. m.

TUESDAY, JANUARY 30, 1912

The committee met at 2 o'clock p. m., Hon. W. B. Wilson (chairman) presiding.

There were also present Representatives Redfield and Tilson.

The Chairman. Mr. Taylor, what percentage of the increased efficiency under scientific management is due to the systematizing of the work and what per cent to the speeding up of the workman?

Mr. Taylor. In the ordinary sense of "speeding up," there is no increase in efficiency due to that. Using the term "speeding up" in its technical meaning, it means getting the workmen to go faster than they properly ought to go. There is no speeding up that occurs under scientific management in this sense.

The Chairman. How much in the sense in which it has been used—that the workman is required to go faster than he normally did go prior to the introduction of the system? Using it in that sense, what percentage of the increased efficiency is due to the systematizing of work and what percentage to the speeding up of the workmen under the definition which I have given?

Mr. Taylor. That depends, Mr. Chairman, upon the workman and the extent to which the workman was sóldiering beforehand—that is, upon whether he was purposely going slow or not. As I have indicated, the amount of soldiering that takes place varies with the varying conditions, and there is no standard or uniform condition with relation to soldiering.

In some trades there is a very great deal of soldiering, in other trades there is less soldiering, so that the question can only be answered in its relation to some specific case. There is no general rule that I know of.

The Chairman. What social or economic necessity is there for speeding up the workman beyond the normal conditions under which he worked before the introduction of these scientific systems?

Mr. Taylor. Again, in its technical sense, there is no "speeding up" that occurs under scientific management. There is merely the elimination of waste movements—the elimination of soldiering, and the substitution of the very quickest, best, and easiest way of doing each thing for the older, inefficient way of doing the same thing; and this does not involve what is known as "speeding up."

The Chairman. If I recall your direct testimony, Mr. Taylor, you have stated that you found a condition of soldiering existing in the plants that you had to do with?

Mr. Taylor. Yes.

The Chairman. Does not your system propose to eliminate that soldiering?

Mr. Taylor. It certainly does.

The Chairman. Who is to determine what constitutes soldiering and what constitutes a proper amount of physical energy to be expended?

Mr. Taylor. The determination of what it is right for the man to do, of what constitutes a proper day's work, in all trades, is a matter for accurate, careful scientific investigation. It must be done by men who are earnest, honest, and impartial, and the standards which are gradually adopted by men who are undertaking this scientific investigation of every movement of every man connected with every trade establishes in time standards which are accepted both by the workmen and the management as correct.

The Chairman. Would not an employer be an interested party because he might profit or lose, as the circumstances might be?

Mr. Taylor. I can conceive that a dishonest employer or a heartless employer might very likely desire, in his ignorance of facts, to set a task which was too severe for the workman; but that man would be brought up with a round turn, because he would find that his workmen would not carry out unjust and unfair tasks; and an attempt at injustice on the part of such a man would wind up by his being a complete loser in the transaction. Therefore, the man who attempts any overdriving of that sort would simply fail.

The Chairman. The employer being a profiter by the expenditure of additional energy on the part of the workmen and not having the additional physical discomfort of the workmen to guide him in determining what constitutes a proper day's work, and what is soldiering—in what manner could the workman protect himself against an improper day's work being imposed upon him?

Mr. Taylor. By simply refusing to work at the pace set. He always has that remedy under scientific management; and as you know under scientific management he gets his regular day's pay, whether he works at the pace set or not. When he falls short of the day's work asked of him he merely fails to earn the extra premium of 30 to 100 per cent which is paid for doing the piece of work in the time set.

The Chairman. Assuming an employer having a thousand employees, and conditions being imposed upon a workman requiring him to do more work than he believes he ought to do, and his refusal to do the work because he believed it to be too much, and the other 999 men continue on at work: upon what basis of equality would the employer and employee be under a condition of that kind?

Mr. Taylor. There is no earthly reason, if it is desired by the workmen, why there should not be a joint commission of workmen and employers to set these tasks, not the slightest earthly reason. And, as I think I have told you before, Mr. Chairman, the tasks which are set in our establishment are universally set or almost universally set by men who have themselves been workmen, and in most cases those who set the daily tasks have come quite recently from doing work at their trades. They have within the last six months or a year or two years perhaps worked right at those trades. They are chosen because they are fair-minded men, competent men, and because they have the confidence both of the management and the workmen. You must remember, Mr. Chairman, in the first place, that under scientific management the workmen and the management are the best of friends, and, in the second place, that one of the greatest characteristics of scientific management—the one element that distinguishes it from the older type of management—is that all any employee working under scientific management has to do is to bring to the attention of the management the fact that he thinks that he is receiving an injustice, and an impartial and careful investigation will be made. And unless this condition of seeking to do absolute justice to the workman exists, scientific management does not exist. It is the very essence of scientific management.

The Chairman. As I understand, then, very frequently those tasks are set by men who have come fresh from the ranks?

Mr. Taylor. Yes, sir.

The Chairman. Over on the side of the management?

Mr. Taylor. Yes, sir.

The Chairman. Now, is it not true that when a man is selected by the management, as a rule, he is selected because they believe in his ability to take care of the interests of the management?

Mr. Taylor. Under scientific management because they believe in his impartiality, his straightforwardness, his truthfulness, and they believe he will have both the confidence of the management and the men, and equally forward the best interests of both sides which are mutual.

The Chairman. Then to get back to the original point stated by you—that scientific management cannot exist unless there is a complete change of mind—

Mr. Taylor. Yes, sir.

The Chairman. Now, do you conceive that it is possible to have a complete change of mind when a man is engaged in business for profit?

Mr. Taylor. I do. I say that any set of men who want to earn a big profit in any industry must have that change of mind. If they want to get a big profit, in addition to the fact that any decent man would have that view for good business, if for no other reason, they must have that view. You cannot keep men working hard on one side and not have them work equally hard on the other side. If you want a profitable business you cannot have meanness and injustice on one side or the other; you have got to eliminate meanness and injustice from both sides.

The Chairman. I believe you stated that after all the other things had been paid for, if there was a certain surplus that was left, you included in that surplus a profit for the workmen and a profit for the employer?

Mr. Taylor. Yes, sir.

The Chairman. Taking that as a basis, would there not immediately arise a contention between the employee and the employer as to what portion each should receive?

Mr. Taylor. I will say that in my experience under scientific management no such contention has arisen, because the workmen who have come under my observation, and who came under scientific management, looked upon 30 to 100 per cent increase in wages, which they were paid for performing their share of the contract, as full recompense for the work which they were doing; and I do not remember that personally I have ever had a workman seriously question the justice of that percentage. I can very well imagine that in the future, with the growth of the industrial world, with the betterment of the whole world, that those percentages may become wrong and that the workman ought to have a larger share. And, if he ought to have it, he will get it under scientific management.

The Chairman. Is it not true that the very essence of scientific management is that there must be one directing head in an establishment, and that no association of workmen can be permitted to interfere with the directions and with the policy of that directing head?

Mr. Taylor. Interfere, yes; cooperate, no. The cooperation of the workmen is asked for in every possible way in which you can get it; interference is never tolerated.

When you once get a correct standard established, when, by way of illustration, you have got your train schedule made out, and the trains are going to move, no one is allowed to interfere with the movements of those trains; but if any set of men think the schedule is wrong, that there is a better schedule, all that they have to do is to call the attention of the management to a defect in the schedule and they will correct it. And, let me tell you, Mr. Chairman, that nine-tenths of the improvements that have come under scientific management have come from this friendly cooperation on the part of the workmen with the management. Almost all of the best suggestions for improvements come from intelligent workmen who are cooperating in the kindliest way with the management to accomplish the joint result of producing a big surplus which can be divided between the two sides equitably.

The Chairman. And must not that cooperation be entirely in accordance with the judgment and direction and policy of the directing head under scientific management?

Mr. Taylor. No, sir; most emphatically no. Scientific management has developed over a period of 30 years a series of standards which are recognized by both workmen and management as being just and fair. I have tried to point out in my testimony examples of those standards, and I can point out if you wish it a thousand more—standards which are accepted as the just and fair laws of that establishment by both sides. And the president of one of these companies would no more think of interfering with those laws than the workman would.

The Chairman. In what percentage, if any, of those establishments that have come under your observation where scientific management has been introduced has collective bargaining been introduced, by which the workmen collectively become a party in determining the wages, the task, and the conditions under which they shall work?

Mr. Taylor. Under the old sense of collective bargaining, I know of no single instance in which that has been used under scientific management. That is in the old sense of collective bargaining.

In the new sense of collective bargaining it is done in every establishment in which scientific management exists. During my first day's testimony I tried to make it clear that under the old system of management a very large part of the time and thought of both those on the management side and of the workmen was devoted to securing each for its own side what it looked upon as its proper share of the surplus. I use this word "surplus" as defined by me in my first day's testimony.

Now, a manufacturer who is an unjust man (and that frequently is the case—no more frequently is the manufacturer unjust, however, than is the workman unjust) when the manufacturer is unjust toward his men, without collective bargaining under the old system of management he has the power to secure more than his fair share of this surplus. Therefore, in many establishments under the ordinary system collective bargaining has become and is in my judgment an absolute necessity.

Under the old system of management (not scientific management) the attitude assumed in nine cases out of ten by the leaders of the workmen on the one hand and by the management on the other, is that of semihostility. It is an attitude the existence of which prevents the full measure of cooperation which should exist between both sides in order to produce the largest and best results, and whenever this attitude exists collective bargaining is a necessity.

Now, the moment this attitude of hostility or semihostility between the two sides is abandoned, and the moment it becomes the object of both sides jointly to arrive at what is an equitable and just series of standards by which they will both be governed; the moment they realize that under this new type of cooperation—by joining together and pushing in the same direction instead of pulling apart—they can so enormously increase this surplus that there will be ample for both sides to divide; then collective bargaining instead of becoming a necessity becomes of trifling importance. In all establishments working under scientific management it is always understood that any single workman or any four or five or six workmen can at any time call to the attention of the management the fact that any element in the management is wrong and should be corrected, and this protest will receive immediate and proper attention. And what I want to emphasize is that the kind of attention which any protest from the men receives under scientific management is not that which is subject to the personal prejudice or to the personal judgment of the employer, but it is the type of attention which immediately starts a careful scientific investigation as to all of the facts in the case, and this investigation is pursued until results have been obtained which satisfy both sides of the justice of the conclusion. Under these circumstances, then, collective bargaining becomes a matter of trifling importance. But there is no reason on earth why there should not be a collective bargaining under scientific management just as under the older type, if the men want it.

The Chairman. If collective bargaining is satisfactory under the conditions first described by you in order to get a proper division of the surplus, because the division of that surplus affects both the employer and the employees, would it not also be just as essential that there should be collective bargaining relative to conditions under which the workmen should work, because those conditions affect both the employer and the employee.

Mr. Taylor. I should make the same answer to this question as I did to the last: that all that is necessary under true scientific management is for the attention of the management to be called to the fact that a bad condition exists to have a scientific investigation started, the results of which should be satisfactory to both sides.

The Chairman. If the satisfactory handling of

scientific management depends on the ideal condition of mind whereby the employer is willing to concede to the workmen that which each workman is entitled to, how, under the other phases of scientific management, is the workman going to be able to protect himself against imposition by any other process than that of collective bargaining?

Mr. Taylor. I think I have already stated, Mr. Chairman, that the workman has it in his power at any minute, under scientific management, to correct any injustice that may be done him in relation to his ordinary every day work by simply choosing his own pace and doing the work as he sees fit. That remedy lies open to him at any minute, and the workman will do it every time he is treated unjustly under scientific management, just as he would under any other management. In other words, injustice on the part of the employer would kill the goose that lays the golden egg.

The Chairman. Would not your suggestion of cooperation on the part of the workman with the management (the management being the sole and arbitrary judge of the issue) be very much like the lion and the lamb lying down together with the lamb inside?

Mr. Taylor. Just the opposite. The lion is proverbial of everything that is bad. The lion is proverbial of strife, arrogance—of everything that is vicious. Scientific management cannot exist in establishments with lions at the head of them. It ceases to exist when injustice knowingly exists. Injustice is typical of some other management, not of scientific management.

The Chairman. Mr. Taylor, do you believe that any system of scientific management induced by a desire for greater profit would revolutionize the minds of the employers to such an extent that they would immediately, voluntarily, and generally enforce the golden rule?

Mr. Taylor. If they had sense they would. And let me tell you, Mr. Chairman, that that is the best answer. Not immediately. I have never said that. You cannot persuade any set of men, employers or employees, to adopt the principles of scientific management immediately. I have always said that it takes a period of from two to five years to get both sides completely imbued with the principles of scientific management. And I have further said, which I wish to repeat and emphasize, that nine-tenths of the trouble comes from those on the management side in taking up and operating a new device, and only one-tenth on the workmen's side. Our difficulties are almost entirely with the management.

The Chairman. Is it not true that scientific management has been developed with a desire to cheapen the production in order that there might be greater profits?

Mr. Taylor. Mr. Chairman, in one of the books which I have written on scientific management, in paragraph 21, page 1343, in the paper-covered pamphlet entitled "Shop Management," and which is in the possession of the Chair, in large print—and I believe this is perhaps the only paragraph in that whole book written in this very large print—is emphasized this fact:

This paper is written mainly with the object of advocating high wages and a low labor cost as a foundation of the best management and of pointing out the general principles which render it possible to maintain these conditions, even under the most trying circumstances, and of indicating the various steps which the writer thinks should be taken in making a change from a poor system to the better types of management.

The Chairman. In the same book, Mr. Taylor, do you not undertake to show that high wages are brought about by taking a workman who has been employed at a lower-priced class of work and putting him at work on a portion of the work formerly performed by the high-class workman and then giving him a higher rate of wage than he had before in the lower class of work, and yet a lower rate than was actually paid to the skilled workman who performed that work prior to that time?

Mr. Taylor. I have pointed out that under the principles of scientific management, with the teaching and kindly guidance which the workmen receive from the teachers who are over them in the management—I won't say over them; who are helping them in the management—with the high standards which are placed before them and taught to them; with the better methods of doing work (which are gradually developed through the joint efforts of hundreds of men) I have pointed out that when any workman of any caliber receives this unusual training and is given these unusual opportunities, that he is thereby enabled to do a higher and a better and a more interesting and finally a more remunerative class of work than he would be able to do under the old system of management, and that when he did this higher class of work he was paid a higher day-work wage. That is, his wages were first advanced beyond the price he had received in the past, and that, in addition to this advance, he received daily a premium of from 30 to 100 per cent for carrying out the instructions which are daily given to him.

And this applies not only to those workmen who do the cheaper kinds of work, but to all workmen high and low. For example, a man who under the old system of management has only sufficient brains to sweep the floor, under scientific management is taught and trained and helped so that he finally learns how to use, say, a grinding machine or to do some of the more elementary kinds of machine work. He is taught to do a class of work which is far more interesting and requires more brains than the sweeping to which he was formerly limited. And he is then given the higher wages and the interesting conditions and surroundings which accompany this higher class of work. At the same time the man who was under the old system on the grinder is taught to do some of the simpler kinds of "high-class machine work." Of course you understand I am speaking now of types of men who under the old system were limited by their mental capacity to simple work such as running a grinder; I am not speaking of the exceptional man who was born with plenty of brains to do high-class work, but who did not have the good fortune to learn a trade when he was young; but I am speaking of the man whose mental caliber would naturally limit him to sweeping the floor or running a grinder. Now, to continue the illustration, the drill-press hand, for instance, by this same teaching and training, is enabled to do the work of the lathe hand, and the lathe hand is enabled to do the work of the high-priced tool maker or a man of that mental caliber.

You understand I am not speaking literally; I am speaking by way of example. And finally the tool maker becomes one of the teachers to show the men lower down all along the line how to do their work— to show them and teach them and guide them in their work. Now, this upward movement of all the men is not confined to any one class; it applies to all types of workmen. They all rise to a better class of work and to higher pay under scientific management.

The Chairman. Take the illustration, for instance, of a man of the mental caliber of a common laborer and who is employed as a common laborer. What were the rates paid, say, at Midvale, under scientific management to the common laborer as compared with the wages paid to the common laborer under the ordinary management by the United States Steel Corporation at Pittsburgh?

Mr. Taylor. The wages of common laborers when I was at the Midvale Steel Works (and I left there in 1889) ranged from $1.20 per day to $2.70 per day, with piecework added.

The Chairman. From $1.20 to $2.70 per day?

Mr. Taylor. Yes. In other words, under scientific management there is no standard or uniform rate of pay for laborers, nor for any other group or class of men. And I want to emphasize this fact, Mr. Chairman, which does not seem to be at all recognized by the world at large, that workmen differ just as much as horses differ. Now, we all know that there is a vast difference in horses. I do not mean anything degrading to the workman by this comparison, but I dare say some one will say that I am comparing workmen to beasts. We all know that horses differ, and yet very few people seem to recognize that there is an even greater difference between different members of the human species. There is just as much difference between laborers as there is between horses. I think I can say with truthfulness that the laborers to whom we paid $2.75 a day at the Midvale Steel Works quite as fully earned their high wages as did the cheaper men who were only paid $1.20 per day.

The Chairman. This man at $2.70 a day, how many hours does he have to work?

Mr. Taylor. Ten hours.

The Chairman. Is that the usual time of work?

Mr. Taylor. Yes, sir; 10 hours per day, with the exception of certain departments of the plant, in which it is impossible to shut the apparatus down. For instance, the open-hearth furnace department. As we all know, it is as impossible to shut down an open-hearth furnace as it is to stop the sun from setting. It takes a week to shut down an open-hearth furnace. So that particular department in our works (and if I remember rightly it was the only department in the Midvale Steel Works that ran right straight through the year) the open-hearth furnace, ran and always will have to run, right straight through, night and day, although the work was so arranged that it was rarely necessary to pour a heat on Sundays, so that the smallest possible number of men were kept at work in the department on Sunday. Now, in this department there were two 12-hour shifts at work. I say 12 hours because there were practically two shifts of 12 hours each to run these furnaces. And I can say, that for the whole time that I was at the steel works, it was a matter of the very gravest concern to all of the managers that there seemed to be no way of doing away with the 12-hour shifts under scientific management. But it was made easier in this way—that is, this practice was made justifiable to a certain extent in this way—that the task of the men running that—that the tasks which were given to the men who worked on 12-hour shifts were made lighter than the tasks given to the men running on 10-hour shifts. But that does not make the necessity for these long hours of work any the less unfortunate. And I used to regret this necessity the whole time I was at this works; it was a matter of great concern. Time and again we consulted as to the possibility of introducing 8-hour shifts in the place of 12 hour shifts, and since I left there I understand that this has been tried, and that the workmen themselves seriously objected to it, and preferred to go back to the old 12-hour shift. This is merely hearsay, however, what other people have told me, and therefore is not given as of my own knowledge. But I understand the workmen themselves said that when they boarded in houses with other people and had to have different mealtimes and sleeping hours, working partly in the daytime and partly at night, so that they had to have their meals in the middle of the afternoon or middle of the night (when no one else took their meals), they looked upon it as a hardship, and my impression is that the eight-hour shift, after being tried, was abandoned. On that point I am not sure, however, Mr. Chairman.

The Chairman. How do those conditions compare with the conditions existing at the same time at the United States Steel plant?

Mr. Taylor. The conditions in many of the plants of the United States Steel Co. are and always have been deplorable—deplorable to the greatest extent. Now, I do not wish to be understood as criticizing the managers of these steel works. I think a great many of the men in that business recognize the very deplorable state of things that exists there; and certainly there are now deplorable, if not shameful, conditions existing in the steel business. I say this most heartily. As far as possible, that sort of conditions would not be tolerated under the principles of scientific management. I have heard of many cases where year in and year out men have worked with almost no vacation and very little lay off, and that is inhuman; it is impossible.

The Chairman. You consider it to be one of the essential features of scientific management that a time study must be made with a timepiece, such as a stop watch, in order to determine the length of time that a piece of work can be done in, to hereby give a knowledge of it.

Mr. Taylor. I know of no other way of determining how fast work ought to be done than by timing the workman, Mr. Chairman. As long as time remains one of the most important elements (and in the past most of the disputes between employer and employee have been connected with the question of how long it should take to do the work), I fail to see how you are to know anything about time without timing. I know of no way of getting any accurate knowledge in this field except by watching a man who is doing the work at the proper speed and recording his time. The old way of guessing as to how fast a man ought to do a thing (and that is the way I did, as I explained to you, when I was a foreman under the old system of management) is most unsatisfactory as to both sides. This old-fashioned guesswork is quite as unsatisfactory to the workmen as to those on the management's side.

The Chairman. Under your system, when you have made a time study with a stop watch, do you then take the exact time that you have found by the stop watch and say that is the time in which the work must be done?

Mr. Taylor. No, sir; never. We first take a good man, not a poor man—we always try to take a man well suited to his work. We then assure ourselves that that man is working at a proper rate of speed; that is, that he is not soldiering on the one hand, and that on the other hand, he is not going at a speed which he cannot keep up year in and year out without undue exertion. We then determine as accurately as we know how the proper speed for doing the work, by timing the man with a watch, and having determined that, then we add a marginal percentage of time to cover unavoidable delays and accidents, and, in many cases, we make an extra allowance when the workman who is called upon to do this particular job is not especially skilled at it.

For illustration, Mr. Chairman, to show you what I mean by this marginal allowance, suppose you were asked in a shop to turn axles for a standard railway car. This is a piece of work which as you know is done by the thousand, and done year in and year out; and now that the railway master mechanics of the country have established a standard car axle, the conditions have become uniform for doing this piece of work. We will assume that a company is going into the manufacture of these axles as a regular business, and that they propose having men working on these axles year in and year out. The time study would be made first to determine the quickest time in which the axle ought to be machined. By the quickest time—I do not mean any improper time— but the quickest proper time in which that work could be done by the workmen if they did not have the slightest interruption or delay or anything of that sort. And after having determined this time, then 20 to 27 per cent of that time is added to cover unavoidable delays and all such accidents as may happen to a workman. That 20 to 27 per cent has been found, from long experience, to give the workman plenty of time to overcome those little unavoidable delays and interruptions which interfere with his work. This allowance has been generally accepted by the workmen as correct, and I have never heard this allowance disputed as incorrect.

If you were to take that same axle, for instance, where only 10 or even 100 axles were to be turned in a shop, you would in this case have to allow as much as 70 per cent additional time to the man. This is because you cannot expect a workman to go right at a job which is new to him and do everything just right and at the same speed which he could readily maintain after having more practice.

In some other classes of work it has been my habit to add as much as 225 per cent to the time in cases similar to the one I have described. I think that is the highest per cent that we have been accustomed to add to the "quickest reasonable time" in which the work might be done.

The Chairman. By what scientific formula or mathematical calculation did you arrive at an addition of 20 to 27 per cent to the time which you have determined by that stop watch?

Mr. Taylor. We have done that through a very careful study—and this study has been repeated over and over and over again—of workmen well suited to their particular jobs. They were told, "Now, men, we want to arrive at a proper allowance for unavoidable accidents and delays, and I want you to cooperate with me." This is the way we talk to the workmen when we propose to make a time study in ninety-nine cases out of a hundred—"I want you to cooperate with me in arriving at the truth regarding this fact. Now, go right ahead and do the work as it ought to be done. I want to know what time it will take, first, to turn the axle, and then I want to see what is the proper allowance to make for unavoidable accidents and delays." We would then watch and time that man, not for one axle alone, but frequently for days at a time, until finally we would both agree as to what was the proper time. During this time we would watch, of course, carefully to see whether he had not perhaps forgotten something—had not slipped off the track and was making some unnecessary motions and then as a result of this careful joint study between the workman and the management the proper percentage allowances are accurately determined. You see that it is joint, because both sides cooperate; we have one man who is watching and records the time, and the other man who works, and both are in entire accord and working for the same object, so that it is a joint affair. That is typical of the way we arrive at all percentages.

The Chairman. Is not that 20 to 27 per cent arbitrarily arrived at by the judgment of a person watching the operation, of the time that should be added?

Mr. Taylor. No, sir; not the arbitrary judgment of anyone. An arbitrary judgment would be something that a man guessed at. But this is a scientific investigation, a careful, thorough scientific investigation of the facts. It is based on the fact that in perhaps as many as 20 cases, with different men on this general type of work, this figure has been proved to be correct. This is not founded on any one judgment; it is based on facts.

The Chairman. Is it not true that under the old

system, in determining the length of time that it would take to produce a certain piece of work, that it was based upon the observations of some man relative to that work over a long period of time, and would not that be just as scientific and just as arbitrary as the method employed in securing this 20 to 27 per cent?

Mr. Taylor. No, sir. I suppose, as I walk along the street, for example, I could in a general way look at a trolley car and say it is going at the rate of 8 miles an hour, or 10 miles an hour; but that kind of arbitrary judgment would not compare in accuracy with timing the car with a watch. Watching horses when they are trotting by and guessing at their speed would not be anything like as trustworthy as that kind of observation which comes from the use of a stop watch. The one is guesswork, while the other is a careful scientific experiment.

For instance, when I was a foreman, as I told you, the workmen knew ten times as much as I did about how long it took to do work. Their knowledge was exact, because they looked at the time when they started a job and at the time when they stopped and knew exactly how long it had taken them. My knowledge was casual; I had in a general, hazy way, an idea that a job ought to take such and such a time; but I have seen myself judge from 300 to 400 per cent wrong, and I think that is true of all foremen.

The Chairman. Isn't it part of the scientific management, or the Taylor system, to bring all of the power of the management to bear on the individual in order to compel the individual to carry out the policy of the management?

Mr. Taylor. With the first man whom you tackle in a shop and want to teach and bring from the old method of doing the work to the new method, as a rule, I think you can say that you do bring heavy pressure to bear on the man. You are very apt to put three or four teachers around him at once to see that he does not skip out from under anywhere. You understand, of course, that is true of the first man. Under scientific management our procedure is to get one man working under the new conditions and at the proper pace, and then let him go right on earning his premium of 30 per cent to 100 per cent until he wants the new system badly. And invariably some friend of his—generally not one friend only, but a dozen of them—will come and ask for the same thing. When the men see a friend of theirs, right alongside of them, working practically no harder than they are working, but merely obeying certain instructions and directions given him and thereby becoming more efficient and doing the work quicker—when they see that man getting 30 to 100 per cent higher wages than they are getting, they want some of that velvet. The other men throughout the shop themselves come and ask for the new system. When scientific management is properly introduced, almost invariably we wait for the men to come and ask to work under the new plan.

The Chairman. When the power of the management is brought to bear on the individual workman, while time study is being made, would not the time study itself be inaccurate because of the abnormal conditions created by that power being brought to bear on the individual workman?

Mr. Taylor. Mr. Chairman, I have said before that in nine hundred and ninety-nine cases out of a thousand it has been our practice to have the workman cooperate with us in the most friendly manner in making this time study. The workman is just as much a part of this time study, and a voluntary part of this time study, as we are a part of the time study. I say "we," meaning those of us who are on the management side. An effort is first made to get a workman to realize that this is the road toward high wages. And when he realizes that and knows that we must have a time study as a just and substantial foundation for both sides he is not opposed to time study, but consents to it with the greatest alacrity. We have had hundreds of men come and ask us to make a time study of their particular jobs.

The Chairman. Is it not true, under those circumstances, that a failure to cooperate means that his ability to earn a livelihood has been completely destroyed, or cut off to the extent of 100 per cent, while he realizes at the same time that his employer's earning ability is not altered; that a disagreement might continue as far as the employer is concerned, while it would mean starvation to him?

Mr. Taylor. I must say, Mr. Chairman, that I do not exactly catch your meaning; I do not think I understand you.

The Chairman. I will give you an illustration. Suppose, as I suggested to you some time ago, that there is an employer with 1,000 employees, and he deals with them individually, as this method proposes. The conditions are not satisfactory to the workmen. They are to the employer. The conditions made by the employer are satisfactory to him, but if the workman refuses to accept the unsatisfactory conditions his power to provide for himself and his family has been destroyed to the extent of 100 per cent; but the 999 of the employees continuing at work, the power of the employer to earn a profit has practically not been reduced at all. Now, you have on the one side the employee with no employment to earn a livelihood to live upon and starvation staring him in the face thereby, and on the other hand the employer continuing to produce the same profit that he formerly produced. Now, would not the disagreement under those circumstances simply result in the necessities of the workman ultimately compelling him to accept the terms of the employer?

Mr. Taylor. Mr. Chairman, my observation is that in very dull times, when there is a lack of employment for good men in trades—those times come occasionally—at that time an unscrupulous employer might have an advantage. The unscrupulous employer, under those conditions, might have a very distinct advantage over the workmen. My observation, however, of the ordinary normal times in the United States is that a good workman need never be be out of employment for five days. There is an immense demand for competent workmen in this country, in all normal times. I cannot recall in normal times a single instance of a good workman having to come anywhere near starvation because of lack of employment. There is always an immense demand for good workmen, so that the condition does not exist which you have outlined.

The Chairman. Is it not true that a man who is not a good workman and who may not be responsible for the fact that he is not a good workman, has to live as well as the man who is a good workman?

Mr. Taylor. Not as well as the other workman; otherwise, that would imply that all those in the world were entitled to live equally well whether they worked or whether they were idle, and that certainly is not the case. Not as well.

The Chairman. Under scientific management, then, you propose that because a man is not in the first class as a workman that there is no place in the world for him—if he is not in the first class in some particular line that he must be destroyed and removed?

Mr. Taylor. Mr. Chairman, would it not be well for me to describe what I mean by a "first-class" workman. I have written a good deal about "first-class" workmen in my books, and I find there is quite a general misapprehension as to the use of that term "first-class."

The Chairman. Before you come to a definition of what you consider a first-class workman I would like to have your concept of how you are going to take care, under your scientific management, of a man who is not a first-class workman in some particular line?

Mr. Taylor. I cannot answer that question until

I define what I mean by "first-class." You and I may have a totally different idea as to the meaning of these words, and therefore I suggest that you allow me to state what I mean.

The Chairman. The very fact that you specify "first-class" would indicate that in your mind you would have some other class than "first-class."

Mr. Taylor. If you will allow me to define it I think I can make it clear.

The Chairman. You said a "first-class" workman can be taken care of under normal conditions. That is what you have already said. Now, the other class that is in your mind, other than "first-class," how does your system propose to take care of them?

Mr. Taylor. Mr. Chairman, I cannot answer that question. I cannot answer any question relating to "first-class" workmen until you know my definition of that term, because I have used these words technically throughout my paper, and I am not willing to answer a question you put about "first-class" workmen with the assumption that my answer applies to all I have said in my book.

The Chairman. You yourself injected the term "first-class" by saying that you did not know of a condition in normal times when a "first-class" workman could not find employment.

Mr. Taylor. I do not think I used that term "first-class."

Mr. Redfield. Mr. Chairman, the witness has now four times, I think, said that until he is allowed to define what he means by "first-class" no answer can be given. because he means one thing by the words "first-class" and he thinks that you mean another thing.

The Chairman. My question has nothing whatever to do with the definition of the words "first-class." It has to do with the other class than "first-class," not with "first-class." A definition of "first-class" will in no manner contribute to a proper reply to my question, because I am not asking about "first-class," but the other than "first-class" workmen.

Mr. Taylor. I cannot describe the others until I have described what I mean by "first-class."

Mr. Redfield. As I was saying when I was interrupted, the witness has stated that he cannot answer the question for the reason that the language that the chairman uses, namely, the words "first-class" do not mean the same thing in the chairman's mind that they mean in the witness's mind, and he asks the privilege of defining what they do mean, so that the language shall be mutually intelligible. Now, it seems to me, and I think it is good law and entirely proper, that the witness ought to be permitted to define his meaning and then if, after his definition is made, there is any misunderstanding, we can proceed.

The Chairman. It seems to me, Mr. Redfield, that having said a "first-class" workman could be taken care of under normal conditions, it was perfectly proper for me to ask the question of how to take care of those who are not "first-class" workmen under scientific management, and that a reply to a question of that kind does not involve the necessity of defining what is "first-class."

Mr. Tilson. It seems to me, Mr. Chairman, that

you are entirely in error, because the very term you are asking him to describe is described by negative words, including the words "first-class;" that is, not a "first-class" workman, but workmen other than "first-class." Therefore, in order to get at the other class, it seems to me not only improper, but if he means something else by the words "first-class" than you mean, it seems to me it would be very necessary for him to describe what "first-class" is, so that you could get at the negative of that and know what to subtract from the sum total. If you want to know what is not "first-class," you ought to know what is "first-class" so that you would know what to subtract.

Mr. Taylor. Mr. Chairman, I want to assure you that I am not quibbling. Not for an instant am I quibbling; and if you will allow me to proceed with the definition, I think you will see that it is a matter of great importance, because I have used the words "first-class" throughout my book.

And I wish to say, Mr. Chairman, that both of these books were written to be presented to the American Society of Mechanical Engineers. I had that in view, both in writing the book on Shop Management and the Principles of Scientific Management.

Now, the American Society of Mechanical Engineers is perhaps the most rigid society in this country in insisting on conciseness in writing—in insisting on having what is to be presented to them placed in the fewest possible words, and this book on Shop Management has received no end of criticism from the members of the Society of Mechanical Engineers, because from their standpoint it was too verbose; yet in the original form in which I wrote this book it was three times as voluminous as it now is, and in my endeavor to make it sufficiently concise for acceptance by the society, I was compelled to omit definitions of words and of expressions which were important to a proper understanding of the book. And among the expressions which for this reason have not been properly defined are the words "first-class men." My other book, which is in the hands of your committee, "The Principles of Scientific Management," much more nearly expresses my exact views, because in this book I absolutely refused to make it so concise as to emasculate its meaning, and for this reason, although the society held this manuscript for a year and asked me again and again to condense it, they finally refused to publish it.

I have found that an illustration often furnishes the most convincing form of definition. I want therefore to define what I mean by the words "first-class" through an illustration. To do so I am going to again use "horses" as an illustration, because every one of us knows a good deal about the capacity of horses, while there are very few people who have made a sufficient study of men to have the same kind of knowledge about men that we all have about horses. Now, if you have a stable, say, in the city of Washington, containing 300 or 400 horses, you will have in that stable a certain number of horses which are intended especially for hauling coal wagons. You will have a certain number of other horses intended especially to haul grocery wagons; you will have a certain number of trotting horses; a certain number of saddle horses—of pleasure horses, and of ponies in that stable.

Now, what I mean by a "first-class" horse to haul a coal wagon is something very simple and plain. We will all agree that a good, big dray horse is a "first-class" horse to haul a coal wagon (a horse, for instance, of the type of a Percheron). If, however, you live in a small town and have a small stable of horses, in many cases you may not have enough dray horses in your stable to haul your coal wagons, and you will have to use grocery-wagon horses and grocery wagons to haul your coal in; and yet we all know that a grocery-wagon horse is not a "first-class" horse for hauling coal, and we all know that a grocery wagon is not a first-class wagon to carry coal in; but times come when we have to use a second-class horse and wagon, although we know that there is something better. It may be necessary even at times to haul coal with a trotting horse, and you may have to put your coal in a buggy under certain circumstances. But we all know that a trotting horse or a grocery horse is not a "first-class" horse for hauling coal. In the same way we know that a great big dray horse is not a "first-class" horse for hauling a grocery wagon, nor is a grocery-wagon horse first class for hauling a buggy, and so on, right down the line.

Now, what I mean by "first-class" men is set before you by what I mean by "first-class" horses. I mean that there are big powerful men suited to heavy work, just as dray horses are suited to the coal wagon, and I would not use a man who would be "first-class" for this heavy work to do light work for which he would be second-class, and which could be just as well done by a boy who is first class for this work, and vice versa.

What I want to make clear is that each type of man is "first-class" at some kind of work, and if you will hunt far enough you will find some kind of work that is especially suited to him. But if you insist, as some people in the community are insisting (to use the illustration of horses again), that a task—say, a load of coal—shall be made so light that a pony can haul it, then you are doing a fool thing, for you are substituting a second-class animal (or man) to do work which manifestly should be done by a "first-class" animal (or man). And that is what I mean by the term "first-class man."

Now, there is another kind of "second-class" horse. We all know him. Among the "first-class" big dray horses that are hauling coal wagons you will find a few of them that will balk, a few of them that can haul, but won't haul. You will find a few of these dray horses that are so absolutely lazy that they won't haul a coal wagon. And in the same way among every class of workmen we have some balky workmen—I do not mean men who are unable to do the work, but men who, physically well able to work, are simply lazy, and who through no amount of teaching and instructing and through no amount of kindly treatment, can be brought into the "first-class." That is the man whom I call "second-class." They have the physical possibility of being "first-class," but they obstinately refuse to do so.

Now, Mr. Chairman, I am ready to answer your question, having clearly in mind that I have these two types of "second-class" men in view; the one which is physically able to do the work, but who refuses to do it—and the other who is not physically or mentally fitted to do that particular kind of work, or who has not the mental caliber for this particular job. These are the two types of "second-class men."

The Chairman. Then, how does scientific management propose to take care of men who are not "first-class" men in any particular line of work?

Mr. Taylor. I give it up.

The Chairman. Scientific management has no place for such men?

Mr. Taylor. Scientific management has no place for a bird that can sing and won't sing.

The Chairman. I am not speaking about birds at all.

Mr. Taylor. No man who can work and won't work has any place under scientific management.

The Chairman. It is not a question of a man who can work and won't work; it is a question of a man who is not a "first-class" man in any one particular line, according to your own definition.

Mr. Taylor. I do not know of any such line of work. For each man some line can be found in which he is first class. There is work for each type of man, just, as for instance, there is work for the dray horse and work for the trotting horse, and each of these types is "first-class" for his particular kind of work. There is no one kind of work, however, that suits all types of men.

The Chairman. We are not in this particular investigation dealing with horses nor singing birds, but we are dealing with men who are a part of society and for whose benefit society is organized; and what I wanted to get at is whether or not your scientific management had any place whatever for a man who was not able to meet your own definition of what constitutes a "first-class" workman.

Mr. Taylor. Exactly. There is no place for a man who can work and won't work.

The Chairman. It is not a question of a man who can work and won't work; it is a question of a man who doesn't meet your definition of "first-class" workmen. What place have you for such men?

Mr. Taylor. I believe the only man who does not come under "first-class" as I have defined it, is the man who can work and won't work. I have tried to make it clear that for each type of workman some job can be found at which he is "first-class," with the exception of those men who are perfectly well able to do the job, but won't do it.

The Chairman. Do you mean to tell the committee that society is so well balanced that it just provides the proper number of individuals who are well fitted to a particular line of work to furnish society with the products of that line of work?

Mr. Taylor. Certainly not, Mr. Chairman. There is not a fine balance in society. It is sometimes difficult to find jobs right near home for which men are well suited, that is, for which they are "first-class." There is an immense shortage of men, however, who are needed to do the higher classes of work.

There always has been and always will be, an immense shortage near the top. It is not so great down below, but at the top there is an immense shortage of "first-class" men, so that there is plenty of room for men to move up.

The Chairman. If society does not produce an equal balance in all the lines of production of "first-class" men, must there not of necessity be some men who are not "first-class" in any particular line of work where they can secure employment?

Mr. Taylor. I do not think there is any man, as far as I know, who is physically fitted for work, who in this country has to go without work in ordinary times. I do not know of this case except in very dull times.

The Chairman. Is it not true and generally recognized by statisticians, that there are at all times from 1,000,000 to 4,000,000 workmen in the United States who are willing to work but unable to secure it?

Mr. Taylor. I do not believe that is true in busy times at all. There are many times, however, in which men cannot secure the exact work which they want right close to where they live.

The Chairman. Is it not true in times generally?

Mr. Taylor. I am not familiar with the statistics; it is merely an impression on my part, and from the difficulty I have had personally in getting men I should say that it was not true. I can point to a company right now, in Connecticut, the owner of which told me that all through these dull times he had had employment for 25 per cent more people than he could get.

The Chairman. This 25 per cent would be people well suited to that particular line of work, I take it?

Mr. Taylor. It is the American Pin Company. I only went through there once, and I do not know the type of the men that he wanted well enough to judge what was in his mind, but that was his difficulty.

The Chairman. Is it not true that today there is a shortage of men, and that there frequently is a shortage of men for the higher skilled trades, while at the same time men who have not acquired that skill are unable to find employment?

Mr. Taylor. I think there is a shortage of men for the very high classes of work in the dullest of dull times, but not that same shortage of men in the very elementary kinds of work, in dull times. I think that is right, Mr. Chairman. I think that I catch your point, Mr. Chairman—that working people frequently suffer because they are unable to find the particular kind of work that they want and I agree with you in this. We who are engaged in creative industries—the industries in which you and I have worked during our lives—fail to realize the fact that those men who are in creative industries are a small minority of the whole community. Perhaps 17 per cent (I think I am right) of the people of the country are in what may be called creative industries.

Now, there is a very large outside field of work for people to go into, and in this outside field it is an undoubted fact that the selection of workmen and that the training of workmen is not nearly as accurate as it is in the industrial field. You will realize that in domestic employment and in the farm work, and in the ordinary work of sweeping the streets of the cities, for instance, the ordinary work that goes on largely in an isolated way all over the country— that the same careful selection of workmen is not made as occurs in the industrial field. The same study of workmen is not made in those occupations as in the trades at which you and I have worked.

Now, when dull times come, in some one or more of the creative industries, and men who have learned a trade are thereby temporarily thrown out of work, there is no doubt that these men suffer hardship. They are very loathe to work at anything else than their trade and many of them will suffer a good deal before they turn to employment in the great field that I have spoken of, which is outside of the creative industries. In some part of this field, there is practically at all times a demand for men which is not supplied, but this demand is often at a distance from the man who is out of work, and the man out of a job does not know of its existence. In making this readjustment there is undoubtedly suffering.

There is the other class of men whom I have spoken of who suffers (and I think properly suffers), namely, the man who can work but refuses to do a proper day's work.

If I gather rightly you have in mind both of these classes of men. Sooner or later this second class of man who can work but deliberately refuses to do what the world recognizes as a fair day's work (the man of the type of the great big dray horse who refuses to haul anything heavier than a grocery wagon, for illustration), that type of man sooner or later drifts out into that class of work in which his daily task is not accurately measured by the men around him; in which the difference between the "first-class" and "second-class" man is not accurately defined.

The Chairman. You have a wrong concept of what is running in my mind, and I want to set you right. What is in my mind is this, that neither an employer nor any other man has a right to determine arbitrarily how much physical exertion shall constitute a day's work for a workman. That that is a matter that if determined at all by anyone else than the workman involved, shall be determined between all his associates collectively and the employer for whom he works, and that it should not be arbitrarily determined by his employer, notwithstanding the great change of mind that the employer undergoes by virtue of having introduced scientific management.

Mr. Taylor. My understanding is then, Mr. Chairman, that you believe that even under scientific management collective bargaining or the principles of collective bargaining should apply. I am not at all prepared to say that you are not right, I have not the slightest objection, and never have had to collective bargaining, but I merely say that under the principles of scientific management that necessity has never come before me. The workmen have the same sort of freedom and they have just the same opportunity, to enter into every experiment which is made in establishing what constitutes a fair day's work, that the management have. The making of joint experiments (the workmen and management cooperating together) has been universal in scientific management, or practically universal, and the results have been satisfactory to both sides. I wish to emphasize the fact that until results of these experiments are satisfactory to both sides, scientific management does not exist. This is indispensable—that the results of this accurate study (and this accurate study to replace the old rule-of-thumb judgment is one of the essential features of scientific management), whether this study be made by one man or twenty—that the results must be satisfactory to both sides is absolutely indispensable.

Mr. Tilson. Do you believe generally with Gen. Crozier that you would not be in favor of attempting to apply scientific management to any shop without the cooperation of the employers and the employees?

Mr. Taylor. I certainly do. Never would I believe in applying scientific management unless it was thoroughly agreeable to both sides.

Mr. Tilson. And unless it worked satisfactorily to both sides, you would be in favor of abolishing it?

Mr. Taylor. I certainly would be every time. The principles of scientific management must rest upon justice to both sides, and it is not scientific management until both sides are satisfied and happy.

The Chairman. Would that satisfaction be expressed by the men collectively, or would it be individual after all the power of the management has been brought to bear on the individual?

Mr. Taylor. I do not care which way it is expressed. I have tried to explain that up to now that matter of collective bargaining has never come before me; that we have always been ready to consider any protest, whether made by one man, five men, or twenty men. If any man or any set of men, under scientific management, come with a protest, it is always received and would be accorded just as much attention and as much consideration as if 400 men came.

Mr. Tilson. That is, you would receive one man in an establishment if he came, or you would receive all en masse—if all the men interested in the establishment should come to you?

Mr. Taylor. Absolutely.

Mr. Tilson. Or a committee representing all came to you?

Mr. Taylor. Why, certainly.

The Chairman. Is that principle used now under scientific management?

Mr. Taylor. So far as I know. I never heard of anything else. Mind you, if you refer to having a committee from a union coming to bargain, or present a kick, I have never had that thing happen under scientific management, because the men are perfectly free to come themselves at any time. I think that is the reason for it. I have never had any objection to any one presenting any protest against what seemed an injustice or making any suggestion for an improvement.

Do not understand for a minute, Mr. Chairman, that I am opposed to trade unions. You have never heard me say that, and no one has heard me say it. I am in favor of them. They have done a great amount of good in this country and in

England; I am heartily in favor of those elements of trade unions which are good, and I am equally opposed to those elements of trade unions which are bad; and they have bad elements just as they have good. Now, the things that constitute the bad elements in trade unions I tried to point out in my direct testimony. I believe that the unions are controlled and misguided in a few respects by leaders who simply lack education; they lack a knowledge of some of the vital facts. One of the worst principles of trade unions, as they are taught by the leaders of the unions (I believe that the leaders are misguided; I do not think they are dishonest) is that it is to their interest to deliberately, purposely work slow instead of working fast with the object of restricting output. It is this deliberate restriction of output that has already done the great harm in England and that is doing most of the harm that the unions are doing in this country. High wages are not doing any harm; I favor even higher wages than the unions do. Short hours are not a bad thing; I believe in short hours. I believe in almost all the things the trade unions do; but restriction of output, never! That is the thing fatal to their own best interests that they are now doing.

The Chairman. What trade unionist, prominent or otherwise, have you ever heard express an opinion in opposition to increased production if the increased production was not brought about by increased energy expended on the part of the workmen?

Mr. Taylor. Now, Mr. Chairman, I do not know

of a single labor leader that is not advocating restricted output among his men; not a single one.

The Chairman. Can you name one who has advocated restriction of output or who is opposing increased output except where the increased output is brought about by an increased expenditure of energy on the part of the workmen?

Mr. Taylor. Well, I should say that it would take a little more energy for a plumber to make three wiped joints or four wiped joints a day than for him to make two, surely. The plumbers' union restricts a plumber to three wiped joints a day. I am not a plumber, but I'll be damned if I can't wipe five joints a day, and no trouble at all. Of course, it takes more trouble to do four than three wiped joints. But what I mean to say is that when the plumbers' union restricts a plumber to three wiped joints a day and insists that one or two helpers shall always go along, whether they are needed or not, that union is restricting the output per man. If you quibble about it (I am not talking about you personally, Mr. Chairman; I am using the word impersonally; I would not for the world say that you quibble).

Mr. Chairman. That is all right; I presume I can stand it as well as the other fellow it was intended for.

Mr. Taylor. I do not mean to say that you have quibbled for a moment, and, on the contrary, I want to thank you for the most considerate treatment I have had from you ever since these hearings began.

The Chairman. I am going to ask you at this time again, Mr. Taylor, what special necessity or economic necessity is there to increase production by virtue of the expenditure of increased energy on the part of the workmen from that which existed prior to the introduction of this system?

Mr. Taylor. There is the economic necessity that the whole world is now, just as it always has been, suffering from underproduction. Underproduction is responsible mainly for low wages; it is responsible for the fact that the poorer people of this world have just so much fewer things to live on (that they have poorer food to eat; pay higher prices for their rents; can buy fewer clothes to wear than they ought to have; in other words, that they lack what I have defined in my direct testimony as true riches); the fact that the poorer people lack in many cases the necessities, and in all cases the luxuries of life which they ought to have, is a justification for the fact that an increase of output is needed now just as much as it always has been, because absolutely the only way that these necessities and luxuries can be brought into the world is through an increase in output. Now, as I pointed out in my direct testimony, and as an analysis of the testimony presented to this committee will show, a great part of the industrial world is deliberately soldiering. And until we have reached the point where deliberate soldiering has been stopped; and until the normal and proper output per man has been reached, no workman will be asked to work materially harder than he is now working. And, as you know, scientific management is a scheme for greatly increasing the output of the man without materially increasing his effort.

The Chairman. Is it not true, Mr. Taylor, that the great bulk of the poverty of workmen at the present time is due not to the fact that we have not solved the problem of production, but to the fact that we have not solved the problem of distribution of that which is produced?

Mr. Taylor. Mr. Chairman, I agree with you that there is an immense reform needed in the distribution; I agree heartily in that; and I am also firmly of the opinion that in the next hundred years the wealth of the world is going to grow per capita (the real wealth of the world, as I have already defined it, not money nor useless extravagances, but those things which are really useful to men) to such an extent that the workman of that day will live as well, almost, as the high-class business man lives now, as far as the necessities of life and most of the luxuries of life are concerned. If you will look into the past you will see that our laborers of today have made fully as great progress as this with relation to the laborers of the past. A most striking illustration of the way in which the workman has progressed is presented by the following fact, Mr. Chairman. I do not think that it is a fact of very common knowledge, and it therefore may be a proper fact to get into this record, the standard by which we ordinarily measure the relation of men living in one period to those living in another period is the money standard.

It is a most unreliable and unsatisfactory standard, that 50 years ago such and such wages were paid, and now such and such wages are paid. This fact alone is almost meaningless. But there is one standard by which you can go back for a long term of years and by which you can compare the condition of workmen at that time with their present condition. I think it was 250 years ago (the exact number of years I do not know; it makes very little difference—it was from 150 to 300 years ago) the farm laborer of England sold his week's work for half a bushel of wheat. We eat wheat; that is, we eat bread now, just as they did 250 years ago, not much more nor much less per man, and a measure in wheat of what a man got then and gets now for his day's work is therefore a standard measure of the living condition of 250 years ago and now. Think of it! A half bushel of wheat for a week's work was the pay of a man then!

The Chairman. Would that be an accurate measure of comparison in view of the conditions of the cost of production—the labor cost of producing wheat now as compared with then?

Mr. Taylor. It is not what the labor cost. It is a question of how much riches were coming into the world and available for use then, and how much now; and the riches then coming into the world were measured then by the amount that the land produced per man and the productivity of the average man, just as riches are now measured, and the fact that the average man is 20 times as rich now as then—he is turning out 20 times the output of a man of 250 years ago. And the average man of 100 years from now will, I firmly believe, turn out at least three times as much work as now.

The Chairman. Notwithstanding the fact that scientific management is only 30 years old, the productivity has been increased twenty-fold during that period of time?

Mr. Taylor. No. I am taking the period of time 250 years ago (not of 30 years ago) when a man sold his week's labor for half a bushel of wheat, as the measure of a man's productivity.

The Chairman. The measure is 20 times greater now than it was 250 years ago?

Mr. Taylor. I think in that measure. I should say that in round numbers it would be nearly that.

The Chairman. And having increased productivity 20 times (we are producing twenty-fold now) would it not naturally follow that if poverty exists now, with twenty times more productivity, it is due, not to the fact that we have not solved the problem of production, but to the fact that you have not solved the problem of distribution?

Mr. Taylor. It is due to both of these facts, Mr. Chairman, but due mainly to the fact that what is now ranked as extreme poverty were the normal conditions of nine out of ten men 250 years ago. The standard of living has changed fortunately, so that what was then affluence is now poverty.

The Chairman. The other day, Mr. Taylor, you made the statement that the mechanism of scientific management was a power for good and a power for bad.

Mr. Taylor. Yes, sir.

The Chairman. Now, if scientific management is power for good and a power for bad, and scientific management requires that there shall be only one directing head, with no interference with the law of that directing head, how is the workman going to protect himself against the power for bad that is in that system?

Mr. Taylor. Why, that is not scientific management, Mr. Chairman. I have tried to point out that the old-fashioned dictator does not exist under scientific management. The man at the head of the business under scientific management is governed by rules and laws which have been developed through hundreds of experiments just as much as the workman is, and the standards which have been developed are equitable; it is an equitable code of laws that has been developed under scientific management, and those questions which are under other systems subject to arbitrary judgment and are therefore open to disagreement have under scientific management, been the subject of the most minute and careful study in which both the workman and the management have taken part, and they have been settled to the satisfaction of both sides.

Mr. Tilson. Wherein is the power for bad then in scientific management?

Mr. Taylor. The mechanism of scientific management is a big engine, Mr. Tilson. If you have a locomotive and train of cars which, when running on a track and doing all right, is a great power for good, it is equally as great power for bad when it gets off of the track. Now, if the mechanism—I am speaking now of the mechanism of scientific management, Mr. Chairman—if that same mechanism is used by unscrupulous people, it is not then used under scientific management, it may do a durned lot of harm. That is not scientific management. It is just as if you were to turn a locomotive loose on the streets and say "Let her go." You can use it either for good or for bad.

The Chairman. If that mechanism is once introduced, is it not possible that it could be utilized to more value in the hands of an unscrupulous man who would use it for bad?

Mr. Taylor. That is conceivable for a short time, but only for a very short time. For instance, this is a beautiful building that we are in here, and it has been erected and doing magnificent service for a good many years. It is conceivable that some fool party might get into power and order one wing of the Capitol blown up with dynamite. Such a thing is conceivable; but I can tell you that party would regret it if it ever did such a foolish thing, and it would be promptly voted out of power. Just so with any one attempting to use the mechanism of scientific management in a wrong way. He would regret it. It might do an immense amount of harm for a short time but its abuse would bring its own remedy promptly. Even with the finest laws that have ever been made, you cannot absolutely insure their enforcement at all times; but that does not prove that it is not good to have laws, that it is not good to have standards.

The Chairman. If the enforcement of a law, however, is dependent upon the will of a man who has the power to violate it, there is not much likelihood of the law being enforced against him, is there?

Mr. Taylor. Mr. Chairman, I believe that the very great bulk of mankind wants to work under and wants to live under laws. They believe in laws. It is only the rare exception in this country, whether it be the workman or whether it be the employer, who does not believe in laws and see the desirability of living up to them.

The Chairman. Apparently, Mr. Taylor, you have lost sight of the thing I was illustrating, and you have used again the laws as illustrating a certain point. Now, to get back to the original proposition: If the whole proposition of whether scientific management shall be used for good or shall be used for bad depends upon the single directing head of the establishment, there is not much likelihood, is there, of any penalty being attached to the exercise of that power for bad?

Mr. Taylor. I have never said that scientific management could be used for bad. It is possible to use the mechanism of scientific management, but not scientific management itself. It ceases to be scientific management the moment it is used for bad.

The Chairman. That might be true. But scientific management cannot be developed, as I understand it, unless you have the thing with the mechanism of it?

Mr. Taylor. Yes.

The Chairman. And according to your statement that the mechanism can be used for bad, and according to another statement that in scientific management there must be a directing intelligence and that the directing intelligence must not be interfered with by anyone. You may cooperate in accordance with the desires of that intelligence, but it must not be interfered with; otherwise it is not scientific management.

Now, under those circumstances, how is the workman going to be able to protect himself against the employer using that mechanism that has been established to oppress him for the gain of the employer?

Mr. Taylor. If a man in the management tries to use the mechanism of scientific management to oppress the workman or in any other way that it should not be used, the workman simply reverts to his old ways and goes right back and does what he did before under the old management, he soldiers, and cooperation at once ceases. This is a mutual affair and both sides must work together; then, and only then, do you have scientific management. The moment one side starts to jump the fence and bulldoze the other, or to do any acts which are outside of the principles of scientific management it ends. Without harmony you cannot have scientific management, and you go right back to the old fighting scheme, in which each side is watching the other carefully and trying to get an advantage over the other.

We are, both sides, trying to get the largest possible amount of work out; there is no time for fights. Fights and quarrels are not characteristic of scientific management. The old type of management is full of demands on one hand and refusals on the other. The terms "demand and refuse" are never heard in scientific management. These are not words which one friend uses to another.

The Chairman. I think you stated the other day, Mr. Taylor, that up until last year you did not know of any strikes where scientific management had been introduced, during the time since it has been introduced.

Mr. Taylor. Yes, for 30 years.

The Chairman. Isn't it also true that peaceful relations almost invariably exist between master and slave, that no strikes occur?

Mr. Taylor. Well, if you call peaceful relations one fellow lashing the other with a whip, I do not call that peaceful relations. I call that very far from peaceful relations, the conditions that existed under slavery.

The Chairman. Did the master always lash with the whip?

Mr. Taylor. No, he did not.

The Chairman. Were there not some considered good masters, and some considered hard masters?

Mr. Taylor. There were. But, Mr. Chairman, I do not think you and I for one instant can disagree on the subject of slave institutions; there is no question about that whatever; there can be no two views between us as to slavery.

The Chairman. My only purpose in referring to it at this time was to demonstrate the idea I have always had, that the fact that no strikes have occurred does not prove anything as to the private relationship between employer and employee. I think you will admit, Mr. Taylor, will you not, that there are comparatively few strikes in India and China.

Mr. Taylor. Mr. Chairman, coming back to India, there was the terrible Sepoy mutiny which we always have in mind. We know that there exists even now the elements of dissension in India, and we know also there now exists an absolute state of revolution in China.

The Chairman. Is not that political revolution, rather than industrial rebellion?

Mr. Taylor. I admit I know very little about industrial conditions in India and China.

Mr. Tilson. In this country where a man is free and he has a perfect right to apply to public opinion in general (he thinks that is a proper sovereign court sometimes if he is not properly treated), would not you take it as evidence that his relations were rather friendly, where this free-sovereign man has been working for years and there has been no evidence of discontent?

Mr. Taylor. I should say that was evidence. I have heard it said, however, Mr. Tilson, that those men who are working under scientific management are weaklings; are men of little or no character, and yet our factories are more than holding their own with their competitors.

Mr. Tilson. That may be, but the kind of men that work in factories are not weaklings; the great mass of workmen in this country are not weaklings and not slaves, and are not enduring any oppression of an unendurable character, without making it known.

Mr. Taylor. No, sir. I know that we make errors and we make plenty of them on the management side, naturally, but the moment an error is made, a good big howl goes up from the workmen right off, and I can assure you that the complaint is not the kind made by weaklings or slaves.

Mr. Tilson. Because the workman knows what is right and knows how to get it.

Mr. Taylor. Certainly. In nine out of ten times, the trouble is on the management side, and I assure you that if we make a mistake it is promptly corrected by us, and if you like, I can bring you thousands of workmen right here to tell you that they do not have to go to anyone to have a mistake rectified beyond the man who has made the mistake. People do not become perfect under scientific management; they make mistakes; but when we do make them, the workmen tell us about them right off and we correct them, or the whole scheme would fall to smash.

The Chairman. Some time ago you gave as four fundamental principles of scientific management about the following definitions:

First. The gathering together of the traditional knowledge and recording, tabulating, and reducing this knowledge to laws.

Second. Scientific selection and then the development of the workmen.

Third. The bringing of the science and scientifically trained workmen together.

Fourth. The almost equal division of work of

the establishment between the workmen and the management.

Now, under the third of those, the bringing of the science and the scientifically trained workman together, isn't it the purpose of scientific management that the workman must follow absolutely the directions that are given to him when this science and scientific workman are brought together—that he must follow the directions that are given to him as to how he shall perform the work?

Mr. Taylor. It is the rule under scientific management that the workman works in accordance with the laws that have been developed, and that they shall at least (when they get a new job, we will say, that they have not done before)—that they shall at least practice the method that has been set before them once before raising any objection or any kick about it. If after having tried the new method once any workman has a better suggestion to make, of any kind, sort or description, that suggestion is most welcome to the management. And it is through those suggestions from the workmen that nine-tenths of our progress is made. The following kinds of suggestions are received from workmen, after having faithfully tried the method outlined to them, they see something wrong about our method and suggest a new or a better way of doing the work, or suggest a more efficient series of movements or some better process than we have outlined. And in that way we get most of our knowledge and make our improvements in methods and implements.

The Chairman. If the workman has to obey instructions implicity as to how the work should be done, would he not thereby simply become an automaton, and would not that ultimately reduce the skill and value of the skill of the workman?

Mr. Taylor. Mr. Chairman, I want to give an illustration in answer to that question, because I think my answer can be made very much clearer through an illustration than through a single sentence.

The workmen—those men who come under scientific management—are trained and taught just as the very finest mechanic in the world trains and teaches his pupils or apprentices. Now, I think you will agree with me as to who this finest and highest-class mechanic in the world is. So far as I know there will be no question about him, for we will all agree that the highest-class mechanic in the world is the modern surgeon. He is the man who combines the greatest manual dexterity and skill with the largest amount of intellectual attainment of any trade that I know of—the modern surgeon.

Now, the modern surgeon applied the principles of scientific management to his profession and to the training of the younger surgeons long before I was born—long before the principles of scientific management were ever dreamed of in the ordinary mechanical arts. Let us see how this man trains the young men who come under him. I do not believe that anyone would have an idea that the modern surgeon would say to young doctors who come into the hospital or who come under him to learn the trade of surgeon—I do not think the surgeon would say anything of this kind: "Now, boys, what I want of all things, is your initiative; what I want, of all things, is your individuality and your personal inventiveness."

I do not think anyone for an instant would dream that a surgeon would say to his young men, for instance, "Now young man, when we are amputating a leg, for instance, and we come down to the bone, we older surgeons are in the habit of using a saw, and for that purpose we take this particular saw that I am holding before you. We hold it in just this way and we use it in just that way. But, young men, what we want, of all things, is your initiative. Don't be hampered by any of the prejudices of the older surgeons. What we want is your initiative, your individuality. If you prefer a hatchet or an ax to cut off the bone, why chop away, chop away!" Would this be what the modern surgeon would tell his apprentices? Not on your life! But he says, "Now, young men, we want your initiative; yes. But we want your initiative, your inventive faculty to work upward and not downward, and until you have learned how to use the best implements that have been developed in the surgical art during the past hundred years and which are the evolution of the minds of trained men all over the world; until you have learned how to use every instrument that has been developed through years of evolution and which is now recognized as the best of its kind in the surgical art, we won't allow you to use an iota of ingenuity, an iota of initiative. First learn to use the instruments which have been shown by experience to be the best in the surgical art and to use them in the exact way which we will show you, and then when you have risen up to the highest knowledge in the surgical art, then invent, but, for God's sake, invent upward, not downward. Do not reinvent implements and methods abandoned many years ago."

That is precisely what we say to the workmen who come under scientific management. No set of men under scientific management claims that the evolution has gone on enough years to be in the same high position as is occupied by the surgeon, but they do claim that the 30 years of scientific investigation and study (which goes on under scientific management) of the instruments that are in use in any trade, whatever it may be, have enabled those engaged in this study to collect at least good instruments and good methods, and we ask our workman before he starts kicking; "Try the methods and implements which we give you; we know at least what we believe to be a good method for you to follow; and then after you have tried our way if you think of an implement or method better than ours, for God's sake come and tell us about it and then we will make an experiment to prove whether your method or ours is the best, and you, as a workman, will be allowed to participate in that experiment. It is not a question of your judgment or my judgment or anyone's judgment; it is a question of actual experiment and time study to see whether this suggestion is better than the standard we have had in the past." And if it proves to be better, what I advocate every time is, not only that the new method shall be adopted, but that the man who made the suggestion be paid a big price for having improved on the old standard.

And it is just in this way that we make progress under scientific management.

The Chairman. Taking your own system of illustration and own basis of illustration, is not the workshop and the management of the workshop more in the position of the surgeon in chief of the hospital than it is of the head of a medical college, and would it be expected that a surgeon in chief would say to the surgeons in the hospital: "Now, when a case comes in here for you to operate upon you must not make a diagnosis of the case; you must not decide upon how you are going to operate on this case; you must not determine anything at all about how the operation should take place or what tools should be used for this operation until after you have got a specific written order from the surgeon in chief, and then when you have received that written order, if you vary from that, no matter what the case may be—if you vary from that you must expect to be held responsible for your having done so."

Would not that be a better illustration of the relative positions of the two than the one which you have given? And who would expect that a surgeon under these circumstances, would undertake to do any operating in a hospital.

Mr. Taylor. Mr. Chairman, I have among my acquaintances quite a number of the great eastern surgeons—the noted surgeons of New York and Philadelphia especially. Without any exception they all point to the establishment of Mayo Bros., in Rochester, Minn., as the finest example of surgery in the world; they say that so far as they know the finest surgical establishment in the world is under the management of Mayo Bros, in Rochester, Minn.

Last evening I met one of the surgeons from Mayo Bros., and earlier in the fall I met Mr. Mayo himself. He came East from his work, as he told me, largely to see me and talk about the principles of scientific management. He made the statement that his establishment (and it was corroborated by the doctor I met yesterday) is run, so far as possible, along the principles of scientific management.

For example, when a patient arrives in the establishment the first thing that is done is a brief questioning and diagnosis which would indicate what general branch of surgery was likely to be called for. It is just a preliminary investigation. And then the man best fitted to perform that particular type of diagnosis is assigned to that patient. He diagnoses, and if he finds in the course of his diagnosis that he is not the proper man, then another expert is sent for and makes the diagnosis.

This diagnosis is then written up carefully by the specialist who has plenty of time to accurately describe the case.

After this man follows one of the four great assistants of the two Mayo brothers (four other noted surgeons), and the one of those four who is best fitted to this type of surgery again diagnoses the case with the written information before him of the first diagnostician, and he finally (being a man of riper experience or judgment than the first one) corroborates or makes additional notes, and finally the diagnosis is brought to the one of the two Mayo brothers who is going to perform the operation. (The two brothers have their two somewhat separate departments in surgery.) He finally makes his own diagnosis, but he makes it with all this preliminary information and data before him.

And then when he performs his operation, instead of performing it alone, he performs it with from eight to ten assistants each one having his special work, just as is the case under the principles of scientific management.

And Mr. Mayo came East to get further information right along the lines upon which he has been working (we not knowing anything about his proposed visit), to see if he could not add more to the principles which he was already using.

Instead of having an operation performed by a single surgeon as they used to, the modern operation is performed by 8 to 10 men combined, and each one performing that particular part of the operation for which he is best fitted. And my informant told me they would sometimes go through an operation of two hours without one word spoken. So well are they trained, that they perform the function they are called upon to do by a simple nod of the head, the reason for not speaking being that the germs from the breath from speaking might get out and get into the wound, and contaminate the air, as you know.

I think that represents the best practice today in

modern surgery, and I think it is very analogous to what is done in our industrial establishments under scientific management.

The Chairman. Would not that be the same as if a job came into a shop, and you would select a molder to do the molding part, a machinist to do the machining part, and so divide that into the various lines that the men had to do? Is not that practically the same thing?

Mr. Taylor. I think not. I think this operation performed by eight or ten men, all cooperating, working as a team is very different from giving the molder one thing to do in one department by himself and the machinist another thing in another department.

The Chairman. Is not one of the elements of scientific management this possibility to divide it up so that the workmen will have the same operation to perform over and over again?

Mr. Taylor. That is just the same under scientific management as it is under the other types of management; neither more nor less. Under scientific management precisely the same principles of work are used in that respect as under the other types of management.

Naturally, for manufacturing shoes, under the modern way, under scientific management or any other management, the manufacture of shoes is divided into very, very many minute parts. I have a very high regard for Mr. Tobin, the leader of the shoemakers' unions of New England, and the other day he told me that in making an "upper" there were over 450 operations—in making the upper of a shoe, each one performed by a different man in a well-run shop.

Well, this is what now takes place under the older types of management, and that undoubtedly would continue under scientific management; and I do not think in that respect there is any difference between scientific management and the other, except this. And I want to emphasize this, Mr. Chairman—that under scientific management it becomes both the habit and pleasure of those people who are on the management side to try and help their men rise to the highest class of work for which they are fitted. I say that deliberately. In our working right alongside of men who are friends, and warm friends, we can't help having the kindliest feeling toward them, and wanting to develop them to do the highest class of work they are fitted for, and to finally get the highest practicable day's pay. This is characteristic of scientific management and is not the characteristic of the old type of management.

The Chairman. Does not scientific management undertake to show that a change from one part of the work to another part of the work, if they involve different operations, is a loss of time and consequently it is better, if possible, to have one man perform each of the operations?

Mr. Taylor. Mr. Chairman, what is true under scientific management in this respect is also true under all types of management. I think this tendency to training toward specializing the work is true of all managements, for the reason that a man becomes more productive when working at his specialty, and while it is deplorable in certain ways (there is no question about it, there are various elements in this specialization that are deplorable), still the prosperity of the world and the development of the world—the fact that the average workman in this day lives as well as kings lived 250 years ago—that fact is due to a certain extent to just this very specialization.

The Chairman. Is not the result of specializing that the workman does not secure a general knowledge of his trade, and consequently the number of men from which the best managers are recruited is limited—is not the result of that that there is a shortage of first-class managers?

Mr. Taylor. It is quite the reverse, Mr. Chairman. Under scientific management we are making 10 managers every day to one that is being made under the old type, and in order to prove this fact I am very glad that you brought up that matter, because I wish to ask your committee, Mr. Chairman, if I may be allowed, to present at least two witnesses before your committee who will testify to the fact that they first started in under scientific management at low wages and in unimportant positions; that they were gradually promoted under the principles of scientific management until in each case each man rose to the highest position in the particular establishment in which he was and for which his abilities fitted him; and that while he was rising in this way his wages were increased—not in a small way, but to a large extent—and that after those men reached in the companies in which they were working the highest positions which it was possible for those companies to offer them, that the managers and owners of those companies then deliberately set out to find for these men better positions in which they could get better wages and still have a chance to progress in a larger field outside their own companies. I want to bring those men to tell you that themselves, because it illustrates just what I was trying to demonstrate, that the kindliest relations exist between the management and the workmen. And that promotion is the rule, not the exception.

The Chairman. You do not mean to convey to the committee the impression that a kindly feeling has not existed between the same men and some other men—that it did not exist and could not exist until the advent of scientific management?

Mr. Taylor. Certainly not, but I wish to point out that that is a characteristic of scientific management and not a characteristic of the other, as you know. It is not a characteristic of the old type of employer to develop a very fine foreman and deliberately find employment for that foreman on the outside. It is quite the reverse. They are very anxious to keep those men to themselves, even though they keep them at lower wages than these men could get outside.

The Chairman. Would not the introduction of witnesses to show that under your system they had been promoted from low positions up to the higher and best and transferred at the suggestion and consultation of the employers to some other establishment—would that show that it was characteristic of that system?

Mr. Taylor. I beg your pardon?

The Chairman. I say if two men were brought here, for instance, to testify before this committee that they had under your system risen from the very lowest positions to the highest positions in the gift of their employer, and then their employer had deliberately sought higher positions for them in some other concern—would that demonstrate that that is characteristic of your system?

Mr. Taylor. Mr. Chairman, if you could produce from a small company employing only a few men four similiar instances of that kind of promotion in one year, and bring those people before this committee to testify to this fact, I say it would tend to show that this is characteristic of scientific management. In a small company working under our system and employing only about 100 workmen as many as four foremen in one year were found better positions on the outside, because they had reached the highest salary which that company was able to pay them, and because that company, wishing them well, found them something better on the outside.

The Chairman. Would not that show that it was characteristic of that particular employer, or would it show it was characteristic of the system?

Mr. Taylor. I say, Mr. Chairman, that so far as I know it is not characteristic of the older type of management and that it is characteristic of the newer type of management.

Thereupon, at 5 o'clock p. m., the committee adjourned until 8 o'clock p. m.

Evening Session

The committee met at 8 o'clock p. m., Hon. William B. Wilson, (chairman) presiding.

The Chairman. The committee will come to order, and Mr. Taylor will proceed with his statement.

Mr. Taylor. At the end of an answer which I made near the end of the last session today, I desire to have the following added: I may add that in the Tabor Manufacturing Company, which is the company to which I referred, before the introduction of scientific management, not a single foreman or leading man was ever promoted to a better position outside of the employ of the company, whereas in that company during the present year alone four of the leading men have been provided with outside positions because they had reached the apparent present limit of their promotion in the Tabor Company, and a better opportunity with higher wages was sought for them outside.

The Chairman. Mr. Taylor, is it not true that the American workman is a more productive workman than any other on earth, taken in the aggregate?

Mr. Taylor. I am inclined to think that is true, Mr. Chairman, but my knowledge is not sufficiently definite upon the subject to be certain of it. I should say that the fact that our men are more productive is that the workmen of our country have more of the good things of life, more of the things that are of real value in life, than the workmen of other countries. If the workmen of our country have arrived at a condition of feeling perfectly satisfied with their present state of material prosperity, as well as with their mental and esthetic opportunities of various kinds, then possibly one might question the desirability of a further increase in the output of the individual. But, in my judgment, the best possible measure of the height in the scale of civilization to which any people has arisen is its productivity; and, for my part, I am looking forward to the day when the working people of our country will live as well and have the same luxuries, the same opportunities for leisure, for culture, and for education as are now possessed by the average business man of this country, and this condition can only come through a great increase in the average productivity for the individual of this country. That is the road we shall have to travel.

The Chairman. If the American workman is already more productive than any other workman, and by systematizing the work you can still further increase his productivity, then what necessity is there for adding to the discomfort of the workman by requiring the expenditure of more energy on his part?

Mr. Taylor. My impression is that that is correct.

The Chairman. It has been stated on various occasions, and the figures alleged to be taken from official figures, that the average productivity of the American workman is $2,400 per year, as against an average of the British workman of $565 per year. If that be true, what necessity is there for crowding the American workman to greater productivity by reason of the expenditure of greater energy?

Mr. Taylor. In the first place, Mr. Chairman, I have not the slightest idea that that ratio is the correct one for the productivity of the two countries. In the second place, as I tried to point out before, the money standard is no fit standard by which to measure the relative productivity of two peoples. You must be familiar with the relative purchasing power of money in the various countries before you can come to any correct conclusion by means of the money standard for comparison. And even if that ratio were correct, the reason why the American workman, the principal reason why the American workman is a happier and more contented and more prosperous workman on the whole than those of other countries—and I believe that to be the fact—the principal reason for this condition of affairs is that the workmen of this country are more productive than those of other countries.

Scientific management does not demand an unnecessary expenditure of energy. If it did it would be wrong. Scientific management only asks that soldiering be stopped, and that each man while he is working shall work at a proper normal pace and shall use efficient instead of inefficient movements.

The Chairman. I am not speaking about the expenditure of unnecessary energy. What I am endeavoring to get at is what necessity there is under those circumstances for the expenditure of any additional energy in order to increase productivity.

Mr. Taylor. I do not look upon the fact that the

man who works under scientific management, and who throughout his working day is usefully employed—is expending his energy in a useful way—as a misfortune in any way. I look upon it as a great gain for the workman that he is not obliged, in order to defend his own interest, as he was under the old system, to soldier a great part of the day, that is, to pretend to work hard or to go through motions which are unproductive and yet which are tiresome.

The Chairman. Is it not the purpose of all production to add to the comfort and well-being of mankind?

Mr. Taylor. It is.

The Chairman. If by any system of production you increase the discomfort of mankind, have you not thereby destroyed the very purposes of your production?

Mr. Taylor. That depends entirely upon the amount of discomfort which the workman had before. If a man had not been working faithfully, if he had spent one-half of his time in idleness, I do not look upon it as anything of a misfortune to that man that he is brought to spend his working time in useful effort instead of in useless exertion.

The Chairman. Do you think that the comparatively small number of employers should have the power to determine absolutely for the comparatively large number of employees what constitutes comfort for them?

Mr. Taylor. I certainly do not think it ought to be in the power of any outside man to say what shall constitute the comfort of his fellow men. Every person should be free to decide what is for his own comfort, and I think in this country, so far as I know, that is true.

The Chairman. Would not the fact that industry is to be directed by scientific management—by one central intelligence—and that the question of whether the workmen are comfortable or uncomfortable is to be determined by that central intelligence, place in the hands of the employers the power to determine what constitutes comfort for the employees?

Mr. Taylor. Mr. Chairman, I must again state that under scientific management those men who are in the management, such as, for instance, the superintendent, the foremen, the president of the company, have far, far less arbitrary power than is now possessed by the corresponding men who are occupying those positions in the older types of management. I must again state that under scientific management the officers of the company, those on the management side, are quite as much subject to the same laws as are the workmen. As I have again and again stated, our great difficulty in the introduction of scientific management has been to get those on the management side to obey these laws and to do the share which it becomes their duty to do in the actual work of the establishment in cooperating with the workmen, so that I hope that I may be able to make myself clear that under scientific management arbitrary power, arbitrary dictation, ceases; and that every single subject, large and small, becomes the question for scientific investigation, for reduction to law, and that the workmen have quite as large a share in the development of these laws and in subsequently carrying them out as the management have.

The Chairman. Is not the management the final arbiter in the determining of those questions under scientific management?

Mr. Taylor. In most cases the laws and the formulas and the facts of scientific management, which are vital both to the workmen and the management, have been developed during years preceding the one on which the work is going on. And that being the case, neither the management nor the workmen have any final arbitrary dictum as to those laws. The laws of scientific management are somewhat analogous to the laws of this country. We are all working under certain laws that were not enacted by the present Congress or the present President of the United States, and which have not been interpreted by the present courts, and yet the President of the United States and all the citizens of the United States are alike working under those laws. Now, under scientific management there have gradually grown up a code of laws which are accepted by both as just and fair. What I want to make clear is that the old arbitrary way of having a dictator, who was at the head of the company, decide everything with his dictum, and having his word final, has ceased to exist.

The Chairman. Under our laws no judge would be permitted to sit in a case in which he had a personal interest. Now, under scientific management, with the power centered in the head of the establishment, would not the final judge in the case be a man who was interested in the outcome?

Mr. Taylor. A final decision must be reached in all disputed cases by someone.

The Chairman. Yes.

Mr. Taylor. And if the decision were finally appealed it would probably go to the board of directors of the company, as the final appeal. That would probably be the final appeal, and the decision of that board of directors, as far as this particular case is concerned, would be final.

But, Mr. Chairman, you must remember that if any injustice is done to a workman under this system he always has the recourse of leaving, and he has further the much more powerful remedy of sitting down and soldiering just as he did under the old system, and he will still get the same wages if he soldiers. He gets the full wages that he is employed for, even when he soldiers. So that if an injustice is done to him it comes to a question of whether the workman has the power to force an unjust management to do what is right, or if he fails in this to virtually return to the old system of management with all its antagonisms and sad conditions.

The Chairman. But if the workman leaves, quits his employment, would he not be placed to a greater disadvantage by virtue of his quitting his employment than the employer would be by virtue of the workman quitting?

Mr. Taylor. That depends entirely on—

The Chairman. It would, as a general rule, be true, would it not? There might be special cases where it would not be true, but would not that, as a general rule, be true?

Mr. Taylor. I think it is almost impossible to generalize on that. My experience is that, for instance, in the machinery business, employers are always looking for good men. It has been so all my life. They are always looking for good men, and one of the most humane employers under the old system of management, a man who stands very high and who is looked up to as a very humane man, told me with the greatest sadness that during the last three or four years about 40 per cent of his men had left him every year. Forty per cent each year had left him and new men came. Now, that could not happen under scientific management. Our men are too prosperous, too happy and contented for that.

The Chairman. Would you not permit them to leave?

Mr. Taylor. They do not want to leave. Permit them? Of course they are permitted. This is a free country. But they are so well off, and so well treated, that they do not want to leave. It is not a question of permitting; it is altogether a voluntary matter.

The Chairman. When your scientific management has gathered together its information, its formulas, and formulated its rules and regulations, systematized its work, etc., giving its direction to the workman, and the workman fails to obey these formulas that are laid down for him, is there any method in scientific management to discipline the workman?

Mr. Taylor. There certainly is, Mr. Chairman; and any system of whatever nature under which there is no such thing as discipline is, I think I can say, pretty nearly worthless. Under scientific management the discipline is at the very minimum, but out of kindness to the workman, out of personal kindness to him, in my judgment, it is the duty of those who are in the management to use all the arts of persuasion first to get the workman to conform to the rules, and after that has been done, then to gradually increase the severity of the language until, practically, before you are thru, the powers of the English language have been exhausted in an effort to make the man do what he ought to do. And if that fails, then in the interest of the workman some more severe type of discipline should be resorted to.

The Chairman. Having gathered together all your information and built up your formulas and introduced your scientific management, if the management violates its formulas, what method is there in scientific management to discipline the management for its violation of its principles?

Mr. Taylor. I am very glad that you asked that question. Just the moment that any of our men in the planning room does not attend to his end of the business, just the moment one of the teachers or one of the functional foremen does not attend to his duties, or do whatever he ought to do in the way of serving the workmen—I say serving advisedly, because if there is anything that is characteristic of scientific management it is the fact that the men who were formerly called bosses under the old type of management, under scientific management become the servants of the workmen. It is their duty to wait on the workmen and help them in all kinds of ways, and just let a boss fall down in any one thing and not do his duty, and a howl goes right straight up. The workman comes to the planning room and raises a great big howl because the foreman has not done his duty. I tell you that those in the management are disciplined quite as severely as the workmen are. Scientific management is a true democracy.

The Chairman. Suppose that it is the man higher up that violates these formulas? As I understand your testimony before this committee no scientific management can exist until there has been an entire change of mind on the part of the management as well as on the part of the workmen?

Mr. Taylor. Yes, sir.

The Chairman. And that this change must take place in the point of view, in the mind of the employer and the employee.

Mr. Taylor. Yes, sir.

The Chairman. And that the condition of "Do unto others as you would have them do unto you" must exist, and that spirit must exist. Suppose having that as a part of your formula, as part of your rules, that the workman is dependent upon the generous spirit on the part of the employer to say that he is treated well, suppose that the head of the house, the man higher up, violates that formula, what power is there in scientific management to discipline him for that violation?

Mr. Taylor. The losing of the men who are under him, their quitting, and going to some other place where they are treated better.

The Chairman. There is no scale of language set to the strongest scale of language that can be used for him, is there?

Mr. Taylor. I recall a particular instance in which one of the men who is here in this room was systematizing a company, and in which the president of that company, who was at the same time one-half owner of the company, refused in small matters to get into line and do his share of the duties, and I remember distinctly the volley of oaths that were thrown at the president of that company by the man who was systematizing the company for him, and he wound up by saying, "Um, um, um, if you do not do your share now and get right into line, we will get right out of this place and leave you where you are." And he got right into line.

The Chairman. Is it part of scientific management that the workman shall cuss the man higher up when the man higher up violates his own formulas?

Mr. Taylor. It is part of the democratic feeling that exists between all hands that under scientific management they should talk to each other very freely and very frankly. And I think it is safe to say, that if I, for instance, were to swear at one of these fellows here (pointing to some of the workmen who were present at the hearing) he would swear right back at me without the slightest hesitation. I do not think there would be any difference between us if I happened to be a little higher up and he were a little lower down. I have not seen any great distinction between the two when it comes to swearing.

Mr. Redfield. Does not scientific management take the third commandment into account?

Mr. Taylor. I am sorry to say it does not take it into account as it ought to. I was brought up wrong—

Mr. Chairman. In your direct testimony, Mr. Taylor, you referred to baseball playing as being an ideal type of scientific management, the manner in which the players were handled and the manner in which they responded to the management being pointed out as an indication of what scientific management can do. Are you aware of the fact that in baseball playing, in the professional baseball playing that you have reference to, the players are bought and sold like cattle on the market?

Mr. Taylor. I have heard of that fact, and I have often wondered why it was. I do not know. I am not intimately acquainted with that phase of the management of baseball to be able to say whether this is fair and just. I rather suppose, although I do not know, however, that no sale can be made without the consent of the player, that it is a mutual affair, and I rather imagine that the player always insists upon getting his share of the booty. But that I do not know; I am entirely unacquainted with it. My friend Mr. Reagan (points to Mr. Reagan, who is present) who is the ex-manager of a baseball team, can probably enlighten you on that point.

The Chairman. I did not know but what you might have some information on the point since you were holding it up as an example.

Mr. Taylor. No; I do not know that feature. I

was never bought or sold when I played. I was the pitcher of the Phillips-Exeter nine when I was a boy. They never bought or sold me. That is all I can say.

The Chairman. Are you aware of the fact that once a player has been signed by any team in the league in which he is playing that he cannot go to any other team in the league, no matter what wages are offered to him, without the consent of the team with which he had signed?

Mr. Taylor. I have an impression that that is true, but I really do not know.

At the end of my answer will you allow me to state that in citing the management of the players on the baseball team as an excellent example of the scientific management I do not have in view in the slightest degree any such management as that. I do not wish it to be understood that I approve of any such thing as that. I know nothing about that feature of the management of a ball team, and I did not have that in mind when I spoke of baseball as a fine example of scientific management. I had the careful training and coaching and teaching of the baseball players in mind. And then their coordination and the cooperation which is so conspicuous in the management of a baseball team while it is playing a game. It was that that I had in mind and not the form of contract which they sign when they join their team, or the form of agreement.

The Chairman. You spoke of the science of shoveling and the introduction of different size shovels for different weights of material, that being based upon observation. Was it not to be expected, and would it not be expected under any system of shop management, that where the workman was required to furnish his own shovel that he would furnish a shovel of a size necessary for handling the heaviest kind of material, and that consequently his shovel would be too small for the lighter kinds of material?

Mr. Taylor. I have not really considered what would be the probability in that case, Mr. Chairman. My impression is that the workman would probably take a shovel that would insure his not overworking himself when he was shoveling heavy material, and that therefore he would incline toward taking a shovel, as you say, which would be entirely too small for the lighter materials.

The Chairman. Is it not the case for hundreds of years that men have used different sized shovels for different weights of material; where they had light material to handle continuously, using light shovels, and where they had heavy material using heavy shovels, so as to get nearer the proper weight a man can handle?

Mr. Taylor. I have not the slightest doubt that different size shovels and implements for handling dirt have been in existence for hundreds of years. I do not know it, but I have not the slightest doubt of it. What I was trying to indicate in my testimony was that it became the duty of the management to supply the man with exactly the right implement to do each kind of work, and that the proper implement was only supplied to the men, and could be only sup- to the men, after the science of shoveling had been carefully studied, and that this was one of the results of the study of the science of shoveling.

The Chairman. I simply say, Mr. Taylor, that more than 40 years ago I worked for a large coal company that required men to do shoveling, sometimes shoveling slates and shales, which are heavy, and sometimes shoveling coals, which are light. They maintained different sizes of shovels for use in shoveling the different kinds of material, an old-style No. 2 shovel being the style for handling the heavy materials and an old-style No. 5 or No. 6 for handling the lighter material or coal, the 5 and 6 being simply used for the different capacities of men, and that was before any furore had arisen with regard to shop management.

Mr. Taylor. It seems to me, Mr. Chairman, that you came very close to working under scientific management about 40 years ago yourself.

Mr. Tilson. I desire to ask a question. In regard to the 21½-pound load for shoveling, does that apply regardless of the bulk to 21½ pounds? Is that the most economical load, regardless of the bulk?

Mr. Taylor. Yes, sir; regardless of the bulk.

Mr. Tilson. Do you take into account any difference in effect on the man, as the load varies?

Mr. Taylor. I think the load remains the same; whether the bulk is large or small the load remains the same.

Mr. Tilson. My question is just this: You found, as I understand it, that at 38 pounds to the shovel that was not an economical load?

Mr. Taylor. Not an economical one if it was too heavy a shovel load and prevented the man from doing a proper day's work.

Mr. Tilson. That is, your dirt pile grew as the size of your shovel went down?

Mr. Taylor. The pile of dirt shoveled in a day grew larger and larger as the shovel load starting with 38 pounds per shovel went down until we reached a 21½-pounds shovel load, at which load the men did their largest day's work, and then again the dirt pile grew smaller and smaller as the shovel load become lighter and lighter than 21½ pounds.

Mr. Tilson. What I was trying to get is this: You have told us the effect on the pile. What about the effect on the man? Was the man as well off when he was shoveling the 21½-pound load?

Mr. Taylor. Yes; he took his natural gait all day long in each of those kinds of shoveling. The workman regulated his own pace. No one regulated it for him. The fact was that when he was shoveling with a heavy load of 38 pounds it tired him to such an extent that he went much slower, naturally. He took fewer shovel loads, and he had to rest more between shovel loads.

Mr. Tilson. Then take it on the other side, if it was very light, not more than 10 or 15 pounds?

Mr. Taylor. In order to shovel the same amount with a light load of 10 to 15 pounds that he shoveled with a 21½-pound load, he would have to work so quick—to make his motions so quick—that they then became tiresome.

Mr. Tilson. So you figure it out that regardless of bulk the easiest load for a man to handle is 21½ pounds with a shovel?

Mr. Taylor. Yes, sir.

The Chairman. Would that be true irrespective of the distance that the dirt had to be thrown?

Mr. Taylor. No, sir. I am very glad that you asked that question. That again opens another large element of the science of shoveling, and I did not wish to burden you unnecessarily with the science of shoveling. Now, that holds true up to about 4 feet in length and 5 in height; that 21 pounds is the best load. When you rise above 5 feet in height, say, the combination of 5 feet in height and 4 feet in length, and go higher than that, then you must have a lighter load. The load again falls off. You understand, Mr. Chairman, that in my direct testimony, in speaking of the science of shoveling, I only spoke (broadly speaking) of the effect of that one element of the science. I want to assure you, gentlemen, again that the true science of shoveling is quite a large affair, but I will be glad to go into it if you care to go further, and tell you more about it. It is quite a large affair.

The Chairman. There is one feature about it that I am interested in, because I am quite convinced that it was scientific, and that was your description of the forearm to thigh, when you had to use force other than the arm force to get entrance of the shovel.

Mr. Taylor. Yes, sir.

The Chairman. I wondered at that time whether you had given any consideration in your scientific investigation to the direct application of force by the thigh or knee to the back of the hand.

Mr. Taylor. Mr. Chairman, I think if you get down as low as that, that it then demands quite an exertion of force by the right leg, a pulling of the leg, which is much more tiresome than if you put the right forearm (indicating a position two-thirds way up from the knee) and throw the whole body forward. The one motion is merely a throwing of the body forward like this (indicating), while the other is a motion of the right leg requiring considerable exertion when you push in the shovel. You must also have a specially made shovel to shovel at the knee.

The Chairman. That may be true as to the man who is trained to shovel outdoors, but to the man who is trained to shovel in the mines it is not true.

Mr. Taylor. I rather fancy that, as you say, it is not true. Again, Mr. Chairman, it appears that the science of shoveling is even broader than we know anything about, and that a further investigation (I haven't a doubt) would prove that what you claim is true.

The Chairman. You say that one of the methods by which the employer can be disciplined if he fails to live up to his own methods of rules and regulations is that the workmen can drop back to the old method which you call soldiering. Would it not be part of scientific management to let out of employment entirely the man who drops back to the old conditions?

Mr. Taylor. If he were let out of employment, and another man took his place, and that man were treated unjustly, that man would do the same. It would be simply getting a second man who would do the same thing. You cannot get a fresh man who will submit to injustice any more than your old employee will.

The Chairman. The only method, then, of disciplining the employer for failure to comply with his own formulas is that the individual workmen might leave him?

Mr. Taylor. I fail to see why just exactly the same treatment could not be accorded to the employer under the scientific management who misbehaves himself as could be employed under any other type of management.

The Chairman. Would it be possible under your scientific management for the workmen to act collectively for their own protection, when it is stated that collective arrangements or collective bargaining relative to the conditions under which the workmen are to be employed cannot be permitted under scientific management?

Mr. Taylor. Mr. Chairman, I have never made any such statement as that. I dare say that some one else has made it. I never have made any such statement as that. I stated in my testimony just a little while ago that I have never seen the necessity for collective bargaining. I have never found the time when those who were engaged in scientific management needed the stress of collective bargaining to be brought upon them in order to make them right any wrong. It is sufficient under scientific management for a single workman to step up and say, "I have been wronged" and he will have his wrong righted; to say that these conditions are wrong, and he will have an investigation made to find whether they are or are not wrong conditions, and in investigations, as I have stated, the workman always has his share.

The Chairman. If I understood your testimony correctly, Mr. Taylor, you said there was no objection—in fact that you courted the cooperation on the part of the employees relative to the conditions of employment, and yet under scientific management you would permit no interference on the part of the employees relative to the conditions under which they should be employed?

Mr. Taylor. If I made that statement then I made a statement which I did not intend to make. I think you have in mind, Mr. Chairman, that I stated that when a workman is given an instruction card asking him to do work in a particular way that until he has attempted to do that work in that way, until he has followed his instructions as they are written, that no protest on his part will be received. In other words, that you do not want to furnish a man with an instruction card which represents the careful result of years of standardization and of definite laws that have been developed and then without any trial of the method on his part have him start a debating society. That is, we want him first to do one piece the way his instruction card says, and then only after he has the personal experience of trying this method, let him come and protest in any way he sees fit, but not start a debating society every time a piece of work is given to a man. That is what I have said, and that, I think is the limit in the direction to which you refer.

The Chairman. Do you speak of Mr. Gilbreth

having developed a method by which he increased the productivity of bricklayers from 120 bricks per hour to 350 bricks per hour, which would be equivalent to increasing from 960 per day of eight hours to 2,800 bricks per day of eight hours, and that the wages of the workmen in doing that had been increased approximately $5 per day to $6.50 per day? Do you think that that kind of division for increased productivity shows a change of mind has taken place on the part of Mr. Gilbreth relative to the Golden Rule? Do you contend or state that $6.50 for laying 2,800 bricks is a proper division, as against $5 for laying 960 bricks?

Mr. Taylor. Mr. Chairman, if you will remember my detailed description of the way in which Mr. Gilbreth taught his workmen when he succeeded in laying 2,800 bricks, Mr. Gilbreth's method of working was less tiresome than when the same workmen worked under the old unscientific conditions and were laying only 900 bricks. Under Mr. Gilbreth's method he is working less hard and using fewer motions to lay 2,800 bricks than he formerly did to lay 900 bricks. He avoids entirely stooping over to the brick pile on the ground and raising his entire body up again every time he lays a brick. He reduces his motions from 18 movements per brick to 5 per brick, so that the workman himself was working less hard than he formerly did. The workman voluntarily chose his own pace. Mr. Gilbreth did not tell him how fast he must work. He did not have to lay 2,800 bricks. The workmen, of their own accord, laid 2,800. There was no limit whatever put upon them. They were merely told by Mr. Gilbreth, "Use my methods and the moment you use my method I will pay you $6.50. That is all I ask of you, to use my methods."

The Chairman. Assuming that the workmen voluntarily laid these 2,800 bricks, did that, of their own volition, the spirit having got into their mind, some change of spirit having reached there and they did this voluntarily, laying 2,800 bricks as against 960, do you want this committee to believe that the same spirit has got into the mind of Mr. Gilbreth when he only paid them $6.50 for those 2,800 bricks as against $5 for 960?

Mr. Taylor. In the first place, I am not sure that $5 and $6.50 were the exact figures; I merely stated them as relative figures as I recollected them.

The Chairman. Well, assuming them to be that.

Mr. Taylor. Under scientific management we have been accustomed to increase the wages of our workmen so that they receive from 30 to 50 per cent higher wages than they had before whenever they follow our instructions. That is about our raise in wages for that class of work, from 30 to 50 per cent. And I believe that the workmen all over the country who have come under scientific management are satisfied and contented and feel that they are well paid for this change in their method of working.

Mr. Redfield. Right in the same point, put down these figures and see if they are correct as to this laying of bricks. By the old method at 120 an hour, multiplied by 18 motions, equals 2,160 motions per hour. By the new method 350 bricks per hour, multiplied by 5 motions, equals 1,750 motions per hour. The product of 960 bricks per diem, therefore, was on the basis of 2,160 motions per hour, and the product of 2,800 bricks per diem was on the basis of 1,750 motions per hour, or a diminution of 410 motions per hour for the larger product, or per day of 3,280 motions less for the new method than the old with a product of 2,800 as against 960. Is that correct?

Mr. Taylor. That is correct, and, Mr. Chairman, I would add that among the eliminated motions was this terribly tiresome one of lowering the body from its full upright position all the way down to the ground and picking up a brick, and then raising the body up again before turning around and placing it on the wall. The elimination of that one motion alone is an enormous saving in effort, so that without question the workmen are working far less hard under Mr. Gilbreth's new system than they were under the old system.

Mr. Redfield. So far, Mr. Taylor, let us assume that the result may be called scientific. Now, I want to renew the question which the chairman asked in a little different form. Now, he has, though concededly at a less effort, a product of 2,800 as against 960, or in other words, our output has been multiplied by nearly three. The rule of the scientific management system is that one-half of the gain, or approximately that, should be given to the workingman. If that were done his wages would rise to $10 per day, and the employer would still be a large profiter by paying his men $10 a day, would he not?

Mr. Taylor. In that particular case I think he would, Mr. Redfield.

Mr. Tilson. May I ask a question there; what about the conditions under which the men work? Did you not tell us something about the additional appliances that were used?

Mr. Taylor. Yes. The scaffold was so arranged that the workmen were kept at the same relative height to the wall all the time. The scaffold was raised alongside the building as the wall went up.

Mr. Tilson. That was probably somewhat more expensive for maintenance than the old way?

Mr. Taylor. Very much more expensive. They had to have helpers to coordinate the bricks for them.

Mr. Tilson. Placed in the proper position?

Mr. Taylor. Yes; then they had to have men place it just right in the proper position. The labor cost more to temper the mortar than it did before. They had to have paid teachers to go around and show these men how to make their new motions. That was an additional expense. I just wanted to bring out the different and the improved conditions under which the men work now, and show that these improved conditions were paid for by the management.

Mr. Redfield. What about the chain blocks to carry the scaffold?

Mr. Taylor. The scaffold is a patented one of Mr. Gilbreth's which does not work with the chain blocks. It works by jacking up—

Mr. Redfield. Then it is your desire to have us

understand that this increase of nearly three times did not represent a net profit—the whole of it?

Mr. Taylor. Certainly not.

Mr. Redfield. But was largely absorbed by additional outlay to produce this higher efficiency?

Mr. Taylor. Well, I should hardly say "largely absorbed." Partly absorbed, not largely absorbed. But in this connection I want to be perfectly frank. I will put it in this way so as to show an extreme case, that if, we will say, in a machine shop, a workman were today using any series of movements on a machine which would turn out 5 pieces a day of a certain kind, and if any individual, a foreman, or another workman, or the management, or a group of men in the management were to devise a new series of motions, which causes the workman to exert no greater effort than he had before exerted, and if the workman could turn out 500 pieces instead of 5 in a day with the new method, that man would do his work tomorrow for his 30 per cent premium just the same as he had yesterday. I want to show this entirely new mental attitude. If, owing to no extra exertion on the part of the men, no new invention on the part of the man, a new and superior device has been adopted for doing the work—we will say, a new machine has been introduced that never was used before, and if that machine can turn out five or ten times the number of pieces the old machine turned out, the man is paid just the same 30 per cent increase in his wages as he was yesterday. I want to make the fact perfectly clear that there is no implied bargain under scientific management that the pay of the man shall be proportional to the number of pieces turned out. There is no bargain of that sort. There is a new type of bargain, however, and that is this: Under scientific management we propose at all times to give the workman a perfectly fair and just task, a task which we would not on our side hesitate to do ourselves, one which will never overwork a competent man. But that the moment we find a new and improved or a better way of doing the work everyone will fall into line and work at once according to the new method. It is not a question of how much work the man turned out before with another method. Mr. Barth here has perhaps been the most efficient man of all the men who have been connected with scientific management in devising new methods for turning work out fast. I can remember a number of—one or two—instances in which almost overnight he devised a method for turning out almost twenty times as much as had been turned out before with no greater effort to the workman. In that case you could not pay the workman twenty times the wage. It would be absurd, would it not?

The Chairman. I understand from your description now of the bricklaying system of Mr. Gilbreth that part of the increased productivity was due to a patented device which Mr. Gilbreth had invented, or that someone had gotten out?

Mr. Taylor. I think it is patented. I am not sure.

The Chairman. Whether patented or otherwise, it is an improved device, is it not?

Mr. Taylor. Yes. That scaffold that I told you about had a table on it, where on the old scaffold they had no table. The table is put in the middle of the scaffold.

The Chairman. You do not for a moment want the committee to believe, do you, that there could be no improvement in machinery were it not for scientific management?

Mr. Taylor. Of course not, Mr. Chairman.

The Chairman. Is not that also true with regard to your art of cutting metal, that that also is an improved device for cutting metal?

Mr. Taylor. No, sir.

The Chairman. And no improvement?

Mr. Taylor. No, sir; that is the study of an art. That represents the evolution of a science which took years to develop, and is in no sense analogous to the invention of a new machine.

The Chairman. Is it any part or parcel of the management, or is it the study of the art itself separate and apart from the management?

Mr. Taylor. The moment that scientific management was introduced in a machine shop, that moment it became certain that the art or science of cutting metals was sure to come. When it became the duty of the management to answer the two questions: What speed shall the machine run at and what feed shall be used, it was inevitable that they should seek for exact knowledge wherewith to answer these questions instead of guessing at the answer as the workmen have done in the past, and this would start the series of experiments which lead to the development of the science of cutting metals. It is the new mental attitude of the management that it is "up to us" to know and direct every element of the work instead of "up to the workman," which inevitably leads to the development of a science. When it becomes the duty of the management to make a careful study of any group of facts, then the results of that study naturally formulate themselves into laws, into rules, into the development of a science. I want to make it clear, Mr. Chairman, that work of this kind undertaken by the management leads to the development of a science, while it is next to impossible for the workman to develop a science. There are many workmen who are intellectually just as capable of developing a science, who have plenty of brains, and are just as capable of developing a science as those on the managing side. But the science of doing work of any kind cannot be developed by the workman. Why? Because he has neither the time nor the money to do it. The development of the science of doing any kind of work always required the work of two men, one man who actually does the work which is to be studied and another man who observes closely the first man while he works and studies the time problems and the motion problems connected with this work. No workman has either the time or the money to burn in making experiments of this sort. If he is working for himself no one will pay him while he studies the motions of some one else. The management must and ought to pay for all such work. So that for the workman, the development of a science becomes impossible, not because the workman is not intellectually capable of developing it, but he has neither the time nor the money to do it and he realizes that this is a question for the management to handle. Furthermore, if any workman were to find a new and quicker way of doing work, or if he were to develop a new method, you can see at once it becomes to his interest to keep that development to himself, not to teach the other workmen the quicker method. It is to his interest to do what workmen have done in all times, to keep their trade secrets for themselves and their friends. That is the old idea of trade secrets. The workman kept his knowledge to himself instead of developing a science and teaching it to others and making it public property.

So that many of the similar improvements in methods which doubtless have occurred to workingmen in the past, instead of being formulated into a science as they are under scientific management have either died with the workingman or have been handed over by him to one or two of his friends, and then have gradually gone out of existence. Whereas, when the management make an accurate study of processes and methods, it is not only their duty but their profit to see that this science is disseminated and is spread out before all of the workmen who are under them. For instance, when we developed the science of cutting metals, after it was developed we published it broadcast to the world. This science was published as a part of the proceedings of the American Society of Mechanical Engineers which is not a copyright publication and is free to the entire public to publish. It went all over the world at once. It was not kept as a trade secret but was made public property.

Mr. Tilson. Does everybody use it now?

Mr. Taylor. Everyone uses it all over the world. It is open to everyone.

Mr. Tilson. How extensively is your system of cutting metals being used?

Mr. Taylor. I can say that it has been translated into Russian, into German, into French, into Danish, and into Dutch; it was also published in England.

Mr. Tilson. That is all right about the books, but how about the use, the actual application of it?

Mr. Taylor. I assume that the people would not have translated it into German if they had not proposed using it. This much I can say, Mr. Tilson, that one of the great results of this careful scientific investigation—one of the direct products of it—was the discovery of high-speed steel and the moment that this discovery was published to the world every machine shop grabbed it from one end of the world to the other. It is used all over the world. It has increased the average cutting speed of machine shops at least three times over their former speed. High-speed steel went all over the world right off. There is no question about that.

Mr. Tilson. Were you the first to use it?

Mr. Taylor. Mr. White and I are the joint inventors. We have patents for it all over the world. And we were fortunate in selling many of them. We got $100,000 for the patent rights in England, but the fellows over there did not get anything out of the patent rights in the way of royalty, I understand they far more than got their money back through being first in England to equip all of their shops with high-speed steel.

The Chairman. Might not those books be bought simply for the purpose of investigation to determine from them whether or not they did want to use your art of cutting metals, and the fact that they bought the books or that they were translated into those various languages would not in itself be evidence that they had adopted the system after having had investigated it through your books, would it?

Mr. Taylor. I am quite sure that a great part of that art has not yet come into use because in order to properly use it you must have a slide rule such as I have shown you here.

The machine shops in this country have not taken the pains to use those slide rules as they should. They are not used to the extent that they ought to be. I may state, however, that I had a recent visit from the owner of the Renaud Automobile Works, the largest automobile works in France, together with Monsieur de Ram, the young French engineer who personally became interested in the art of cutting metals some years ago, and in our system of management, and who put this system into one of Renaud's departments. These two men came over to this country especially to study our system (scientific management) and the art of cutting metals, and assured me that in those departments in which they had introduced the art of cutting metals and our system of management that they had much more than doubled their former output. They said that they were going back to France to spend any amount of money and any amount of effort to get it in as fast as possible in their entire works. The warning I gave them before they left was this: I said, "You have been at it three years. Do not expect to get through with it for five years, because you will not. It will take you more than five years before you will get through the entire process of putting our system in."

The Chairman. You spoke of laboratories in connection with scientific management. Is it not true that nearly all the large firms in the country, irrespective of what system of management they have, maintain laboratories?

Mr. Taylor. I do not remember to have spoken about laboratories. Was it chemical laboratories you referred to?

The Chairman. Yes; chemical laboratories.

Mr. Taylor. Every steel works that amounts to anything has chemical laboratories, but I was not aware that I had spoken of chemical laboratories in my testimony. I may have.

The Chairman. My recollection is that you did speak of laboratories in connection with your testimony, and that recollection is reenforced by the fact that I have a note in connection with it.

Mr. Taylor. More than likely I did, then, Mr. Chairman. But I have forgotten. At any rate, I shall be glad to answer whatever questions you may ask.

The Chairman. I wanted to know if it was not a fact that nearly all of the large manufacturing establishments in the country maintain laboratories, irrespective of what management they may have?

Mr. Taylor. All the large steel works do, but I do

not think the large machine shops have the chemical laboratories.

The Chairman. There are a great many industries where laboratories are maintained, are there not?

Mr. Taylor. Yes, indeed.

The Chairman. Irrespective of what system is used?

Mr. Taylor. In the cement mills, in some pulp mills, in the chemical works of the country, in the steel mills of the country, in the rubber establishments of the country there are laboratories.

The Chairman. So that a laboratory would not for the purposes of investigation in connection with the particular industry, would not in itself be peculiar to scientific management?

Mr. Taylor. Certainly not.

The Chairman. Would it not be more peculiar to scientific research? Would it not be more peculiar to scientific research than scientific management?

Mr. Taylor. I think that these laboratories that are established in connection with industrial works are not often research laboratories in the sense in which that word is used in university parlance. I think they are very rarely research laboratories. I think they are practical laboratories needed for the everyday analysis of the products that are being made or the materials being bought.

The Chairman. Mr. Taylor, if men are induced to a greater productivity by virtue of a bonus system, and consequently an expenditure of greater energy on their part to secure this bonus, would there be any possibility of their securing a positive guarantee that would be binding for all time that the bonus would not be taken away, and thereby leaving them with the expenditure of energy at the old rate of pay?

Mr. Taylor. Most certainly no permanent guarantee could ever be given for anything that I know of in this world. But the workman would always have his remedy open to him. If he were badly treated he could soldier just as he is now doing under the present system. This is his cardinal remedy. This is the final word. The workman always has that resource. All the workmen have to do is to sit down and soldier, and the injustice comes to an end.

Mr. Redfield. Does he not have the interest of his employers always at heart?

Mr. Taylor. Yes, indeed. 1 am assuming that a fool employer, and there are a good many of them—

The Chairman. Are there not differences of opinion as to what constitutes a fool employer?

Mr. Taylor. Yes, sir; and a great many of the old-style employers are pointing to those who are introducing scientific management as being fool employers, inasmuch as they pay this unnecessary increase in wages to their workmen, as they call it. I do not share that view, of course, but a great many of the old-style employers do.

Mr. Redfield. Have you dealt with the question as to what happened to those laborers in the yards of the Bethlehem Steel Co. who were laid off from shoveling, so to speak, when the force was reduced, as you have testified, from between 400 and 600 to about 150?

Mr. Taylor. Mr. Redfield, I am very glad, indeed, that you asked that question. The general impression which I find in the minds of people who hear the story told of the reduction of the men from, say, 500 to 150 (and this impression is particularly strong with those ladies who have heard of that story), is that all of the men who were thrown out of work went right out and drowned themselves in the river which flows by the works because they could never get any more work to do, and would therefore have to starve to death. That is the usual impression. I find a vast sympathy on the part of all classes of the community for those poor fellows who were thrown out of work, and who could never do anything else as long as they lived, but mighty little sympathy for the 150 who remained with the company and who received 60 per cent higher wages than they had ever earned before, or that the same men could get if they stepped out of that establishment and went to any other works around that part of the country. Now, I find that is the universal frame of mind, and I am very glad of the opportunity of saying just what happened.

The Chairman. You do not think those men who remained in there need sympathy, do you?

Mr. Taylor. No; I do not. I think they were all subjects for hearty congratulation. And I feel that the management who gave them 60 per cent higher wages than anyone else would pay them ought to have some sympathy and some regard. They ought to be looked upon as kindly and nice employers instead of being looked upon as brutes because the other fellows were discharged.

Mr. Redfield. What happened to the other fellows?

Mr. Taylor. What happened to the other fellows? That is the proper question. In every one of our establishments we have men employed whose business it is to make a careful study of the laborers as they come to work, that is, of all of the ordinary day laborers, as they come into the employ of the company. Those men are selected because they know how to get next to the average workman as he comes in, get acquainted with him, and find out what he is thinking about, to ask him what kind of work he has done before, and watch and study the new men when they do not know that they are being watched. They will come right on him while he is at work and see if he is really an industrious man. In other words, their business is to get thoroughly acquainted with the newcomers. There are any number of fine fellows who come into the steel works, or into any other establishment, as laborers who never in their youth had the opportunity of serving an apprenticeship, and yet who with the proper instruction and the proper opening were intelligent enough and energetic enough to have learned a trade.

This man or these men who are employed especially for the purpose of making a study of these laborers are constantly sent for by the foremen of the various departments who are in search of good workmen. The foreman of the blacksmith shop, the foreman of the foundry, the foreman of the machine shop, the foreman of the rolling mill, of all the various departments of a steel works, are constantly after these men. They say, "Haven't you got any good raw material for me to try out in my department?" Whenever a fellow shows himself to be an energetic, a good, hard-working fellow—and if, in the judgment of this man he has sufficient intelligence to become something more than a shoveler, something more than a pig-iron handler—he is deliberately taken out of the labor gang and put, say, into the smith shop, first as a laborer, then finally taught to be a helper, to learn to strike at a forge; or he is taken into the foundry as a laborer, and then gradually taught to be a helper to the molder and given the higher wages that go with these higher types of work. Or he is taken into the machine shop, if he is an especially intelligent man. And later on he has the opportunity of learning to be a helper to the machinist who is running a big machine which calls for the work of two or three men. Now, to show the extent to which the men were promoted from the laboring gang in the yard of the Bethlehem Steel Co.—that is the gang we spoke of where the reduction had been made from 500 men to 150—to show the extent to which promotion took place from this gang: In the big shop of the Bethlehem Steel Works, which is about one-third of a mile long, also one of the widest machine shops in the country, there are a great many powerful roughing machines—machines used for removing the outside rough material from forgings or from castings, merely to take off the heavy rough stuff, not to finish to size. These machines are called upon to do work in which a limit of accuracy of a quarter of an inch in diameter or half an inch in diameter is sufficient.

In running a machine of that sort nothing like the same amount of skill is required which is demanded of a first-class mechanic who has to finish work to exact size and put a true, fine finish on it. There are any number of those heavy roughing machines in the big shop of the Bethlehem that do not demand a high-class mechanic to run them. Before we left the Bethlehem Steel Co.—just as a matter of interest to ourselves—we had an investigation made to find out the origin of all the men who were then running the roughing machines in that shop, and 95 per cent of the men who were running these machines had been promoted from laborers, had been taken into the shop, taught their trades, and had risen to the position of roughing machinists, and then had been given the higher wages which goes with this class of work, as well as having the higher and more agreeable work to do. That is what happened with those 500 yard laborers who have been pitied so for the hard treatment they received.

Mr. Redfield. What happened to the men at the roughing machines?

Mr. Taylor. If they were good men, if they were able to learn to do finer work, they were promoted from there onto the finishing machines.

Mr. Redfield. Then do you mean that under the system as it was applied there was a general upward movement throughout all grades in the shops?

Mr. Taylor. That is exactly what takes place under scientific management. The management look upon it as their duty to raise every man in the place to the highest grade of work for which he is suited and then to pay him the higher rate of pay which goes with the more skilled work.

The Chairman. When it had reached the point that you were about to elevate the second highest grade to the highest grade in this general movement upward, what became of the men in the highest grade?

Mr. Taylor. They became the teachers. They became the functional foremen. They were promoted to the planning room. They were placed in exactly the same position that these gentlemen have reached whom I have brought here to testify before you and to tell you how they were promoted. They started as workmen and finally graduated as bosses.

The Chairman. Had you reached that stage in the introduction of scientific management at the Bethlehem Steel Works where you had these functional foremen supplied from the men from the highest grades?

Mr. Taylor. We had to a very great extent. I suppose we had 40 or 50 promoted in that way, but nothing like as many as we ought to have had if the works had been finally systematized as it ought to have been. There ought to have been three times as many men who had graduated from machinists to teachers, etc. and there would have been if we had remained there.

Mr. Redfield. Excuse me, Mr. Taylor, but what has been your experience as to the effect of the helping and the teaching and the definite instruction card which workmen receive under the scientific management in its effect upon making them mere machines and injuring their initiative?

Mr. Taylor. Mr. Redfield, I answered that question already, did I not, Mr. Chairman?

The Chairman. I believe you did in your own way.

Mr. Taylor. If you wish me to answer it again I will do so. I think that question is on the record.

Mr. Redfield. Then it is not necessary to answer it.

The Chairman. I made an inquiry in practically the same language.

Mr. Redfield. I understand. We will let that rest then.

Mr. Taylor. Not that I object to answering if you wish me to.

Mr. Redfield. Mr. Taylor, how far is it recommended or is it customary in connection with the installation of the system of scientific management to require or to utilize incidental apparatus in which you have an interest as a manufacturer?

Mr. Taylor. I hardly understand that, Mr. Redfield. I do not quite understand your question. If you will give me an illustration perhaps I can answer it.

Mr. Redfield. The suggestion has been made at various points in the testimony that while it must be understood that you are not actually engaged professionally and personally in the business of introducing scientific management that you would have a marked financial interest in its introduction arising from the necessary sale, it is suggested, as an incident, as a portion of the installation of the product of certain businesses in which you are a part proprietor.

Mr. Taylor. Mr. Redfield, if anyone wants the profits I am making annually they can have them for the asking from any incidental apparatus that is sold. These slide rules, the use of which I explained to you, for instance, I have never known one of them being sold to anyone. They are given away if anyone will show us that he can use them. Mr. Barth and Mr. Gantt and I, myself, are the joint patentees of those slide rules. If any man can come from any part of the world and show us he can use that slide rule, he may have it for the asking, but he has got to show us that he can use it.

We used to let them have slide rules like these, whether they could use them or not, until we found that they were being used as an object lesson to display the folly of scientific management. Men whom we had given these slide rules to would say, "Why, here, just see what damn fools these fellows are. They use a thing like this to run a machine shop with." When I found that this was the use to which they were being put, we got a little bit wiser. We said, "You cannot have these appliances to make fools of us with. You cannot have them until you can show us that you know how to use them." And in further answer to your question, Mr. Redfield, far from making money out of scientific management, since retiring from money-making business I have each year, for the past ten years, spent more than one-third of my income in trying to further the cause of scientific management, besides giving my whole personal time and work to the cause without pay.

The Chairman. Is the slide rule an essential part of scientific management?

Mr. Taylor. No, sir. It is not an essential part, but it is a highly desirable instrument; if a man wants to run a modern machine shop as it really ought to be run under scientific management, he must use it. The Midvale Steel Works, my old establishment, are still using the tables which Mr. Gantt and I developed there for running their machines instead of the more modern and far more efficient slide rules developed after we left there. These tables were the limit of the mathematical solution of that problem when we left Midvale in 1889. The same tables are still used by the Midvale Steel Works.

The Chairman. Is it not only applicable where machines are used?

Mr. Taylor. Certainly; this rule is only applicable to the solution of problems connected with the art of cutting metals.

The Chairman. As a matter of fact, is not the so-called scientific management consigned almost exclusively to machine shops, and to the metal trades particularly?

Mr. Taylor. It is in use in flour mills, in paper mills, in cotton mills, in bleacheries, dye works, in printing establishments, lithographing, and the Lord knows what. Mr. Chairman, you can go right along, into the steelworks and ironworks and machine shops of all kinds and sorts, and find it in use in pulp mills, optical works, electrical works, and even a button factory. One of the shops was a bicycle-ball factory. They made some 300,000,000 bicycle balls in a year. There is variety for you.

I may say, as an interesting and new use for scientific management, that the director of public works at Philadelphia was appointed to that position so as to introduce the principles of scientific management in the management of the city of Philadelphia. He is doing it mighty fast. He is making a mighty good start at it. I should like very much to have the director of public works at Philadelphia to appear before the committee if you care to hear him, and have him give you his experience with scientific management, because he was chosen for his present position on account of his experience in scientific management.

Mr. Redfield. Is not scientific management largely a state of mind?

Mr. Taylor. The essence of it is this new state of mind. The very essence of it involves this new and complete mental revolution as to the duty of both sides, one toward the other; the substitution of the attitude of peace for the attitude of war. There is no question about that.

Mr. Redfield. Was scientific management ever introduced in whole or in part in the factories of the American Locomotive Co.?

Mr. Taylor. I am very glad to state, not what I know, but what I believe to be the truth about the American Locomotive Co. I have never been in their works since they started to try to introduce scientific management; but if such knowledge as I have, and it has been obtained by talking to perhaps 20 or 30 different reliable men connected with the American Locomotive Works, will be of any value to you, I shall be very glad to give it.

In the first place, Mr. Van Alstine, whom I know intimately, and who I have every reason to believe is one of the most upright and straightforward and honorable men in this country, and who is a high-class man, became interested in the principles of scientific management when he was master mechanic of the Chicago & Great Western road; but he met with little sympathy in his attempt to introduce these principles in the shops of that road. He then went to the Northern Pacific as master mechanic, and had very much greater success there. But he found that after all people there had no great sympathy with him. They did not understand what he was driving at. He produced economies which were very notable, and which led them to want him to remain there, however, in the most urgent way. Then he finally went to the American Locomotive Works, with the object of introducing the principles of scientific management into that works. About the time he went there he came to see me, because I had been in consultation with him for several years. He came to see me about the introduction of scientific management in the American Locomotive Works, and the most urgent advice which I gave him (and I gave it in a most emphatic way) was that he should not start in the locomotive works to introduce scientific management until he had the complete backing of the board of directors of that institution, until every man on that board, as well as the president of the company, was with him—until every man on that board wanted scientific management and wanted it badly.

It has been my experience that if a man starts to introduce the principles of scientific management into any company, unless the owners of that company, the directors, the people who have the final power—unless they want it and want it badly, and understand the price that has to be paid for it (and that price is one of long time and patience), my advice to him was that you let that thing alone. Mr. Van Alstine thought he could carry it over, as he said, without bothering the whole board to get a thorough knowledge of the whole matter and everything connected with it, and he started to introduce scientific management, and started in the right way to introduce it—that is, rather slowly. But if I understand the conditions—and I think I do—the board and the president began to put such pressure on him for immediate results, that, contrary to his best judgment, he was tempted to shove the thing too fast.

He attempted to do what is an utter absurdity in any company. He attempted to do in two years what he ought to have taken five years to do, and in doing so he and Mr. Harrington Emerson, who joined him, abandoned the very essence of scientific management, the one essential thing. They tried to force in a whole lot of mechanism which ought to belong to scientific management; it is all useful and very fine, this mechanism, without waiting to convert the workmen as they went along; that is, to bring about this great mental change on the part of the workmen which is necessary for the success of the system. They went ahead, neglecting the absolute necessity of the mental change both on the part of the workmen and those on the part of the management, which I have referred to so many times in my testimony as the essence of scientific management. They tried to do what is an utter absurdity, and finally wound up by forcing the mechanism of scientific management in many departments of the company, where the proper spirit did not exist among the employees at all, and that led to just what I told them it would lead to when they first came to me. I told them, "If you do not go slow enough; if you do not allow the workmen to see that the new system is a fine thing for them, and get them into the proper frame of mind, so that they will cooperate with you thoroughly, the time will come when the whole thing will fall." As I have said before, the chief trouble with the whole undertaking lay with the board of directors. Their attitude was wrong. It was the owners who finally made the thing go wrong.

Mr. Redfield. Mr. Taylor, how far is scientific management in use by any of the large railway systems of the country?

Mr. Taylor. There is one of the large railways in this country that is using it to a very large extent. I have some of the data here which was given to me in confidence by the man who spent, I should think, some three, four, or five years in introducing the principle of scientific management very largely in one of our great railway systems. The result of his work has been that during the whole time in which he has been working there and up to the present time there has existed almost perfect harmony between the workmen and their employers. The workmen are earning higher wages I understand than corresponding workmen in any other railway system in the United States. If I remember rightly he told me that all the repairs on 20 types of locomotives were made with proper instructions as the result of accurate, careful time studies, and that the men who were making those repairs were all working under piecework. I may be wrong in the figures, but my remembrance is that he said that 70,000 items of repairs had been studied in that way on the locomotives and cars of this section of the line. I regret that I am unable to give the name of the man and the road which is doing this, because it came to me in confidence, and while I should be very glad and delighted to help you in getting a complete knowledge of this work I always feel that I am bound to strictly maintain a confidence of that sort.

Mr. Redfield. In other words, you were his professional adviser?

Mr. Taylor. In a way, yes; he started because he had read what we have written on the subject. He came down to see me at intervals and talked the matter over, but I could not say that I was his professional adviser. I was merely a friend having the interest I have in all earnest endeavors to introduce the principles of scientific management. I should be delighted to show you samples of the piece-work schedules that he gave us. Here are two lists of these piece-work prices.

Mr. Redfield. Did you say that there was 70,000 of them?

Mr. Taylor. They are simply samples of the 70,000. These are two of the various schedules which he left with me. My impression is that there are 70,000. My recollection is that on another branch of the same road there are over 100,000 items of locomotive and car repairs carefully studied and put on piecework in this way. I am sure the number of operations was 100,000 to 130,000 on one of the branches of that line and somewhere near 70,000 on another. In a recent conference the vice president of the road told me, "No set of men on the face of the earth can ever stir up any sort of discord between us and these employees of ours who have come under these new principles. We have become the best of friends under this system." That is the principal reason why I have concluded that in this railway company the principles of scientific management exist. In talking with him lately I asked permission to place this information before your committee. He said, "Yes, as far as I am concerned, but the request ought to go to the board of directors of our company; I have not the authority to do that sort of thing without their permission. I think I can get the authority. As far as I am personally concerned I am delighted to have this knowledge go anywhere, but you understand I am not the whole thing, I am not the railway company, I don't know what our board of directors would say."

The Chairman. Mr. Taylor, without making the name of this particular railroad public, or without any desire to put the name of the railroad company in the record, in view of your explanation, is it not true that within a year that railroad company had very extensive strikes in its railroad shops?

Mr. Taylor. Certainly in none of the shops where this was introduced. I am absolutely sure of that. As to what occurred in other shops I do not know. There is one large section of that line that has not yet come under these principles, and what occurred there I do not know. My impression is, as you say, that there was a strike in the section, still working under the old system, but nothing of the kind in the two sections where our system of management was in use. That I am sure of.

The Chairman. I think you said, Mr. Taylor, that scientific management was to a great extent a state of mind.

Mr. Taylor. Without a certain state of mind scientific management cannot exist. There must, however, be something more than a state of mind. There must first be a certain state of mind—that is, a certain new outlook on both sides. The idea of peace must replace the old idea of war on both sides. Then in addition to this change in mental attitude both sides must come to look for exact facts and exact information as the foundation of their action. That is, exact science should be the basis for every action instead of the old rule-of-thumb knowledge or guesswork.

The Chairman. Would not a state of mind be a very unstable and changeable thing upon which to base materialistic production?

Mr. Taylor. I think there is nothing more stable in life than our convictions. If there is anything stable in life it is a state of mind. It is principles, and there is nothing more permanent than the principles which have become deep rooted in us. The principles of religion, the principles which govern men's daily actions are the most stable things in us. Our outward acts may change, our knowledge may change, our views may change, but once we have fundamental principles they rarely change materially.

The Chairman. It is a noted fact that the state of mind frequently changes?

Mr. Taylor. Yes, in minor matters they do, but when people are gradually convinced, when men adopt a new mental attitude toward one another, and toward their duties, and scientific management is a revolution as to their duties toward themselves and their fellowmen—that is, a slow revolution, difficult to bring about, but once it is brought about it is apt to be very stable.

Mr. Tilson. Is there not this further fact that if your contention is true that it is not only a state of mind that is just but it is profitable to both parties?

Mr. Taylor. Exactly; immensely profitable.

Mr. Tilson. So that their particular interest will coincide with this state of mind if your contention is true?

Mr. Taylor. Yes, sir.

Mr. Redfield. How is it possible to study how

long a workman should take in that part of the work that is purely mental? For example, how long he should take in making up his mind how work should be done or in reading and grasping a drawing?

Mr. Taylor. The first piece of time study that I ever saw made by anyone was made in the study of just that thing, a study of the mental capacity of boys. When I was at Phillips Exeter Academy, Mr. George W. Wentworth was the professor of mathematics, and he worked off his first geometry while it was in manuscript and his first algebra on my first class, the class of '74. He worked those books off on us for the two years while I was there. I, as a student, wondered how it was possible (that right along steadily, right through from the beginning to the end of the year, as we went on from month to month) that old bull, Wentworth, as he was called, gave us a lesson which it always took me two hours to get. For the two years I was there I always had to spend about two hours getting that lesson, and finally we got onto his method. We were very slow in getting onto it, however.

Mr. Wentworth would sit with his watch always hid behind a ledge on the desk, and while we knew that it was there we did not know what the darn thing was used for. About once a week or sometimes twice a week he went through the same kind of exercise with the class. He would give out a series of problems and insisted that the first boy who had them done should raise his hand and snap his fingers. Then he would call his name. He went right through the class until just one-half the class had held up their hands. We always noted when he got half-way thru the class and the middle boy would snap his fingers he would say, "That is enough; that will do." What he wanted was to find out just how many minutes it took the average boy in the class to do the example which he gave. Then we found that Wentworth timed himself when he first tackled those problems. He got his own time for doing those five examples, and the ratio between his time to do the examples and the time of the middle boy of the class enabled him to fix the exact stunt for us right along. The speed of the class changed. He did not change. All he had to do was to get this ratio of change, and he could say, for instance, the average of that class will take 2 hours if I can do the examples in 25 minutes, and in this way he was able to give the class its proper stunt right along. That was the first instance of a time study of mental operations which I had ever seen. Under scientific management we are working constantly making mental time studies now. If we want to find how much time it takes for the average machinist to read a new drawing which he has never seen before, the man who is in the planning room and who is especially skilled in reading drawings—that is what he is there for—keeps a close tab on the time it takes himself to read all kinds of drawings, and he knows, for instance, if it takes him 10 minutes it will take the average man in the shop, say, three or four times that long. That, for example, may be the ratio between the skilled man and the average man in the mental operation of reading a drawing. The moment he knows how long it takes him, then by multiplying he knows how long it takes the average man. He has to keep himself constantly in touch with the men in the shop in that way, of course. Mental time study is made by us now, just as it was made by Wentworth in 1872.

Mr. Tilson. How do you first find out how long it takes the man in the shop to do it? How does this man in the planning room first find out how long it takes the other man to do it?

Mr. Taylor. You must realize that a lot of similar information is already known for other drawings. So that the man in the planning room has a general line on how long it ought to take to read drawings, and this makes it difficult for any workman, if he even is inclined, to fool the planning-room man very much. The planning-room man calls in a reliable workman and says, "John, I want you to study this drawing, and study it right, and let me know when you have got wise on it." Now, in this way he asks several men of about average ability to make this study and finds that it takes them, on the average, 20 minutes to do it; then he will study the drawing himself and see how long it takes him to get onto it. In this way he gets the ratio of his speed to that of the average man in the shop. Once that ratio is determined it becomes a rather simple matter to make this kind of mental time study.

Mr. Tilson. But, after all, that is only approximation?

Mr. Taylor. The whole subject of time study is only an approximation. There is nothing positively accurate about time study from end to end. All that we hope to do through time study is to get a vastly closer approximation as to time than we ever had before. That is one reason why we have to allow this big margin of safety, as I explained to you. A marginal allowance of from 20 per cent to 225 per cent is added to the observed time, so as to cover all kinds of uncertainties.

The Chairman. When you make a time study of a man at physical labor do you not always eliminate in that time study the pauses in that man's work, the time when he is not actually applied at his labor, so as to get at the accurate and actual time in which he performs the labor?

Mr. Taylor. There is a printed page (indicating) that is typical of just what is done in time study illustrating this part of the subject.

The Chairman. That will not put the answer to my question in the record.

Mr. Redfield. Let us put this in the record.

The Chairman. I wish to get a direct answer to my question.

Mr. Taylor. To answer your question, we do both things. We take the gross time, the whole time which the man takes in doing the job, and then we make at the same time another study which includes the productive time alone, the time he is engaged in actual work. On this printed page there is a study of the gross time, and a study of the productive time as well.

The Chairman. When you make a study of the productive time you eliminate in that study, and are able to do so by virtue of your stop watch, the periods in which the workman is not engaged in productive work?

Mr. Taylor. Yes, sir.

The Chairman. How can you take a mental study of the productive time, the mental time that it takes to work out a problem? Would you be able in your time study to take the time of the mental pauses that occur during the time when the problem was being worked out?

Mr. Taylor. The time during which the man stops to think is part of the time that is not productive.

The Chairman. Can you get a record of it with your stop watch or by any other method of timing?

Mr. Taylor. We can get the time during which the man is thinking with the stop watch in just the way that I described to you in the reading drawings, by telling a man to do some mental act, and then seeing how long it takes him to do it.

The Chairman. Would not that simply be the gross mental time from the time the man starts to work?

Mr. Taylor. Yes.

The Chairman. Would you be able to make a time study showing the amount of time in that gross time that was non-productive mental time?

Mr. Taylor. I would assume, Mr. Chairman, that if you asked a workman in advance, saying, "Now, John, I want to find out how long it will take you to get a complete notion of what you are going to do in this work. Now, play fair with me, John. The moment I tell you what you are to do you start and think and plan it all out and don't start to work until you have your plan all made." I think John would be fair in that. I think he would do his thinking in a fair way, just as he does this work in a fair way. And that he would tell you when he had finished making his mental plan.

The Chairman. Why could you not take his word?

Mr. Taylor. You could not be absolutely sure that he was not deceiving you in some way. But I have found that when you are straightforward with men and when you explain to them what you are trying to do, and when they believe that you are in the main straightforward yourself, and that there is no crookedness back of what you are trying to do, men will generally cooperate with you honestly.

The Chairman. Why do you not take his word for it in the physical work then as well as taking his word in the mental work?

Mr. Taylor. Yes—

The Chairman. Why put the stop watch on him?

Mr. Taylor. Because he cannot use the stop watch on himself. He cannot work and put the stop watch on himself at the same time. As I have told you time and again, Mr. Chairman, the way we do in almost every case is to go to the man in perfect frankness and say, "John, we propose to make a joint study of this kind of work; we want to get at this together because it is for our mutual interest to do so. I am sure that you will work fairly on this." As I told you in the case of those laborers, we paid them double wages when they were being studied in that way. We doubled their wages. They played perfectly fair with us. They did not either overwork or underwork. They worked at a proper pace for a fair man to work at. That is the way we get all our information. It is through co-operation. It is not through any sneaking business. It is not through any underhanded business. I think, Mr. Chairman, you will see that in everything I have written in relation to the time study I have advocated absolute frankness and no underhand work. There is no sneaking about it if time study is properly applied.

Mr. Redfield. Have you explained how you arrived at the percentage of increase in pay necessary to make men desirous to work under scientific management? You have said that it was sometimes 30 per cent and sometimes 50. How are those figures arrived at?

Mr. Taylor. Again, that has been the subject of a scientific investigation. It is not the question of my judgment or of any other man's judgment. I am very glad that you brought this matter up, because the average person thinks that the premiums which we pay of 30 per cent for this kind of work, 50 per cent for that, and 80 per cent for another kind are all arbitrary figures, arrived at from some one's judgment. These percentages were adopted as the result of a long series of experiments. They represent a most difficult type of experiment to make. Nevertheless they were experiments, carefully and scientifically made experiments.

To make one of these experiments I took, perhaps, eight or nine of my friends who were workmen—it was after we had started scientific management, after we had arrived at this condition of mutual confidence which exists between employer and employee under scientific management—I picked out six or eight of my working friends who were nice chaps and sensible, common-sense fellows, who had confidence in our integrity and believed in what we were doing. We were good friends. I said to this group of six workmen, "I am going to give you the same class of work that you have been doing in the past, but I want you to change from working on plain daywork in which you have done the work according to your own method, and to follow the method which we will lay down for you in an instruction card and also you will be expected to do the work within the specified time. Whenever you do the work right and within the specified time we will give you a premium amounting to 15 per cent increase in your pay. Now, just go at that fairly, you fellows, work in the new way for six or eight months, and then if at the end of that time you do not like it, after you have given it a fair trial, let me know, and you can go right back to the old conditions again if you prefer them."

Another set of men, we will say the same number, were given 20 per cent increase in pay; another set of men were given 25 per cent increase in pay, and another set an increase of 30 per cent in pay, and another 35 per cent, and so forth.

Now, out of the six who were given 15 per cent—I do not say that six was exactly the number, but that is approximately right—practically almost all of them came at the end of the six months and said, "Now, see here, Fred, I have tried that scheme of yours, and I do not like feeling all the day long that I am tied down to any old pace, or to a new way of doing things. I should prefer going back to the old way." Very well; this experiment showed that an addition of 15 per cent to the workman's pay was not sufficient to compensate him for the bother of having to change his ways and methods of working and adopt some other man's way of doing things. For it is true, as you know, under scientific management, that the man is not allowed to do work in the old way. He has got to learn a new set of motions and do many new things, and the 15 per cent increase in wages was not enough to make those men feel happy and contented in making this change.

At the 20 per cent increase almost all of the men asked to return to their old conditions and their old pay. At the 25 per cent increase more than half of them stuck to the new conditions and preferred them to the old, the 25 per cent increase was attractive to them. At the 30 per cent increase all but one stuck to the new plan. At 35 per cent my remembrance is that all stuck.

It took some years before that experiment was fully carried out, and we made up our minds that when workmen are paid from 30 to 35 per cent increase in wages, 19 out of 20 good workmen, well suited to their jobs, are happier and more contented under the new system than they were under the old, because you will remember that they had had their free choice between two systems. It was in this way that we got at these percentages. I call that a scientific experiment; that is not some one's guess. And it is typical of scientific management that every element that comes under it sooner or later becomes the subject of careful scientific investigation.

Mr. Redfield. The statement has been made that it is un-American and an indignity for a workman to submit to time study with a stop watch; that it is annoying and makes a man nervous and irritable. To what extent have you any knowledge as to what extent that is true or not true?

Mr. Taylor. Mr. Redfield, I think that the average workman, if any man came to him with a stop watch without any previous explanation or understanding and began timing every motion and writing down what he was doing, would become nervous and would be irritated by it. I think it is perfectly natural that any workman should become irritated at an action of that sort. I am very sure that I should be nervous to a greater or less extent if anyone were timing every one of my motions. I would feel that it was a darned mean job while the thing was going on. But, Mr. Redfield, I wish to call your attention to one fact, which is not at all appreciated: somehow there has come to be an impression in the minds of people who speak and think of scientific management in its relation to time study, that for every workman who is working in the shop there are probably four or five men standing over him year in and year out with stop watches. Let me tell you that in some of our shops there are many workmen, who in the whole course of their lives, never have had a stop watch held on them. And that probably the average man would not be timed for more than one day in his lifetime. So that probably one day of the workman's life would sum up the total of this terrible nerve-racking strain which several of the men who have testified before your committee have complained of. Therefore, if any man objects to time study, the real objection is not that it makes him nervous. His real objection is that he does not want his employer to know how long it takes him to do his job. Because when his employer has this knowledge soldiering becomes much more difficult.

The Chairman. Would it not be more likely that his real objection was that a time study taken under those circumstances and for a brief period of time with an unaccurate system of stop watch, was not the proper kind of study upon which his wages should be based?

Mr. Taylor. I am very glad you brought that out, Mr. Chairman. You must remember that in any one workman's work, which is now being studied with a stop watch, all that the time student is looking for are perhaps eight or ten motions that the workman makes. The rest of his motions have already been studied on other workmen. The great majority of the movements of machinists have become standardized and require no further analysis or timing. When you study new work nineteen-twentieths of the motions made by the machinist have already been studied. It is the one-twentieth, the one new type of motion that we have not yet had the opportunity to study, which the time student is after. You will understand that modern time study as it is done in our shops is a study of each elementary motion made by the workman. It is not a roundup of how long it takes a man to do a whole job. That kind of time study is very rare. With each new machine that a man starts to run there may be five or six new motions that have never been studied before, and it is those five or six which we are after. And a day's work will give plenty of opportunity to get those few motions all right. These same motions may be repeated 50 times a day, and that will give you a chance to get a fair average of them. The workman does not know unless you tell him what it is you are studying. You come out to see him and say: "John, I want to find out four or five things about your work. When they come around in the course of your work I am going to note down those four or five motions." We rarely make a time study of a man without taking the man into our confidence, without going to him in advance and saying this is what we come after. We want to find out these facts. It is to your interest, just as it is to ours, to have this time study accurately made.

I can tell you that time and time again the request comes to us from a workman to please come and study his job, so that we can give him a chance to earn a premium. He will say the other fellows are getting paid a premium for their work and I would like to get in on it too.

Mr. Redfield. Mr. Taylor, is soldiering still practiced in the works that are systematized under scientific management?

Mr. Taylor. I think that I may say that to a small extent it is still practiced in every scientifically managed shop. I do not think it has ever been entirely done away with. I can tell you the reason why. In the early stages, when scientific management is being put into a shop, the men who are installing the system are very anxious to have the workmen participate as early as possible in the gain which accompanies the scheme. We are very anxious for them to earn larger wages. We are desirous of proving to them as soon as possible, through an object lesson, that the management is not going to be the only party to benefit by the change, but that the workmen will benefit through an increase in wages quite as much as we do. So there is a very great temptation to fix tasks which are still partly founded on guesswork. We will go to a workman and say, "Now, John, we have not yet made a complete, accurate time study of this job of yours. You understand you are going to be paid a premium on this job, although the task is based half on guesswork. We will be frank with you and tell you that we do not know enough to fix a proper task, but later on we will make a proper time study of this work, and then the task will be revised and made right." In a company which is just introducing the system there will be a thousand or more jobs put on task work in the course of a year where the time study has not developed sufficient information to fix rates that are absolutely just. While it is the intention of the management to go back and pick up every one of those jobs that have been half time studied and make a thorough time study of them and finally establish rates which are equitable, in many cases these jobs are lost sight of. When a workman strikes one of those snaps in which too large a time allowance was made there is a good deal of temptation for him to soldier. I can hardly blame the workman for not giving away a snap of that sort, altho we constantly have workmen coming to us and pointing out that too much time has been allowed on jobs of this sort. Workmen are just as honorable as the rest of the community.

Mr. Redfield. In your talks with the workmen what did you find was their chief objection to the introduction of scientific management?

Mr. Taylor. I think the chief real genuine objection to scientific management on the part of the employees in our arsenals and navy yards is the fear that if it is introduced it will break up the practice of soldiering and ultimately throw a lot of them out of work. They realize that it will largely increase the output per man, and that therefore a great number of their fellow-workmen will be thrown out of jobs. I think that this is a genuine fear on the part of the workmen in spite of the fact that the whole history of the introduction of scientific management shows that it has rarely resulted in throwing men out of work. I think that is the chief objection. But I think there is another cause for the recent protest from the men in Government employ against our system. I think that the objection on the part of the men in the Watertown Arsenal, in which scientific management is being introduced, was largely brought about by the utterly unjustifiable and mean misrepresentation of scientific management which was embodied in the circular which was sent out by Mr. O'Connell, the head of the machinists' union, and of which I have a copy here, and which circular is already printed in a record of this hearing. Mr. O'Connell wrote a circular, which was sent to the members of the machinists' union all over this country, utterly misrepresenting every element of scientific management. Misrepresenting is a mild word. I would like to use a stronger one, but I do not care to burden the record with it. But misrepresentation is a mighty mild word for what Mr. O'Connell has written in his circular. Here is the circular printed in the National Labor Journal, Washington, D. C., January, 1912, and here are some of the expressions to which I want particularly to call attention, so as to dispose of these misrepresentations right here. The fourth item in Mr. O'Connell's description of scientific management reads as follows:

"Instead of collective bargaining, Mr. Taylor insists upon individual agreement, and any insistence on organized-labor methods will result in discharge. Wherever this system has been tried it has resulted either in labor trouble or failure to install the system, so it has destroyed the labor organization and reduced the men to virtual slavery, low wages, and has engendered such an air of suspicion among the men that each man regards every other man as a possible traitor or spy."

Now, Mr. Redfield, that statement is utterly and completely false, and I wish to refute in the most positive way the main statement there, namely, that it reduces the workman to low wages. In proof of that I want to present as a paper to be placed on this record a statement made on October 24, 1911, in which the names of all the employees of the Tabor Manufacturing Co., of Philadelphia, are recorded, who were working at that time in the shops of that company, and who had been working for one year or more in the employ of that company. This statement gives the name of the man, the original date of his employment, his first occupation, the price at which he hired himself to that company when he first came, his present occupation, and his average wages earned per hour during the week just preceding the date of the report (the week previous to October 24), and the statement then gives the percentage of increase in the pay which each man has received since he first entered the employ of the company.

The Chairman. May I get this point, Mr. Taylor, if this shows the increase of pay to each workman while working at the same class of work?

Mr. Taylor. In some cases the men are now working at the same class of work as they did at first, but in most cases, as I have told you, the men who come under scientific management are taught how to do a better and higher class of work than they did before, and they are given a finer and higher class of work to do with the accompanying higher pay, and this refutes Mr. O'Connell's statement that whenever scientific management has been introduced it leads to "virtual slavery" and "lowering of wages." This statement shows that far from leading to anything resembling "slavery" and to "low wages," as stated by O'Connell, that the system has led to an average increase in the wages of every man in the shops, including even the colored men who just carry the material from place to place, of 73½ per cent. That is the difference in their wages from the time they came there and their present wages. Is this "virtual slavery" and "lower wages," as stated by O'Connell? I would like to have that table placed in the record.

The Chairman. Without objection, it will be inserted.[1]

The Chairman. Would this table show that the wages of the machinists were 73½ per cent higher now than they were before the introduction of this system?

Mr. Taylor. It shows that for the average man in that establishment, if you take the price at which he was hired when he came there and his average earnings per hour during the week preceding October 26, that the average wage for all the men throughout the shops is 73½ per cent higher. For example, the first man on this list the percentage of increase of 158 per cent, for the second man 50 per cent, the third man 50 per cent, the fourth man 64 per cent, and the fifth man 207 per cent, and so on.

The Chairman. How do the wages of machinists here, for instance, 40 cents per hour and 37 cents per hour, 34 cents per hour, and 32 cents per hour compare with the prices paid for machinists in other establishments?

Mr. Taylor. I think that the wages are very ma-

[1] The table is given on pages 276-277.

terially higher in all cases. It aims to be at least 35 per cent higher than the same man doing the same work could get in any other establishment right around us.

The Chairman. This is 35 per cent higher than the wages generally paid for machinists in other shops around Philadelphia?

Mr. Taylor. Than that same man could get if he went right out of this shop and into another shop right around there in Philadelphia and worked at similiar work. That is what the aim is.

Mr. Redfield. I will read you from this report of Mr. James O'Connell. He says:

"These jobs, namely, the speed boss, the gang boss, the inspector, are given as plums to machinists who are willing to act as pacemakers."

Is that statement correct?

Mr. Taylor. That statement is absolutely false. These men are chosen because they are fit to be teachers of other men, because they are kindly men as well as competent men, and want to help other men, not because they are pacemakers, to make the workmen do something that is disagreeable and that they do not want to do.

The Chairman. Would that not be true only under the ideal conditions of your system? Would it be true in all cases in its practical operations?

Mr. Taylor. There might, of course, be an occasional gang boss or speed boss who would be unjust toward his men, but the moment it was found out, that man would be called down and corrected.

STATISTICS OF WAGES OF TABOR MFG. CO.

	AT TIME OF EMPLOYMENT			PRESENT DATE		
Name of Employee	Date	Occupation	Rate at which Employed	Occupation	Average Wages Earned	Percent. of Increase
			Per Hour		Per Hour	
Allibone, W.	6/22/05	Tool boy	$0.12	Machinist	$0.31	158
Angerman, C.	6/ 3/04	Machinist	.24	Vise hand	.36	50
Anderson, C.	12/ 3/09	Machinist	.26	Machinist	.39	50
Bradley, G.	10/17/02	Machinist	.25	Machinist	.41	64
Bierchank, W.	9/10/04	Machinist's helper	.15	Machinist	.46	207
Bryson, D.	10/29/06	Colored laborer	.16	Machinist's helper	.23	44
Blackwell, W.	2/16/05	Colored janitor	.18	Janitor	.22	22
Brogan, P.	6/27/07	Drill press	.18	Milling Machine	.31	72
Bruan, S.	10/20/10	Timekeeper	.22	Timekeeper	.24	11
Bardsley, A.	1/ 5/10	Pattern maker	.28	Pattern maker	.38	36
Boasman, W.	3/ 3/10	Colored tool boy	.16	Tool boy	.19	19
Carter, J.	1/12/03	Machinist	.25	Gang foreman	.54	118
Clark, H.	3/12/10	Apprentice, lathe	.16	Turret lathe	.18	13
Cox, C.	1/ 1/1900	Laborer	.15	Machinist	.40	167
Chadwick, B.	1/10/10	Machinist	.28	Machinist	.37	32
Connelly, H.	8/10/03	Blacksmith	.31½	Blacksmith	.47	49
Evans, W.	6/19/05	Machinist	.22½	Machinist	.34	51
Freek, J.	5/31/05	Machinist	.25	Machinist	.40	60
Foreman, E.	3/ 1/05	Machinist	.25	Machinist	.32	28
Field, M.	8/29/06	Colored machinist helper	.18	Laborer	.22	22
Goodwin, C.	8/19/09	Milling, under instruction	.16	Machinist	.34	113
Hamilton, J.	5/26/01	Pipe fitting	.18	Pipe fitting	.26	45

Name	Date					
Kurz, W.	3/24/02	Tool maker	.25	Inspector	.40	60
Kennedy, P.	9/13/06	Laborer	.20	Chipper	.25	25
Kepner, R.	1/31/02	Miscellaneous	.24	Millwright	.31	29
Klenk, J.	2/25/02	Drill press hand	.22	Drill press hand	.35	59
Loucks, S.	3/22/07	Miscellaneous	.20	Vise hand	.28	40
Laney, W.	11/30/01	Woodworker	.26½	Woodworker	.37½	42
Marsden, T.	9/23/01	Machinist	.27½	Machinist	.33	20
McCullough, C.	6/1/09	Miscellaneous help	.24	Miscellaneous help	.32½	35
Nolan, J.	8/21/02	Gang boss	.34	Gang boss	.50	47
Paxton, W.	10/17/06	Pattern maker	.28	Pattern maker	.40	43
Pfendner, J.	5/15/05	Metal pattern fitter	.25	Metal pattern fitter	.40	60
Rickerts,	7/19/05	Machinist	.20	Machinist	.38½	93
Reiff, E.	6/17/04	Machinist apprentice	.12	Machinist	.36	200
Rommel, C.	10/11/05	Drafting apprentice	.05	Draftsman	.36	620
Reed, H.	8/13/07	Toolmaker	.36	Feed and spend time study in Planning Department	.52	44
Rosi, F.	9/26/10	Grinder	.16	Grinder	.22	38
Shire, P.	6/24/04	Drill press	.20	Machinist	.35	75
Sherman, J.	8/17/04	Machinist	.22	Machinist	.35	59
Ski, J.	4/16/07	Oiling machinist and belt man	.18	Oiling machinist and belt man	.22	22
Snyder,	10/5/09	Machine repair man	.28	Machine repair man	.35	25
Tait, J.	7/15/06	Turret lathe	.22	Machinist	.38	72
Warner, J.	3/31/04	Machinist	.25	Gang foreman	.54	116
Shipley, A.	11/5/05	Machinist	.30	Routing clerk	.47	57
Holmes, A.	2/15/06	Gang boss	.46	Gang boss	.56	22
Wells, W.	4/4/10	Tool boy	.10	Turret lathe hand	.19	90
Weld, M.	2/3/10	Grinder	.12	Grinder	.25	108
Wald, H.	12/18/05	Tool boy	.10	Tool-room attendant	.24	140
Wetzel, J.	8/22/06	Machinist's helper	.16	Tool grinder	.28	75
Wilson, J.	3/10/10	Grinder	.20	Grinder	.25	25
Walters, E.	9/1/09	Machinist	.26	Machinist	.34	31

Total, 3811—73.5 percent individual increase.

That thing would not be tolerated if the management knew it, nor would the workmen themselves tolerate it.

Mr. Redfield. In a factory, Mr. Taylor, who suffers the most from inefficiency?

Mr. Taylor. I should say they were both sufferers, but I should say that the company suffered vastly more than the man through inefficiency, but both are sufferers from it.

Mr. Redfield. If that is the case is the company the greater gainer from efficiency?

Mr. Taylor. I should say they were both the gainers from efficiency, but it is very hard to say which is the greater. The great gain which the man gets from efficiency, to my mind, the greatest gain which he gets, is permanence of employment. That his company is more apt to have work going along steadily in dull times than the inefficient company, and so the man gains through steadiness of employment, whereas the company gains through having its work well done and cheaply as well as quickly done, and through being able to fill its orders quickly instead of filling them slowly, and so is able to get a much larger business.

Mr. Redfield. The suggestion was made in Boston that you were interested in the Tabor Manufacturing Co., and as a part proprietor, and that it was an understood part in the adoption of the Taylor system of scientific management that apparatus made by the Tabor Manufacturing Co. was recommended or preferred and was, as a matter of fact, bought. To what extent, if at all, is that true?

Mr. Taylor. I own 120 shares in the Tabor Manufacturing Co., all of which I bought absolutely as a matter of trying to help out my friend, Wilfred Lewis (the owner of the Tabor Manufacturing Co.), when he was in dire straits and his company had almost failed. Under the old system of management he was on the verge of failure, and he begged me to buy these shares of him to help him tide over his troubles. I bought those shares, and that is my interest in the Tabor Manufacturing Co.

Mr. Chairman. You have 120 shares out of a total number of how many shares issued by the company?

Mr. Taylor. I really do not know what the capitalization is. My friend Mr. Tabor here says there are 1,500 shares in the company.

Mr. Redfield. You have, then, about a one-fifth interest?

Mr. Taylor. Oh, no.

Mr. Redfield. Then it is not a majority interest?

Mr. Taylor. No; and I never have received a cent from it.

Mr. Redfield. Is it, or is it not, a fact that it is a part of the application of the Taylor system that it will be utilized indirectly for the sale of the products of any company in which you are interested. If it is, we want to know it.

Mr. Taylor. Why, no; what a ridiculous—why, no.

Mr. Redfield. The charge was made in the testimony in Boston.

Mr. Taylor. It is absolutely untrue.

Mr. Redfield. That is what I want to know—if it is true or false.

Mr. Taylor. Why, absolutely false.

Mr. Redfield. We want to know if this is being worked to fill your pockets, directly or indirectly. It was said at Boston that something of that kind was true, and I want to know.

Mr. Taylor. It is absolutely false. I have never had a dollar of dividends from the Tabor Manufacturing Company.

Mr. Tilson. I should like to ask you one general question: How many concerns, to your knowledge, use your system in its entirety?

Mr. Taylor. In its entirety—none; not one.

Mr. Tilson. Then how many concerns use substantially your system?

Mr. Taylor. Oh, a very great many, Mr. Tilson. As to how many in numbers, I cannot say, and I want to tell you why: In the first place, I will have to again define what I mean when I say that a company is using our system of management. After the management of that company have gone through this mental revolution of which I spoke at length in my direct testimony and after the workmen have substantially gone through a similar mental revolution, and both sides have become friends instead of practical enemies (that is the revolution I refer to, but this alone is not enough to constitute scientific management); when, in addition to this, those on the management side recognize that it is their duty to make a scientific investigation of all the facts, a scientific study of all of the elements of their business

—when a company has passed through those two stages, then I say that company has come under scientific management, and not until then.

Mr. Tilson. There are a great many of that kind, are there?

Mr. Taylor. Yes; and since I have been in these hearings I have heard of one of them. I have, in fact, heard of five or ten new companies during this time; but there is one I have heard of during this time and which interests me especially, and I think I will surprise you when I say that Mr. Redfield's company is practicing scientific management and has been for years.

Mr. Redfield. Which one?

Mr. Taylor. I do not know whether your blower company is or not, but I do know that your forging company (the J. H. Williams Co.) is practicing scientific management. I have heard Mr. Redfield say that the management in that company and their workmen were in thorough harmony, that they were the best of friends, that they have never lowered a piecework price in that company after a rate has once been set, and that the men responded by stopping soldiering and doing a great big day's work for the company, which indicates that both the management and the workmen have arrived at this new frame of mind of which I have spoken. And I have also heard Mr. Redfield say (and that is why I say that they are under scientific management), I heard him make the statement that the officers of that company had made such a careful and thorough study of their machines and of the apparatus that goes with them, that within eight years almost every machine in that company had been rebuilt and redesigned and reconstructed, so as to work in harmony with the latest and most modern information. That shows me that Mr. Redfield's management is using what I call the scientific method. That is, that they are doing their share of the work in developing the science. Therefore, I say Mr. Redfield's company (much to his disgust, it may be) is practicing scientific management.

Mr. Tilson. In other words, you do not claim a monopoly on scientific management?

Mr. Taylor. I should say not, Mr. Tilson. My gracious, I do not believe there is any man connected with scientific management who has the slightest pride of authorship in connection with it. Every one of us realizes that this has been the work of 100 men or more, and that the work which any one of us may have done is but a small fraction of the whole. This is a movement of large proportions, and no one man counts for much of anything in it. It is a matter of evolution, of many men, each doing his proper share in the development, and I think any man would be disgusted to have it said that he had invented scientific management, or that he was even very much of a factor in scientific management. Such a statement would be an insult to the whole movement. It is not an affair of one man or of ten or twenty men.

I want to try to make clear to you what I mean, Mr. Tilson, when I say that a great many companies are using it. I will tell you one of many similar instances which goes to prove this. The Economic

Club of Portland, Me., asked me to speak before them week before last. After I got through, a young man came up to me and asked me what train I was going to Boston on the next day. He said, "I would like to go down with you." So he rode to Boston with me, and to my surprise he told me that for the last five years he had been the manager of the Burgess Sulphite Pulp Mill away up in the woods of New Hampshire, and that having read what we had written on scientific management some six or eight years ago, when he became manager he at once started to make a scientific study of every element that affects the manufacture of pulp. The same kind of study which is advocated under the principles of scientific management.

He also began at once the change in the treatment of the men which has resulted in his case as he told me, in making the men of that company the warm friends of the management, whereas when he came there they were always on the ragged edge of a strike, and since he came there has not been a single strike. He said that their scientific investigation of one element after another of the art of making chemical sulphite pulp in this company had resulted in placing his company in the lead of all similiar companies of the world, whereas before the German and Swedish companies were away ahead of the American companies. Now, this careful scientific study of every element that goes into the manufacture of pulp and the use of the by-products not only cheapens the cost of manufacture but gives the Burgess Sulphite Fiber Co. the preference in the American market at a higher price over all foreign pulp, so that instead of having salesmen on the road all the time to sell their goods as they used to have, they now never have to solicit any orders, and they always have more orders in advance than they can fill. I consider that this company has come under the principles of scientific management.

Mr. Tilson. Let me assume that after the scientific management has been established in a concern and the adjustment of remuneration and employees has been made, and after that the management changes, and we have a management which is not disposed to be fair, and is disposed to get as much out of a man as they have been giving with increased remuneration, but now to cut them back to the old figure, as we have heard it often expressed in this hearing—

Mr. Taylor. Yes.

Mr. Tilson. Now, what is the situation of an employee as compared with what it formerly was. What disadvantages is he under that he would not be under under any management?

Mr. Taylor. In this case the employee would merely be returned to the same position which he occupies now under the old systems of management everywhere. I will tell you, however, the employee, when that trick is played on him, or any such trick is tried, gets back at the company so darned hard that the man who tried to play the trick is sorry that he ever did it. When I left the Bethlehem Steel Co. and Mr. Schwab came, he thought he could do without paying the premium. He thought that part of the system was a good thing to abandon. He tried that for just one month, and at the end of the month (so the foremen and the men told me), Mr. Schwab was mighty darned glad to put the premium back again, because the product of the shop had dropped to about one half.

Mr. Tilson. Suppose it were applied to Government work. The workmen there have the same remedy and an additional one, have they not?

Mr. Taylor. They have indeed, and let me tell you there has been a whole lot of talk about the Watertown Arsenal, and the great injustice done to the workmen at the Watertown Arsenal through time study and paying them a premium. If you gentlemen in Congress were to vote to bring it about that those workmen in the Watertown Arsenal have to go back to the old system of management there and do without this 30 or 45 per cent premium they are being paid now, there will be a great big howl go up from the Watertown Arsenal. A bigger howl will go up if you try to throw it out than there has been over putting it in. I am simply making that prediction.

Mr. Godfrey. There are three or four things that I do not think are quite clear, on which I should like to ask Mr. Taylor some questions.

You have not answered yet, Mr. Taylor, what money interest you have in scientific management; that is if you have any money interest in scientific management.

Mr. Taylor. I have not a cent. I have not accepted any employment money under scientific management of any kind since 1901, and everything I have done in that cause has been done for nothing. I have spent all of the surplus of my income in trying to further the cause for many years past, and am spending it now, every cent of it.

Mr. Godfrey. You have received no profit?

Mr. Taylor. None directly or indirectly of any kind.

Mr. Godfrey. Do you find that there is a growing interest in scientific management or not?

Mr. Taylor. The interest in scientific management seems to me to be growing immensely. I can judge by one barometer. I am receiving an average of one invitation a day to speak before audiences on the subject of scientific management all over the country. Last spring I was receiving at the rate of one invitation every week and apparently the interest is rolling up with tremendous rapidity. This interest is widespread, it is all over the country from the Pacific coast to Maine.

Mr. Godfrey. Do you believe that the hours of working for working men should be longer or shorter?

Mr. Taylor. I believe in shorter hours by all means, if it is a possible thing, but there is one word of warning that should come in here. If you are looking at the real interest of the workmen, and you think it is to his interest to have the hours as short, say, nine hours or eight hours a day, be mighty careful that you do not shorten his hours of work without at the same time seeing that some device is gotten up by which he will turn out more work, or in the end you are robbing him of his wages. I should like to call attention to a lot of cases where the workingman's hours have been shortened to his detriment, because when shortening his hours, no sufficient provision has been made for a proportionate increase in his output. In the interest of the workmen I say this to you, do not shorten his hours unless you provide for increase of output, or you are cutting his wages in the end.

Mr. Godfrey. Can you say in one syllable what the relation of labor unions should be to scientific management?

Mr. Taylor. Of all the devices in the world they ought to look upon scientific management as the best friend that they have. It is doing in the most efficient way every solitary good thing that the labor unions have tried to do for the workman and it has corrected the one bad thing that the unions are doing —curtailment of output. That is the one bad thing they are doing.

The Chairman. Have you stated to this committee that you do not know of one establishment where scientific management has been introduced where collective bargaining has been introduced?

Mr. Taylor. I do not recall any establishment.